# Forward Error Correction via Channel Coding

Orhan Gazi

# Forward Error Correction via Channel Coding

Springer

Orhan Gazi
Electronic & Communication Engineering Department
Cankaya University
Ankara, Turkey

ISBN 978-3-030-33382-9 ISBN 978-3-030-33380-5 (eBook)
https://doi.org/10.1007/978-3-030-33380-5



This Springer imprint is published by the registered company Springer Nature Switzerland AG
The registered company address is: Gewerbestrasse 11, 6330 Cham, Switzerland

# Preface

Coding and decoding can be considered as a mapping and de-mapping between the elements of two sequence sets. For the two sequence sets A and B, mapping a symbol sequence belonging to set A to a symbol sequence in set B can be considered as coding and the reverse operation can be considered as decoding.

The purpose of mapping depends on the aim of design. If the sequences in set A are shorter than the sequences in set B, such a mapping can be considered as data compression or source coding in short. If the aim of the mapping is to hide the information content of the sequences in set A, then this type of coding can be named as encryption. On the other hand, if the sequences in set B are longer compared to sequences in set A, and they are more robust to noisy transmissions, then such a mapping is named as channel coding.

The purpose of channel coding is to transmit more information bits reliably through a communication channel using a fixed amount of energy in unit time. For this purpose, instead of transmitting information bit sequences directly, a mapping can be performed to longer information sequences, and such a mapping is done mathematically. Since as the length of the information sequence increases, the total number of data words, i.e., information sequences, become a huge number, and it becomes almost impossible to do a manual mapping and de-mapping. The use of channel codes for error correction purpose in communication engineering became a hot subject of the researchers after the publication of Shannon's paper titled as "A Mathematical Theory of Communication" in 1948. In his paper, Shannon drew the limits for the maximum speed of reliable communication, i.e., maximum number of bits transmitted reliable per unit of time, for a definite signal bandwidth and signal-to-noise ratio, and it is also mentioned in the paper that the Shannon's limits can be achieved via the use of channel codes. Since then numerous channel codes have been designed to get the highest reliable speed of communication. Most of the channel codes are designed in a trivial manner until 2009 in which the first mathematically designed channel codes, polar codes, are introduced. Hence, it took almost 60 years for the researches to design the first channel code, achieving Shannon's limits, in a mathematical manner.

The use of channel codes in communication engineering is a must issue. Without the use of channel codes, it is not possible to design energy-efficient communication systems. In this book, preliminary information is provided about the channel coding. Channel codes can be divided into two main categories which are block and convolutional codes. Block codes are nothing but vector subspaces. For this reason, the first chapter of this book is devoted to the fundamental subjects of linear algebra. Without a good knowledge of linear algebra, it is not possible to comprehend the construction and use of channel codes. We suggest the reader not to skip the first chapter of this book. The second chapter deals with the binary linear block codes, and fundamental information is provided about the construction and error correction capability of the binary linear block codes. The preliminary decoding approach, syndrome decoding, is explained in the third chapter along with some well-known binary linear block codes. Cyclic codes which can be considered as a subset of linear block codes are explained in the fourth chapter. Galois fields, which are the background subject for algebraic design of linear block codes, are explained in the fifth chapter. For the understandability of the topics presented in the sixth and seventh chapters, the subjects explained in the fifth chapter should be comprehended very well. In the sixth and seventh chapters, two important cyclic codes, which are BCH and Reed Solomon codes, are explained. BCH codes are binary cyclic codes; on the other hand, Reed Solomon codes are non-binary cyclic codes. These two types of codes were employed in practical communication and data storage systems; for instance, Reed Solomon codes were used in compact disc and digital video broadcasting (DVB) standard DVB-S, and similarly BCH codes were employed in satellite communications, DVDs, disk drives, solid-state drives, two-dimensional bar codes, etc. In the eighth chapter, the second type of channel codes, convolutional codes, are explained with Viterbi decoding algorithm. Convolutional codes are used in some mobile communication standards such as GSM.

This book can be studied in one semester graduate or undergraduate course, or it can be read by anyone interested in the fundamentals of error-correcting codes. We tried to be simple and brief while writing the book. We avoided the use of heavy mathematics in the book and unnecessary long explanations which distracts the readers' attention to focus on the main points. We tried to provide as many simple solved examples as we could do. As a last word, I dedicate this book to my lovely daughter Vera Gazi who was six years old when this book was being written. Her love was an energy source for my studies.

Ankara, Turkey Orhan Gazi
Monday, November 25, 2019

# Contents

# Chapter 1
# Review of Linear Algebra

## 1.1 Group

A group is a set $G$ together with a binary operation $*$ defined on the set elements, and the set elements together with the binary operation satisfy a number of properties. These properties are as follows:

1. Closure

   If a set $G$ is closed under binary operation $*$, then we have

$$\forall a, b \in G \rightarrow a * b \in G. \tag{1.1}$$

2. Associative

   If a set $G$ has associative property under binary operation $*$, then we have

$$(a * b) * c = a * (b * c). \tag{1.2}$$

3. Identity element

   If a set $G$ has identity element $e$ under binary operation $*$, then we have

$$\forall a \in G, a * e = e * a = a. \tag{1.3}$$

4. Inverse element

   If a set $G$ has inverse property under binary operation $*$, then $\forall a \in G$; there exists an element $b$ such that


O. Gazi, *Forward Error Correction via Channel Coding*,
https://doi.org/10.1007/978-3-030-33380-5_1

$$a * b = b * a = e. \tag{1.4}$$

5. Commutative (not must)

If a set $G$ has commutative property under binary operation $*$, then $\forall a, b \in G$; we have

$$a * b = b * a. \tag{1.5}$$

If the properties 1 – 4 are satisfied, then the set is called a group under binary operation $*$. On the other hand, if all the properties are satisfied, then the set is called a commutative group under binary operation $*$.

Now, let's give some examples to the groups.

**Example 1.1** Let's show that the set of integers $Z$ with the ordinary addition operation + form a group, i.e., the pair $(Z, +)$ forms a group. For this purpose, let's verify all the properties of the group $(Z, +)$.

1. Closure

If $a, b \in Z$, then it is obvious that $a + b \in Z$. Thus, closure property is satisfied $\surd$.

2. Associative

If $a, b, c \in Z$, then it is obvious

$$(a + b) + c = a + (b + c) \quad \surd. \tag{1.6}$$

3. Identity element

For + operation, 0 is the identity element such that

$$a \in Z \rightarrow a + 0 = 0 + a = a \quad \surd. \tag{1.7}$$

4. Inverse element

The inverse of the integer $a \in Z$ is $-a$ such that

$$a + (-a) = 0 \quad \surd. \tag{1.8}$$

These four properties are sufficient for the set of integers $Z$ to be a group under ordinary addition operation +. In fact, the set of integers under addition operation satisfies the commutative property also. Let's show that commutative property is also satisfied.

5. Commutative

**Table 1.1** Mod-5 addition table

| ⊕ | 0 | 1 | 2 | 3 | 4 |
|---|---|---|---|---|---|
| 0 | 0 | 1 | 2 | 3 | 4 |
| 1 | 1 | 2 | 3 | 4 | 0 |
| 2 | 2 | 3 | 4 | 0 | 1 |
| 3 | 3 | 4 | 0 | 1 | 2 |
| 4 | 4 | 0 | 1 | 2 | 3 |

If $a$, $b \in Z$, then it is clear that

$$a + b = b + a \quad \surd. \tag{1.9}$$

Hence, we can say that the set of integers under addition operation is a commutative group, and this group includes infinite number of elements. We can categorize it as infinite group.

Now let's give an example for a group including finite number of elements.

**Example 1.2** The finite set $G$ and the operation $\oplus$ are given as

$$G = \{0, 1, 2, 3, 4\}$$

$$\oplus \rightarrow \text{Mod} - 5 \text{ addition operation.}$$

Determine whether $G$ is a group under the defined operation $\oplus$ or not.

**Solution 1.2** Let's check each property of the group for $G$ using the operation $\oplus$. Since $G$ is a finite set, we should consider all possible pairs while checking the closure property, and we have to inspect all the triples while checking the associative property. To make it practical, we can make a table as Table 1.1 for Mod-5 addition operation considering the elements of $G$, and check the group properties in a quick manner using the table.

1. If we inspect the table, we see that closure property is satisfied $\surd$.
2. From the table, it is seen that associative property is satisfied $\surd$.
3. Identity element is 0 $\surd$.
4. Identity element is available at every row. This means that every element has an inverse. For example, we can find the inverse of 2 as 3. This is illustrated in Fig. 1.1.
5. Commutative property is also satisfied $\surd$.

Hence, we can say that the finite set

**Fig. 1.1** Finding the inverse of 3

| ⊕ | 0 | 1 | 2 | 3 | 4 |
|---|---|---|---|---|---|
| 0 | 0 | 1 | 2 | 3 | 4 |
| 1 | 1 | 2 | 3 | 4 | 0 |
| 2 | 2 | 3 | 4 | 0 | 1 |
| 3 | 3 | 4 | 0 | 1 | 2 |
| 4 | 4 | 0 | 1 | 2 | 3 |

**Table 1.2** Mod-4 addition table

| ⊕ | 1 | 2 | 3 |
|---|---|---|---|
| 1 | 2 | 3 | 0 |
| 2 | 3 | 0 | 1 |
| 3 | 0 | 1 | 2 |

$$G = \{0, 1, 2, 3, 4\}$$

is a group under Mod-5 addition operation. That is, the pair $(G, \oplus)$ forms a group.

**Example 1.3** The finite set $G$ and the operation $\oplus$ are given as

$$G = \{1, 2, 3\}$$

$$\oplus \rightarrow \text{Mod} - 4 \text{ addition operation.}$$

Determine whether $G$ is a group under the defined operation $\oplus$ or not.

**Solution 1.3** In Table 1.2, we show the Mod-4 addition operation results considering the elements of $G$.

If we inspect the table, we see that closure property is not satisfied. Since $2 \oplus 2 = 0$ and $0 \notin G$, then closure is violated. If one property is not satisfied, then no need to check the rest of the properties. We can say that the given set $G$ is NOT a group under Mod-4 addition operation.

**Example 1.4** The finite set $G$ and the operation $\oplus$ are given as

$$G = \{1, 2, 3, 4\}$$

**Table 1.3** Mod-4 multiplication table

| $\otimes$ | 1 | 2 | 3 |
|---|---|---|---|
| 1 | 1 | 2 | 3 |
| 2 | 2 | 0 | 2 |
| 3 | 3 | 2 | 1 |

$$\oplus \rightarrow \text{Mod} - 5 \text{ addition operation.}$$

The set $G$ is not a group under Mod-5 addition operation. Closure property is not satisfied, since $1 \oplus 4 = 0 \notin G$.

**Definition** The number of elements in $G$ is denoted by $|G|$ and it is named as the order of the group $G$.

**Example 1.5** The finite set $G$ and the operation $\otimes$ are given as

$$G = \{1, 2, 3\}$$

$$\otimes \rightarrow \text{Mod} - 4 \text{ multiplication operation.}$$

Determine whether $G$ is a group under the defined operation $\otimes$ or not.

**Solution 1.5** In Table 1.3, we show the results of Mod-4 multiplication operation considering the elements of $G$.

If we inspect the Table 1.3, we see that closure and inverse group properties are not satisfied, since $2 \otimes 2 = 0 \notin G$, and the element 2 does not have an inverse, i.e., we cannot find an element $x$ such that $2 \otimes x = 1$ where 1 is the identity elements of the Mod-4 multiplication operation.

**Example 1.6** The finite set $G$ and the operation $\otimes$ are given as

$$G = \{1, 2, 3, 4\}$$

$$\otimes \rightarrow \text{Mod} - 5 \text{ multiplication operation.}$$

Determine whether $G$ is a group under the defined operation $\otimes$ or not.

**Solution 1.6** In Table 1.4, we show the Mod-5 multiplication operation results considering the elements of $G$.

If we inspect the Table 1.4, we see that all the properties of a group are satisfied. Then, we can say that $G$ is a group under Mod-5 multiplication operation.

**Table 1.4** Mod-5 multiplication table

| $\oplus$ | 1 | 2 | 3 | 4 |
|---|---|---|---|---|
| 1 | 1 | 2 | 3 | 4 |
| 2 | 2 | 4 | 1 | 3 |
| 3 | 3 | 1 | 4 | 2 |
| 4 | 4 | 3 | 2 | 1 |

**Example 1.7** The set $G$ and the operation $\otimes$ are given as

$$G = \{\text{Set of } 2 \times 2 \text{ matrices with nonzero determinants}\}$$

$$\otimes \rightarrow \text{Matrix multiplication operation.}$$

Determine whether $G$ is a group under the defined operation $\otimes$ or not.

**Solution 1.7** The set $G$ owns all the four properties of a group, i.e., "closure, associative, identity element, and inverse element" properties are satisfied. Thus, $G$ is a group. However, commutative property is not satisfied. Since $\boldsymbol{A}$ and $\boldsymbol{B}$ square matrices are of the same size, then we can write for some matrices $\boldsymbol{A}$ and $\boldsymbol{B}$ that

$$\boldsymbol{A} \otimes \boldsymbol{B} \neq \boldsymbol{B} \otimes \boldsymbol{A}.$$

Although $G$ is a group, it is not a commutative group.

**Summary** If $G$ and $\oplus$ are given as

$$G = \{0, 1, \ldots, M - 1\}$$

$$\oplus \rightarrow \text{Mod} - \text{M addition operation,}$$

then it can be shown that $G$ is a group under Mod-M addition operation.

If $G$ and $\oplus$ are given as

$$G = \{1, \ldots, M - 1\}$$

$$\oplus \rightarrow \text{Mod} - \text{M addition operation},$$

then it can be shown that $G$ is NOT a group under Mod-M addition operation.

If $G$ and $\otimes$ are given as

$$G = \{1, \ldots, M - 1\}$$

$$\otimes \rightarrow \text{Mod} - \text{M multiplication operation},$$

then it can be shown that if M is a prime number, then $G$ is a group under Mod-M multiplication operation; otherwise, $G$ is NOT a group under Mod-M multiplication operation.

**Example 1.8** The finite set $G$ and the operation $\otimes$ are given as

$$G = \{1, 2, \ldots, 15\}$$
$$\otimes \rightarrow \text{Mod} - 16 \text{ multiplication operation}.$$

Determine whether $G$ is a group under the defined operation $\otimes$ or not.

**Solution 1.8** According to the information provided in the previous summary part, for

$$G = \{1, \ldots, M - 1\}$$

to be a group under Mod-M multiplication operation, M should be a prime number. For our example, $M = 16$ which is not a prime number. Thus, $G$ is not a group under Mod-16 multiplication operation.

**Example 1.9** For a set to be a group which properties should be owned by the set, list the properties.

**Solution 1.9** Fundamentally, there are four basic properties, and these properties are "closure, associative, identity element, and inverse element." If the set has these four properties, then the set is a group. Besides, if the "commutative" property is also available, then the set becomes a commutative group.

**Example 1.10** Let $G$ be a group under $*$ operation, and $a$, $b \in G$. Show that

$$(a * b)^{-1} = b^{-1} * a^{-1}. \tag{1.10}$$

**Solution 1.10** Let $x = a * b$. If the inverse of $x$ is $x^{-1} = b^{-1} * a^{-1}$, then we have

$$x * x^{-1} = a * b * b^{-1} * a^{-1}$$

where, employing the associative property for the right-hand side, we get

$$a * \left( \underbrace{b * b^{-1}}_{=e} \right) * a^{-1} \rightarrow a * e * a^{-1} \rightarrow a * a^{-1} \rightarrow e \rightarrow x * x^{-1} = e.$$

Thus, the inverse of $x = a * b$ is $x^{-1} = b^{-1} * a^{-1}$.

**Exercise** Let $G$ be a group under $*$ operation, and $a, b \in G$. Verify that if

$$(a * b)^{-1} = a^{-1} * b^{-1} \tag{1.11}$$

then $G$ is a commutative, i.e., abelian, group.

**Exercise** Let $G$ be a commutative group under $*$ operation, and $a, b \in G$. Show that $a^2 * b^2$ commute, i.e., $a^2 * b^2 = b^2 * a^2$. Note that $a^2 = a * a$ and $b^2 = b * b$.

### *1.1.1 Subgroup*

A subgroup $(H, *)$ is a subset of a group $(G, *)$ having all the group properties. Note that the same operation $*$ is considered by subgroup and group.

**Example 1.11** The finite set $G$ and the operation $\oplus$ are given as

$$G = \{0, 1, 2, 3, 4, 5\}$$

$$\oplus \rightarrow \text{Mod} - 6 \text{ addition operation.}$$

$G$ is a group under Mod-6 addition operation. Find a subgroup of $G$.

**Solution 1.11** We can form a subgroup of $G$ as

$$H = \{0, 2, 4\}.$$

If we check the properties “closure, associative, identity element, and inverse element,” we see that $H$ has all these properties under Mod-6 addition operation. Note that every subset of $G$ is not a subgroup. For instance, $L = \{1, 3, 4, 5\}$ is a subset of $G$ such that

$$G = H \cup L.$$

However, $L$ is not a subgroup of $G$. Closure property is not satisfied since $1 \oplus 5 = 0 \notin L$.

**Example 1.12** The finite set $G$ and the operation $\otimes$ are given as

$$G = \{1, 2, 3, 4, 5, 6\}$$

$$\otimes \rightarrow \text{Mod} - 7 \text{ multiplication operation.}$$

$G$ is a group under Mod-7 multiplication operation. Find a subgroup of $G$.

**Solution 1.12** We can form a subgroup of $G$ as

$$H_{g1} = \{1, 2, 4\}.$$

Another subgroup of $G$ can be formed as

$$H_{g2} = \{1, 6\}.$$

**Definition** Let $G$ be a group under the operation $*$, and $a \in G$. Then, $a^n$ is defined as

$$a^n = \underbrace{a * a * \ldots * a}_{n \text{ times}} \tag{1.12}$$

and $a^{-n}$ is calculated as

$$a^{-n} = \underbrace{a^{-1} * a^{-1} * \ldots * a^{-1}}_{n \text{ times}} \tag{1.13}$$

where $a^{-1}$ is the inverse element of $a$ considering the operation $*$, i.e., $a * a^{-1} = e$ where $e$ is the identity element.

**Definition** Let $G$ be a group. If there exists an element $a$ of $G$ such that $a^n$, $n \in Z$ generates all the elements of $G$, then the group $G$ is a cyclic group. In this case, the element $a$ is named as a generator element of $G$.

Similarly, let $H$ be a subgroup of $G$. If there exists an element $b$ of $H$ such that $b^n$, $n \in Z$ generates all the elements of $H$, then the subgroup $H$ is a cyclic subgroup. In this case, the element $b$ is named as a generator element of $H$.

**Example 1.13** The finite set $G$ and the operation $\oplus$ are given as

$$G = \{0, 1, 2, 3, 4\}$$

$$\oplus \rightarrow \text{Mod} - 5 \text{ addition operation.}$$

$G$ is a group under the Mod-5 addition operation. Determine whether $G$ is a cyclic group or not.

**Solution 1.13** For $G$ to be a cyclic group, we need to find an element $a$ of $G$ such that $a^n$, $n \in Z$ generates all the elements of $G$. In fact, one can find more than one generator element for $G$. One such generator element is 2, since using $2^n$ we can generate all the elements of $G$ as in

$$2^2 \rightarrow 2 \oplus 2 = 4$$
$$2^3 \rightarrow 2 \oplus 2 \oplus 2 = 1$$
$$2^4 \rightarrow 2 \oplus 2 \oplus 2 \oplus 2 = 3$$
$$2^5 \rightarrow 2 \oplus 2 \oplus 2 \oplus 2 \oplus 2 = 0.$$

Similarly, using 3, we can generate all the elements of $G$ as in

$$3^2 \rightarrow 3 \oplus 3 = 1$$
$$3^3 \rightarrow 3 \oplus 3 \oplus 3 = 4$$
$$3^4 \rightarrow 3 \oplus 3 \oplus 3 \oplus 3 = 2$$
$$3^5 \rightarrow 3 \oplus 3 \oplus 3 \oplus 3 \oplus 3 = 0.$$

In fact, using the elements 1 and 4, we can generate all the other elements of $G$. Thus, we can say that all the elements of $G$ are generator elements except for 0.

**Definition** Let $G$ be a group, and the identity element of $G$ be $e$. Let $a$ be an arbitrary element of $G$. If there exists $m \in Z$ such that

$$a^m = e, \tag{1.14}$$

then $m$ is said to be the order of $a$.

**Example 1.14** The finite set $G$ and the operation $\otimes$ are given as

$$G = \{1, 2, 3, 4\}$$
$$\otimes \rightarrow \text{Mod} - 5 \text{ multiplication operation.}$$

$G$ is a group under Mod-5 multiplication operation. Find the order of every element of $G$.

**Solution 1.14** For $a \in G$, the order of $a$ is $m_a$, if $a^{m_a} = e$. For our example, the identity element is $e = 1$. We can determine the order of each element as in

$$1^1 = 1 \rightarrow m_1 = 1$$
$$2^4 = 1 \rightarrow m_2 = 4$$
$$3^4 = 1 \rightarrow m_3 = 4$$
$$4^2 = 1 \rightarrow m_4 = 2.$$

**Theorem** If $G$ is a noncyclic group, then $G$ can be written as the union of subgroups.

## *1.1.2 Cosets*

Let $G$ be a group and $H$ be a subgroup under the operation $*$, and $a \in G$. The left and right cosets of $H$ are the sets obtained using

$$a * H = \{a * h | h \in H\} \quad \text{and} \quad H * a = \{h * a | h \in H\}.$$

For a commutative group, the left and the right cosets are the same of each other.

**Example 1.15** The finite set $G$ and the operation $\oplus$ are given as

$$G = \{0, 1, 2, 3, 4, 5\}$$
$$\oplus \rightarrow \text{Mod} - 6 \text{ addition operation.}$$

$G$ is a group under Mod-6 addition operation. A subgroup of $G$ can be determined as

$$H = \{0, 2, 4\}.$$

Find the cosets of $H$.

**Solution 1.15** To find a coset of $H$, we will find an element $a$ of $G$ such that $a \notin H$, and if there are any cosets formed, $a$ should not also be an element of the previously determined cosets. Considering this, we can form the cosets as

$$1 \in G, 1 \notin H \rightarrow C = 1 \oplus H = \{1, 3, 5\}.$$

There is only a single coset of $H$, and the group $G$ can be written as

$$G = H \cup C.$$

**Example 1.16** The finite set $G$ and the operation $\oplus$ are given as

$$G = \{0, 1, 2, 3, 4, 5, 6, 7, 8, 9\}$$
$$\oplus \rightarrow \text{Mod} - 10 \text{ addition operation.}$$

$G$ is a group under Mod-10 addition operation. A subgroup of $G$ can be determined as

$$H = \{0, 5\}.$$

Find the cosets of $H$.

**Solution 1.16** To find a coset of $H$, we will find an element $a$ of $G$ such that $a \notin H$, and if there are any cosets formed, $a$ should not also be an element of the previously determined cosets. Considering this, we can form the cosets as

$$1 \in G, 1 \notin H \rightarrow C_a = 1 \oplus H = \{1, 6\}$$
$$2 \in G, 2 \notin H, 2 \notin C_a \rightarrow C_b = 2 \oplus H = \{2, 7\}$$
$$3 \in G, 3 \notin H, 3 \notin C_a, 3 \notin C_b \rightarrow C_c = 3 \oplus H = \{3, 8\}$$
$$4 \in G, 4 \notin H, 4 \notin C_a, 4 \notin C_b, 4 \notin C_c \rightarrow C_d = 4 \oplus H = \{4, 9\}.$$

The given group $G$ can be written as the union of subgroup and cosets, that is

$$G = H \cup C_a \cup C_b \cup C_c \cup C_d.$$

**Example 1.17** The set of integers $Z$ under addition operation + is a group. A subgroup of $Z$ can be formed as

$$H = 4Z \rightarrow H = \{\ldots - 8, -4, 0, 4, 8, \ldots\}.$$

Find the cosets of $H$.

**Solution 1.17** The cosets of $H$ can be formed as in

$$C_a = 1 + H \rightarrow C_a = \{\ldots, -7, -3, 1, 5, 9, \ldots\}$$
$$C_b = 2 + H \rightarrow C_b = \{\ldots, -6, -2, 2, 6, 10, \ldots\}$$
$$C_c = 3 + H \rightarrow C_c = \{\ldots, -5, -1, 3, 7, 11, \ldots\}.$$

The group $Z$ can be written as

$$Z = H \cup C_a \cup C_b \cup C_c.$$

**Lemma** The subgroup $H$ and its cosets $C_a$, $C_b$, ..., have the same number of elements. In other words, the subgroup and its cosets have the same order.

**Theorem** Let $G$ be a finite group and $H$ be a subgroup of $G$. The order of $H$, i.e., the number of elements of $H$, divides the order of $G$.

***Proof*** A group can be written as the union of subgroups and its cosets. That is, the group $G$ can be written as

$$G = H \cup S_a \cup S_b \cup \ldots S_i \tag{1.15}$$

where any two sets are disjoint with each other, and they have equal number of elements. Let's say that there are $N - 1$ cosets, and the order of a coset is $K$, then we can write

$$|G| = |H| + |S_a| + |S_b| + \ldots + | S_i | \rightarrow |G| = KN \rightarrow K = \frac{|G|}{N} \tag{1.16}$$

which means that $|G|$ is a multiple of $K$.

**Example 1.18** The finite set $G$ and the operation $\otimes$ are given as

$$G = \{1, 2, 3, 4, 5, 6\}$$

$$\otimes \rightarrow \text{Mod} - 7 \text{ multiplication operation.}$$

$G$ is a group under Mod-7 multiplication operation. A subgroup of $G$ can be formed as

$$H = \{1, 2, 4\}.$$

Determine the cosets of $H$.

**Solution 1.18** The coset of $H$ can be formed as in

$$C_a = 3 \otimes H \rightarrow C_a = \{3, 5, 6\}$$

The given group $G$ can be written as the union of subgroup and coset, that is

$$G = H \cup C_a.$$

## 1.2 Fields

Let $F$ be a set and assume that two operations denoted by $\oplus$ and $\otimes$ are defined on the set elements. For the set $F$ to be a field under $\oplus$ and $\otimes$ operations, the following properties are to be satisfied:

1. $F$ is a commutative group under $\oplus$ operation.
2. $F$ is a commutative group under $\otimes$ operation.
3. If $a$, $b$, $c \in F$, then we have

$$a \otimes (b \oplus c) = (a \otimes b) \oplus (a \otimes c) \quad (1.17)$$

which implies the distributive property of $\otimes$ over $\oplus$.

For a commutative group, we need to check five properties. If the items 1), 2), and 3) are considered, it is clear that to determine whether a set is a field or not under the defined operations $\oplus$ and $\otimes$, we need to check 11 properties in total.

Since fields are sets with some special properties, fields as sets can contain finite or infinite number of elements.

Note that if the set contains "zero" as one of its elements, we ignore the "zero" element while searching for the inverse property.

### 1.2.1 Binary or Galois Field

The smallest finite field is the binary field $F_2$ defined as (Table 1.5)

$$F_2 = \{0, 1\}$$

$$\oplus \rightarrow \text{Mod} - 2 \text{ addition operation}$$

$$\otimes \rightarrow \text{Mod} - 2 \text{ multiplication operation.}$$

If we check the field properties, we see that $F_2$ is a commutative group under mod-2 addition and mod-2 multiplication operations, and multiplication operation has the distributive property over addition operation.

**Table 1.5** Mod-2 addition and multiplication tables

| $\oplus$ | 0 | 1 |
|---|---|---|
| 0 | 0 | 1 |
| 1 | 1 | 0 |

| $\otimes$ | 0 | 1 |
|---|---|---|
| 0 | 0 | 0 |
| 1 | 0 | 1 |

Binary field is sometimes denoted by $GF(2)$ or by $GF_2$.

**Example 1.19** The finite set $F$ and the operations $\oplus$ and $\otimes$ are given as

$$F = \{0, 1, 2, 3, 4\}$$

$$\oplus \rightarrow \text{Mod} - 5 \text{ addition operation}$$

$$\otimes \rightarrow \text{Mod} - 5 \text{ multiplication operation.}$$

Determine whether $F$ is a field or not under the defined operations $\oplus$ and $\otimes$.

**Solution 1.19** If we check the field properties, we see that

1. $(F, \oplus)$ form a commutative group.
2. $(F - \{0\}, \otimes)$ form a commutative group.
3. It can be shown that for the given field elements and defined operations, we see that $\otimes$ has distributive property over $\oplus$.

Thus, we can say that $F$ is a field under Mod-5 addition and Mod-5 multiplication operations.

**Example 1.20** Determine whether the set of integers

$$Z = \{\ldots, -3, -1, -1, 0, 1, 2, 3, \ldots\}$$

under ordinary addition "+" and multiplication "·" operations is a field or not.

**Solution 1.20** It can be shown that $Z$ is a commutative group under "+" operation. However, $Z$ is not a commutative group under "·" operation, since every element does not have an inverse. For instance, the inverse of 2 is 1/2 which is not an element of $Z$. So, inverse property is violated for "·" operation. Thus, $Z$ is not a field under ordinary addition "+" and multiplication "·" operations.

**Example 1.21** Determine whether the set of real numbers $R$ under ordinary addition "+" and multiplication "·" operations is a field or not.

**Solution 1.21** The set of real numbers is a commutative group under ordinary addition "+" and multiplication "·" operations, and multiplication operation has distributive property over addition operation. Then, we can say that the set of real numbers is a field under ordinary addition "+" and multiplication "·" operations.

**Example 1.22** The finite set $S$ and the operations $\oplus$ and $\otimes$ are given as

$$S = \{0, 1, 2, 3\}$$

$$\oplus \rightarrow \text{Mod} - 4 \text{ addition operation}$$

$$\otimes \rightarrow \text{Mod} - 4 \text{ multiplication operation.}$$

Determine whether $S$ is a field or not under the defined operations $\oplus$ and $\otimes$.

**Solution 1.22** If we check the field properties, we see that

1. $(S, \oplus)$ form a commutative group.
2. $(S - \{0\}, \otimes)$ does not form a commutative group. Inverse property is violated. The inverse of 2 is not available.

Thus, we can say that $S$ is NOT a field under Mod-4 addition and Mod-4 multiplication operations.

### *1.2.2 Prime Fields*

If the set

$$F = \{0, 1, \ldots, M - 1\},$$

where $M$ is a prime number, satisfies all the properties of a field under the operations

$$\oplus \rightarrow \text{Mod} - \text{M addition}$$

$$\otimes \rightarrow \text{Mod} - \text{M multiplication,}$$

then this set is called a "prime" field.

**Example 1.23** Find a prime field under Module-7 arithmetic operations.

**Solution 1.23** The required prime field can be formed as

$$F = \{0, 1, 2, 3, 4, 5, 6\}$$

$$\oplus \rightarrow \text{Mod} - 7 \text{ addition}$$

$$\otimes \rightarrow \text{Mod} - 7 \text{ multiplication.}$$

Indeed, we can show that

1. $(F, \oplus)$ is a commutative group.
2. $(F - \{0\}, \otimes)$ is a commutative group.
3. $\otimes$ has distributive property over $\oplus$.

The Mod-7 additions and Mod-7 multiplications are shown in Table 1.6.

**Table 1.6** Mod-7 multiplication and addition tables

| ⊕ | 0 | 1 | 2 | 3 | 4 | 5 | 6 |
|---|---|---|---|---|---|---|---|
| 0 | 0 | 1 | 2 | 3 | 4 | 5 | 6 |
| 1 | 1 | 2 | 3 | 4 | 5 | 6 | 0 |
| 2 | 2 | 3 | 4 | 5 | 6 | 0 | 1 |
| 3 | 3 | 4 | 5 | 6 | 0 | 1 | 2 |
| 4 | 4 | 5 | 6 | 0 | 1 | 2 | 3 |
| 5 | 5 | 6 | 0 | 1 | 2 | 3 | 4 |
| 6 | 6 | 0 | 1 | 2 | 3 | 4 | 5 |

| ⊗ | 1 | 2 | 3 | 4 | 5 | 6 |
|---|---|---|---|---|---|---|
| 1 | 1 | 2 | 3 | 4 | 5 | 6 |
| 2 | 2 | 4 | 6 | 1 | 3 | 5 |
| 3 | 3 | 6 | 2 | 5 | 1 | 4 |
| 4 | 4 | 1 | 5 | 2 | 6 | 3 |
| 5 | 5 | 3 | 1 | 6 | 4 | 2 |
| 6 | 6 | 5 | 4 | 3 | 2 | 1 |

## 1.3 Vector Spaces

Before giving the definition of vector space, let's explain the meaning of a vector.

**Vector**

Let $(F, \oplus, \otimes)$ be a field where $\oplus$ and $\otimes$ are the two operations defined on the field elements. A vector $\overline{v}$ is a sequence of field elements in any number. The number of elements in vector $\overline{v}$ is called the length of the vector.

**Example 1.24** The field$(F, \oplus, \otimes)$ is given as

$$F = \{0, 1, 2, 3, 4\}$$

$$\oplus \rightarrow \text{Mod} - 5 \text{ addition}$$

$$\otimes \rightarrow \text{Mod} - 5 \text{ multiplication.}$$

Using the field elements, we can define some vectors as in

$$\boldsymbol{u} = [2\,2\,3\,1\,1\,2\,3\,2\,3\,4\,2\,1\,1\,3]$$

$$\boldsymbol{v} = [2\,3\,0\,4\,4\,3\,1\,2\,3\,4\,2]$$

$$\boldsymbol{k} = [1\,2\,2\,3\,4].$$

Let $(F, \oplus, \otimes)$ be a field under $\oplus$ and $\otimes$ operations. Using the field elements, let's define two vectors $\boldsymbol{u}$ and $\boldsymbol{v}$ of equal length as

$$\boldsymbol{u} = [u_1\ u_2 \ldots u_k]$$

$$\boldsymbol{v} = [v_1\ v_2 \ldots v_k].$$

**Vector Addition**
The sum of the vectors $\boldsymbol{u}$ and $\boldsymbol{v}$ is performed as

$$\boldsymbol{u} \oplus \boldsymbol{v} = [u_1 \oplus v_1\ u_2 \oplus v_2 \ldots u_k \oplus v_k].$$

**Example 1.25** The field$(F, \oplus, \otimes)$ is given as

$$F = \{0, 1, 2, 3, 4\}$$
$$\oplus \rightarrow \text{Mod} - 5 \text{ addition}$$
$$\otimes \rightarrow \text{Mod} - 5 \text{ multiplication}.$$

Using the field elements, we can define some vectors as in

$$\boldsymbol{u} = [2\ 1\ 3\ 1\ 4]$$

$$\boldsymbol{v} = [4\ 4\ 3\ 1\ 3].$$

The sum of the vectors $\boldsymbol{u}$ and $\boldsymbol{v}$ is performed as

$$\boldsymbol{u} \oplus \boldsymbol{v} = [(2 \oplus 4)\ \ (1 \oplus 4)\ \ (3 \oplus 3)\ \ (1 \oplus 1)\ \ (4 \oplus 3)] \rightarrow \boldsymbol{u} \oplus \boldsymbol{v} = [1\ \ 0\ \ 1\ \ 2\ \ 2].$$

**Multiplication of a Vector by a Scalar**
Let $a \in F$. $a \otimes \boldsymbol{u}$ is performed as

$$a \otimes \boldsymbol{u} = [a \otimes u_1\ a \otimes u_2 \ldots a \otimes u_k]. \tag{1.18}$$

**Example 1.26** The field$(F, \oplus, \otimes)$ is given as

$$F = \{0, 1, 2, 3, 4\}$$
$$\oplus \rightarrow \text{Mod} - 5 \text{ addition}$$

$$\otimes \rightarrow \text{Mod} - 5 \text{ multiplication}.$$

Using the field elements, we can define a vector as

$$\boldsymbol{u} = [2\ 1\ 3\ 1\ 4].$$

We can multiply the vector $\boldsymbol{u}$ by the scalar $2 \in F$ as in

$$2 \otimes \boldsymbol{u} = [(2 \otimes 2)\ \ (2 \otimes 1)\ \ (2 \otimes 3)\ \ (2 \otimes 1)\ (2 \otimes 4)] \rightarrow 2 \otimes \boldsymbol{u} = [4\ \ 2\ \ 1\ \ 2\ \ 3].$$

**Dot Product of Two Vectors**
The dot product of $\boldsymbol{u}$ and $\boldsymbol{v}$ is performed as

$$\boldsymbol{u} \otimes \boldsymbol{v} = [u_1 \otimes v_1\ u_2 \otimes v_2 \ldots u_k \otimes v_k]. \tag{1.19}$$

**Example 1.27** The field$(F,\ \oplus, \otimes)$ is given as

$$F = \{0, 1, 2, 3, 4\}$$
$$\oplus \rightarrow \text{Mod} - 5 \text{ addition}$$
$$\otimes \rightarrow \text{Mod} - 5 \text{ multiplication}.$$

Using the field elements, we can define a vector as

$$\boldsymbol{u} = [2\ 1\ 3\ 2\ 3]$$
$$\boldsymbol{v} = [3\ 1\ 3\ 4\ 4].$$

$\boldsymbol{u} \otimes \boldsymbol{v}$ is calculated as

$$\boldsymbol{u} \otimes \boldsymbol{v} = [(2 \otimes 3)\ \ (1 \otimes 1)\ \ (3 \otimes 3)\ \ (2 \otimes 4)\ \ (3 \otimes 4)] \rightarrow \boldsymbol{u} \otimes \boldsymbol{v} = [1\ \ 1\ \ 4\ \ 3\ \ 2].$$

## *1.3.1 Vector Spaces*

Let $(F,\ \oplus, \otimes)$ be a field under $\oplus$ and $\otimes$ operations. Using the field elements, let's define some vectors and let's denote the set of vectors generated using the field elements by

$$\mathcal{V} = [\boldsymbol{v}_1\ \boldsymbol{v}_2 \ldots \boldsymbol{v}_M] \tag{1.20}$$

where $\boldsymbol{v}_i$, $i = 1 \ldots M$ is defined as

$$\boldsymbol{v}_i = [v_1\ v_2 \ldots v_N]. \tag{1.21}$$

The set of vectors $\mathcal{V}$ is called a vector space if the following properties are satisfied by the elements of $\mathcal{V}$:

1. $(\mathcal{V}, \oplus)$ is a commutative group.
2. For every $\alpha$, $\beta \in F$ and $\boldsymbol{v}_i, \boldsymbol{v}_j \in \mathcal{V}$, we have

$$(\alpha \otimes \boldsymbol{v}_i) \oplus (\beta \otimes \boldsymbol{v}_j) \in \mathcal{V} \tag{1.22}$$

which is named as the closure property.

3. For every $\alpha$, $\beta \in F$ and $\boldsymbol{v}_i \in \mathcal{V}$, we have

$$(\alpha \oplus \beta) \otimes \boldsymbol{v}_i = (\alpha \otimes \boldsymbol{v}_i) \oplus (\beta \otimes \boldsymbol{v}_i). \tag{1.23}$$

4. For every $\alpha \in F$ and $\boldsymbol{v}_i, \boldsymbol{v}_j \in \mathcal{V}$, we have

$$\alpha \otimes (\boldsymbol{v}_i \oplus \boldsymbol{v}_j) = (\alpha \otimes \boldsymbol{v}_i) \oplus (\alpha \otimes \boldsymbol{v}_j). \tag{1.24}$$

5. The operation $\otimes$ has associative property, i.e.,

$$(\alpha \otimes \beta) \otimes \boldsymbol{v}_i = \alpha \otimes (\beta \otimes \boldsymbol{v}_i). \tag{1.25}$$

**Example 1.28** The field$(F, \oplus, \otimes)$ is given as

$$F = \{0, 1, 2\}$$

$$\oplus \rightarrow \text{Mod} - 3 \text{ addition}$$

$$\otimes \rightarrow \text{Mod} - 3 \text{ multiplication}.$$

Using the field elements, we construct vectors, and using these vectors, we form a vector set as in

$$\mathcal{V} = [(0\,0\,0)\ \ (1\,1\,1)\ \ (2\,2\,2)].$$

Determine whether $\mathcal{V}$ is a vector space or not.

**Solution 1.28** For the given vector set $\mathcal{V}$, we check the properties of a vector space as in

1. $(\mathcal{V}, \oplus)$ is a commutative group.
2. For every $\alpha$, $\beta \in F$ and $\boldsymbol{v}_i, \boldsymbol{v}_j \in \mathcal{V}$, we have

$$(\alpha \otimes \boldsymbol{v}_i) \oplus (\beta \otimes \boldsymbol{v}_j) \in \boldsymbol{\mathcal{V}}.$$

For instance,

$$1, 2 \in F \text{ and } (1\ 1\ 1\ ), (2\ 2\ 2) \in \boldsymbol{\mathcal{V}} \rightarrow (1 \otimes (1\ 1\ 1\ )) \oplus (2 \otimes (2\ 2\ 2)) = (2\ 2\ 2) \in \boldsymbol{\mathcal{V}}.$$

3. For every $\alpha$, $\beta \in F$ and $\boldsymbol{v}_i \in \boldsymbol{\mathcal{V}}$, we have

$$(\ \alpha \oplus \beta) \otimes \boldsymbol{v}_i = (\alpha \otimes \boldsymbol{v}_i) \oplus (\beta \otimes \boldsymbol{v}_i).$$

For instance,

$$1, 2 \in F \text{ and } (1\ 1\ 1\ ) \in \boldsymbol{\mathcal{V}} \rightarrow (1 \oplus 2) \otimes (1\ 1\ 1\ ) = (1 \otimes (1\ 1\ 1\ )) \oplus (2 \otimes (1\ 1\ 1\ )).$$

4. For every $\alpha \in F$ and $\boldsymbol{v}_i, \boldsymbol{v}_j \in \boldsymbol{\mathcal{V}}$, we have

$$\alpha \otimes (\boldsymbol{v}_i \oplus \boldsymbol{v}_j) = (\alpha \otimes \boldsymbol{v}_i) \oplus (\alpha \otimes \boldsymbol{v}_j).$$

For instance,

$$\begin{aligned} 2 \in F \text{ and } (1\ 1\ 1\ ), (2\ 2\ 2) \in \boldsymbol{\mathcal{V}} &\rightarrow 2 \otimes ((1\ 1\ 1\ ) \oplus (2\ 2\ 2)) \\ &= (2 \otimes (1\ 1\ 1\ )) \oplus (2 \otimes (2\ 2\ 2)). \end{aligned}$$

5. The operation $\otimes$ has associative property, i.e.,

$$(\alpha \otimes \beta) \otimes \boldsymbol{v}_i = \alpha \otimes (\beta \otimes \boldsymbol{v}_i).$$

For instance,

$$1, 2 \in F \text{ and } (2\ 2\ 2\ ) \in \boldsymbol{\mathcal{V}} \rightarrow (1 \otimes 2) \otimes (2\ 2\ 2\ ) = 1 \otimes (2 \otimes (2\ 2\ 2\ )).$$

Since all the vector space properties are satisfied by $\boldsymbol{\mathcal{V}}$, we can say that the set of vectors $\boldsymbol{\mathcal{V}}$ is a vector space.

***N*-Tuple**
A number vector consisting of $N$ elements is also called $N$-tuple.

### 1.3.2 Subspace

Let $\mathcal{V}$ be a vector space defined on a scalar field ($F$, $\oplus$, $\otimes$). A subset of $\mathcal{V}$ denoted by $\boldsymbol{W}$ is called a subspace if the condition

$$\forall\alpha,\beta \in F \text{ and } \boldsymbol{w}_i, \boldsymbol{w}_j \in \boldsymbol{W} \rightarrow (\alpha \otimes \boldsymbol{w}_i) \oplus \left(\beta \otimes \boldsymbol{w}_j\right) \in \boldsymbol{W} \tag{1.26}$$

is satisfied for $\forall \boldsymbol{w}_i$, $\boldsymbol{w}_j \in \boldsymbol{W}$.

**Example 1.29** The field($F$, $\oplus$, $\otimes$) is given as

$$F = \{0, 1, 2\}$$

$$\oplus \rightarrow \text{Mod} - 3 \text{ addition}$$

$$\otimes \rightarrow \text{Mod} - 3 \text{ multiplication.}$$

Using the field elements, we construct vectors, and using these vectors, we form a vector set as in

$$\mathcal{V} = [(0\,0\,0)\ (1\,0\,2)\ (2\,1\,1\,)\ (2\,0\,1)\ (1\,2\,2\,)\ (0\,1\,0)\ (2\,2\,1)\,].$$

It can be shown that $\mathcal{V}$ is a vector space.
A subset of $\mathcal{V}$ is given as

$$\boldsymbol{W} = [(0\,0\,0)\ \ (2\,1\,1\,)\ \ (1\,2\,2\,)].$$

Determine whether $\boldsymbol{W}$ is a subspace or not.

**Solution 1.29** For any two elements $\boldsymbol{w}_i$, $\boldsymbol{w}_j$ of $\boldsymbol{W}$, it can be shown that

$$(\alpha \otimes \boldsymbol{w}_i) \oplus \left(\beta \otimes \boldsymbol{w}_j\right) \in \boldsymbol{W}.$$

For instance,

$$\boldsymbol{w}_1 = (2\,1\,1\,)\ \ \boldsymbol{w}_2 = (1\,2\,2\,) \rightarrow (1 \otimes \boldsymbol{w}_1) \oplus (2 \otimes \boldsymbol{w}_2) = (1\,2\,2) \in \boldsymbol{W}.$$

Thus, we can say that $\boldsymbol{W}$ is a subspace.

**Example 1.30** The field ($F$, $\oplus$, $\otimes$) is given as

$$F = \{0, 1\}$$

$$\oplus \rightarrow \text{Mod} - 2 \text{ addition}$$

$$\otimes \rightarrow \text{Mod} - 2 \text{ multiplication.}$$

Using the field elements, we construct vectors, and using these vectors, we form a vector set as in

$$\mathcal{V} = [(0\,0\,0)\ \ (1\,0\,0)\ \ (0\,1\,0\,)\ \ (0\,0\,1)\ \ (1\,1\,0\,)\ \ (1\,1\,1)\ \ (0\,1\,1)\ \ (1\,0\,1)].$$

It can be shown that $\mathcal{V}$ is a vector space.

A subset of $\mathcal{V}$ is given as

$$\boldsymbol{W} = [(0\,0\,0)\ \ (1\,0\,0)\ \ (0\,1\,0\,)\ \ (1\,1\,0\,)].$$

Determine whether $\boldsymbol{W}$ is a subspace or not.

**Solution 1.30** For any two elements $\boldsymbol{w}_i$, $\boldsymbol{w}_j$ of $\boldsymbol{W}$, it can be shown that

$$(\alpha \otimes \boldsymbol{w}_i) \oplus (\beta \otimes \boldsymbol{w}_j) \in \boldsymbol{W}.$$

For instance,

$$\boldsymbol{w}_1 = (1\,1\,0\,)\ \ \boldsymbol{w}_2 = (1\,0\,0\,) \rightarrow (1 \otimes \boldsymbol{w}_1) \oplus (1 \otimes \boldsymbol{w}_2) = (0\,1\,0) \in \boldsymbol{W}.$$

Thus, we can say that $\boldsymbol{W}$ is a subspace.

### 1.3.3 Dual Space

Let $\boldsymbol{W}$ be a subspace of the vector space $\mathcal{V}$. The dual space of $\boldsymbol{W}$ denoted by $\boldsymbol{W_d}$ is a subspace of $\mathcal{V}$ such that

$$\text{for} \forall \boldsymbol{w}_i \in \boldsymbol{W} \text{ and} \forall \boldsymbol{w}_j \in \boldsymbol{W_d} \text{ we have } \boldsymbol{w}_i \otimes \boldsymbol{w}_j = \boldsymbol{0} \tag{1.27}$$

where $\boldsymbol{0}$ is the zero vector defined as

$$\boldsymbol{0} = [0\,0\ldots0]_{1\times \mathrm{N}}.$$

### 1.3.4 Linear Combinations

Let $\mathcal{V}$ be a vector space defined on a scalar field $(F, \oplus, \otimes)$ and $\boldsymbol{v}_1, \boldsymbol{v}_2, \ldots, \boldsymbol{v}_k$ be a set of vectors in $\mathcal{V}$. Let $a_1, a_2, \ldots, a_k$ be some scalars in the field $F$. The linear combination of the vectors $\boldsymbol{v}_1, \boldsymbol{v}_2, \ldots, \boldsymbol{v}_k$ is calculated using

$$(a_1 \otimes \boldsymbol{v}_1) \oplus (a_2 \otimes \boldsymbol{v}_2) \oplus, \ldots, \oplus (a_k \otimes \boldsymbol{v}_k). \tag{1.28}$$

Linear combination can also be represented via matrix multiplication using either

$$[a_1\ a_2 \ldots a_k] \otimes \begin{bmatrix} \boldsymbol{v}_1 \\ \boldsymbol{v}_2 \\ \vdots \\ \boldsymbol{v}_k \end{bmatrix} \tag{1.29}$$

or

$$[\boldsymbol{v}_1\ \boldsymbol{v}_2 \ldots \boldsymbol{v}_k] \otimes \begin{bmatrix} a_1 \\ a_2 \\ \vdots \\ a_k \end{bmatrix}. \tag{1.30}$$

**Linear Independence and Dependence**

Let $\mathcal{V}$ be a vector space defined on a scalar field $(F, \oplus, \otimes)$ and $\boldsymbol{v}_1, \boldsymbol{v}_2, \ldots, \boldsymbol{v}_k$ be a set of vectors in $\mathcal{V}$. Let $a_1, a_2, \ldots, a_k$ be some scalars in the field $F$.

If

$$(a_1 \otimes \boldsymbol{v}_1) \oplus (a_2 \otimes \boldsymbol{v}_2) \oplus, \ldots, \oplus (a_k \otimes \boldsymbol{v}_k) = \boldsymbol{0} \tag{1.31}$$

is satisfied for only $a_1 = a_2 = \ldots = a_k = 0$, then the vectors $\boldsymbol{v}_1, \boldsymbol{v}_2, \ldots, \boldsymbol{v}_k$ are said to be linearly independent of each other.

On the other hand, if there exists a nonzero $a_i$ value for Eq. (1.31), then the vectors $\boldsymbol{v}_1, \boldsymbol{v}_2, \ldots, \boldsymbol{v}_k$ are said to be linearly dependent vectors.

**Example 1.31** The field $(F, \oplus, \otimes)$ is given as

$$F = \{\text{real numbers}\}$$
$$\oplus \rightarrow \text{Mod} - 10 \text{ addition}$$
$$\otimes \rightarrow \text{Mod} - 10 \text{ multiplication}.$$

It can be shown that the set of real number vectors, each having three numbers, is a vector space. Some sets of vectors from the vector space are given as

(a) $\boldsymbol{v}_1 = [3\ 8\ 5]\ \ \boldsymbol{v}_2 = [-2\ 2\ 12]\ \ \boldsymbol{v}_3 = [1\ 10\ 17]$
(b) $\boldsymbol{v}_1 = [-2\ 0\ 0]\ \ \boldsymbol{v}_2 = [1\ 7\ 0]\ \ \boldsymbol{v}_3 = [2\ 5\ 3]$

Determine whether the set of vectors given in (a) and (b) are linearly independent or not.

**Solution 1.31** For part (a),

$$(a_1 \otimes \boldsymbol{v}_1) \oplus (a_2 \otimes \boldsymbol{v}_2) \oplus (a_3 \otimes \boldsymbol{v}_3) = \boldsymbol{0}$$

can be satisfied for only $a_1 = a_2 = 1$ and $a_3 = -1$. This means that $\boldsymbol{v}_3$ can be written as the linear combination of $\boldsymbol{v}_1$ and $\boldsymbol{v}_2$, i.e., we have

$$\boldsymbol{v}_3 = \boldsymbol{v}_1 + \boldsymbol{v}_2.$$

Hence, the vectors of part (a) are linearly dependent vectors.

For part (b),

$$(a_1 \otimes \boldsymbol{v}_1) \oplus (a_2 \otimes \boldsymbol{v}_2) \oplus (a_3 \otimes \boldsymbol{v}_3) = \boldsymbol{0}$$

is satisfied for only $a_1 = a_2 = a_3 = 0$. This also means that one vector cannot be written as the linear combination of the other vectors. Hence, the vectors of part (a) are linearly independent.

### 1.3.5 Basis Vectors

In a vector space, there exist a number of linearly independent vectors from which all other vectors can be generated. These linearly independent vectors are called basis vectors, and the set of basis vectors is referred to as basis. Each vector in the vector space can be expressed as the linear combination of basis vectors.

***Span***
The basis vectors are said to span the vector space.

***Dimension***
The number of vectors in the basis is called the dimension of the vector space.

**Subspace Formation**
If we have the basis vectors of a vector space, then using a subset of the basis vectors, we can generate the elements of a subspace via linear combination of the subset of the basis vectors. In fact, the subset of the basis vectors is the basis of a subspace.

**Standard Basis**
In an $m$-dimensional vector space, the vectors

$$\boldsymbol{b}_1 = [1 \ldots 0] \quad \boldsymbol{b}_2 = [0\ 1 \ldots 0] \cdots \boldsymbol{b}_m = [0 \ldots 1] \tag{1.32}$$

are referred to as the standard basis vectors.

**Number of Bases**

For an $m$-dimensional vector space, we can generate other bases using the standard basis. This can be achieved keeping the linear independence properties of the vectors in the basis.

**Example 1.32** The standard basis of a vector space $\mathcal{V}$ is given as

$$\boldsymbol{B} = [\boldsymbol{b}_1 \;\; \boldsymbol{b}_2 \;\; \boldsymbol{b}_3]$$

where we have

$$\boldsymbol{b}_1 = [1\,0\,0] \;\; \boldsymbol{b}_2 = [0\,1\,0] \;\; \boldsymbol{b}_3 = [0\,0\,1].$$

Using standard basis vectors, we can generate another basis as

$$\boldsymbol{B}_1 = [\boldsymbol{b}_1 + \boldsymbol{b}_2 \;\; \boldsymbol{b}_2 \;\; \boldsymbol{b}_3 + \boldsymbol{b}_2]$$

leading to

$$\boldsymbol{B}_1 = [110 \quad 010 \quad 011].$$

Both $\boldsymbol{B}$ and $\boldsymbol{B}_1$ can generate all the elements of vector space $\mathcal{V}$ which contains eight vectors.

Assume that a vector space is constructed from a field consisting of $q$ numbers, and the vectors contain $n$ numbers. The total number of bases for this vector space can be calculated using

$$\frac{1}{n!}(q^n - 1)(q^n - q)\ldots\left(q^n - q^{n-1}\right). \tag{1.33}$$

For our example, we can calculate the total number of bases as

$$\frac{1}{3!}\left(2^3 - 1\right)\left(2^3 - 2\right)\left(2^3 - 2^2\right) \rightarrow 28.$$

**Orthogonality**

If the dot product of two vectors is zero, then these vectors are said to be orthogonal to each other, i.e., $\boldsymbol{v}_i$ is orthogonal to $\boldsymbol{v}_j$ if we have

$$\boldsymbol{v}_i \cdot \boldsymbol{v}_j = \boldsymbol{0}. \tag{1.34}$$

### 1.3.6 Matrices Obtained from Basis Vectors

We can form a matrix by placing the basis vectors of a vector space as the rows or columns of the matrix. If the basis vectors are used for the rows of the matrix, then the row span of the matrix generates the vector space. On the other hand, if the basis vectors are used as the columns of the matrix, then the column space of the matrix generates the vector space. The number of basis vectors used as the rows or columns of the matrix is referred to as the rank of the matrix.

**Elementary Row Operations**
For a matrix of size $N \times M$, the elementary row operations are outlined as follows:

1. Interchanging any two rows
2. Multiplying a row by a nonzero scalar
3. Adding a row to another row

Elementary row operations do not affect the row space of the matrix.

**Example 1.33** For the given matrix

$$A = \begin{bmatrix} 1 & 0 & 1 & 1 & 0 \\ 0 & 1 & 0 & 1 & 1 \\ 0 & 1 & 1 & 0 & 0 \end{bmatrix}$$

we can perform some elementary row operations as in

$$R2 \leftrightarrow R3 \quad \rightarrow \quad A_1 = \begin{bmatrix} 1 & 0 & 1 & 1 & 0 \\ 0 & 1 & 1 & 0 & 0 \\ 0 & 1 & 0 & 1 & 1 \end{bmatrix}$$

$$R1 \leftarrow R1 + R3 \quad \rightarrow \quad A_2 = \begin{bmatrix} 1 & 1 & 1 & 0 & 1 \\ 0 & 1 & 1 & 0 & 0 \\ 0 & 1 & 0 & 1 & 1 \end{bmatrix}$$

where "$\leftrightarrow$" refers to exchange of two rows and "$\leftarrow$" refers to the assignment of the right argument to the left one.

**Theorem** Let $(F, \oplus, \otimes)$ be a finite field, and assume that that there are $q$ elements in $F$. If we construct an $n$-dimensional vector space $\mathcal{V}$ using field elements, then the number of vectors in the vector space happens to be $q^n$ which is denoted by $|\mathcal{V}|$.

### 1.3.7 Polynomial Groups and Fields

Groups and fields are some sets satisfying a number of properties under some defined operations. We can construct polynomial sets satisfying the properties of a group or field. Let's illustrate the concept by examples.

**Example 1.34** The field $(F, +, \cdot)$ is given as

$$F = \{0, 1\}$$
$$+ \rightarrow \text{Mod} - 2 \text{ addition}$$
$$\cdot \rightarrow \text{Mod} - 2 \text{ multiplication.}$$

Using field elements for coefficients, we can construct the polynomials of the form

$$\alpha, \beta \in F, \alpha x + \beta,$$

and from these polynomials, we can obtain a polynomial set as in

$$S = \{0, 1, x, x + 1\}.$$

(a) Show that $(S, +)$ is a group.
(b) Find the inverse of the elements $x$ and $x + 1$.

**Solution 1.34** (a) Using the elements of $S$ and mod-2 addition operation, we can construct addition Table 1.7.

If the Table 1.7 is inspected, we see that $(S, +)$ is a group.

(b) Using Table 1.7, we can determine the inverses of $x$ and $x + 1$ as $x$ and $x + 1$, respectively.

**Example 1.35** Consider the polynomial set of the previous example again. For the elements of $S$, we define two operations $\oplus$, $\otimes$ as

**Table 1.7** Mod-2 polynomial addition

| $\oplus$ | $0$ | $1$ | $x$ | $x+1$ |
|---|---|---|---|---|
| $0$ | $0$ | $1$ | $x$ | $x+1$ |
| $1$ | $1$ | $0$ | $x+1$ | $x$ |
| $x$ | $x$ | $x+1$ | $0$ | $1$ |
| $x+1$ | $x+1$ | $x$ | $1$ | $0$ |

**Table 1.8** Polynomial tables for ⊕, ⊗ operations

| $\oplus$ | 0 | 1 | $x$ | $x+1$ |
|---|---|---|---|---|
| 0 | 0 | 1 | $x$ | $x+1$ |
| 1 | 1 | 0 | $x+1$ | $x$ |
| $x$ | $x$ | $x+1$ | 0 | 1 |
| $x+1$ | $x+1$ | $x$ | 1 | 0 |

| $\otimes$ | 0 | 1 | $x$ | $x+1$ |
|---|---|---|---|---|
| 0 | 0 | 0 | 0 | 0 |
| 1 | 0 | 1 | $x$ | $x+1$ |
| $x$ | 0 | $x$ | $x+1$ | 1 |
| $x+1$ | 0 | $x+1$ | 1 | x |

$$p(x) \oplus q(x) \rightarrow R_{x^2+x+1}(p(x) + q(x))$$
$$p(x) \otimes q(x) \rightarrow R_{x^2+x+1}(p(x)q(x))$$

where $R_{x^2+x+1}(\cdot)$ means the remainder polynomial after division of the input argument by $x^2 + x + 1$.

Determine whether $(S, \oplus, \otimes)$ is a field or not.

**Solution 1.35** Using the elements of $S$, we can construct tables for defined $\oplus$, $\otimes$ operations as in Table 1.8.

When Table 1.8 is inspected, we see that

1. $(S, \oplus)$ is a commutative group.
2. $(S - \{0\}, \otimes)$ is a commutative group.
3. $\otimes$ has distributive property over $\oplus$.

Hence, we can say that $S$ is a field under the defined operations $\oplus$ and $\otimes$.

## 1.4 Ring

Let $\boldsymbol{R}$ be a set and assume that two operations denoted by $\oplus$ and $\otimes$ are defined on the set elements. For the set $\boldsymbol{R}$ to be a ring under $\oplus$ and $\otimes$ operations, the following properties are to be satisfied:

1. $\boldsymbol{R}$ is a commutative group under $\oplus$ operation.
2. $\otimes$ operation has associative property. That is, if $a, b, c \in \boldsymbol{R}$, then we have

$$a \otimes (b \otimes c) = (a \otimes b) \otimes c. \tag{1.35}$$

3. $\otimes$ operation has left and right distribution properties over $\oplus$. That is, if $a, b, c \in \boldsymbol{R}$, then we have

$$
\begin{aligned}
a \otimes (b \oplus c) &= (a \otimes b) \oplus (a \otimes c) \\
(a \oplus b) \otimes c &= (a \otimes c) \oplus (b \otimes c).
\end{aligned}
\tag{1.36}
$$

In addition, if the commutative property also holds, then the ring is said to be a commutative ring. If the ring has identity element for $\otimes$ operation, then the ring is said to be an identity ring, and it is typically denoted by $\boldsymbol{I_R}$.

Note that for a set to be a ring, we do not require that $\boldsymbol{R}$ is a commutative group under $\otimes$ operation. Inverse and commutative properties may not even hold for $\otimes$ operation.

**Example 1.36** The set of $4 \times 4$ matrices under ordinary addition and multiplication operations form a ring. This ring does not have commutative and inverse properties.

## Problems

1. The finite set $G$ and the operation $\oplus$ are given as

$$
G = \{0, 1, 2, 3\}
$$
$$
\oplus \rightarrow \text{Mod} - 4 \text{ addititon operation.}
$$

Determine whether $G$ is a group under the defined operation $\oplus$ or not.

2. The finite set $G$ and the operation $\otimes$ are given as

$$
G = \{0, 1, 2, 3, 4\}
$$
$$
\otimes \rightarrow \text{Mod} - 5 \text{ multiplication operation.}
$$

Determine whether $G$ is a group under the defined operation $\otimes$ or not.

3. The finite set $G$ and the operation $\otimes$ are given as

$$
G = \{1, 2, 3, 4\}
$$

$$\otimes \rightarrow \text{Mod} - 5 \text{ multiplication operation.}$$

Show that $G$ is a group under Mod-5 multiplication operation. Find a subgroup of $G$, and determine all the cosets using the subgroup.

4. The set of integers $Z$ under addition operation + is a group. A subgroup of $Z$ can be formed as

$$H = 6Z \rightarrow H = \{\ldots - 12, -6, 0, 6, 12, \ldots\}.$$

Find the cosets of $H$.

5. The finite set $S$ and the operations $\oplus$ and $\otimes$ are given as

$$S = \{0, 1, 2, 3, 4, 5, 6, 7, 8, 9, 10\}$$
$$\oplus \rightarrow \text{Mod} - 11 \text{ addition operation}$$
$$\otimes \rightarrow \text{Mod} - 11 \text{ multiplication operation.}$$

Determine whether $S$ is a field or not under the defined operations $\oplus$ and $\otimes$.

6. The field($F$, $\oplus$ , $\otimes$) is given as

$$F = \{0, 1\}$$
$$\oplus \rightarrow \text{Mod} - 2 \text{ addition}$$
$$\otimes \rightarrow \text{Mod} - 2 \text{ multiplication.}$$

Using the field elements, we construct vectors, and using these vectors, we form a vector set as in

$$\mathcal{V} = [0000 \;\; 1011 \;\; 0101 \;\; 1110\,].$$

Determine whether $\mathcal{V}$ is a vector space or not.

7. The dimension of a vector space constructed using the elements of the binary field is 4. Write all the elements of the vector space, and find three different bases of this vector space.
8. The dimension of a vector space constructed using the elements of the binary field is 5. Find a basis of this vector space other than the standard basis, and using the basis, determine a subspace of the vector space.

9. Using the elements of the prime field $F_3 = \{0, 1, 2\}$, construct a basis of a vector space whose dimension is 3, and determine all the elements of the vector space.
10. The basis of a vector space constructed using the binary field is given as

$$\boldsymbol{B} = [00101\ 01010\ 10010\ 10101].$$

Find the elements of the vector space, and find a subspace of this vector space.

11. Explain the difference between linear independence and orthogonality.

# Chapter 2
# Linear Block Codes

## 2.1 Binary Linear Block Codes

In this chapter, we will only inspect the binary block codes, and for the simplicity of writing, we will use the term "linear block codes" for the place of "binary linear block codes" otherwise indicated. Now, let's give the definition of binary linear block codes, i.e., linear block codes.

**Definition** The binary field $F$ is given as

$$F = \{0, 1\}$$

$$\oplus \rightarrow \text{Mod} - 2\,\text{addition operation}$$

$$\otimes \rightarrow \text{Mod} - 2\,\text{multiplication operation.}$$

Let $\mathcal{V}$ be an $n$-dimensional vector space constructed using the elements of $F$ and $\boldsymbol{C}$ be $k$-dimensional subspace of the vector space $\mathcal{V}$. The subspace $\boldsymbol{C}$ is called a linear block code, and the elements of $\boldsymbol{C}$ are denoted as code-words.

Note that from now on while considering the vector spaces, we will assume that the vector spaces are constructed using the binary field. We will not explicitly mention that the vector spaces are constructed using the binary field.

**Example 2.1** A vector space of dimension 3 has the standard basis

$$\boldsymbol{B} = \{001, 010, 100\}.$$

Find the elements of vector space $\mathcal{V}$ generated by $\boldsymbol{B}$. How many elements do $\mathcal{V}$ have?

Find a subspace, i.e., code, using the generated vector space.

**Solution 2.1** The basis vectors span the vector space. In other words, the elements of the vector space can be obtained by taking the linear combinations of the basis



O. Gazi, *Forward Error Correction via Channel Coding*,
https://doi.org/10.1007/978-3-030-33380-5_2

vectors. Since there are 3 elements in the basis, it is possible to generate $2^3 = 8$ vectors by taking the linear combinations of the basis vectors. Thus, the vector space generated by $\boldsymbol{B}$ has 8 vectors. The elements of the vector space can be generated as in

$$
\begin{aligned}
0 \times 001 &\rightarrow 000\\
1 \times 001 &\rightarrow 001\\
1 \times 010 &\rightarrow 010\\
1 \times 100 &\rightarrow 100\\
1 \times 001 + 1 \times 010 &\rightarrow 011\\
1 \times 001 + 1 \times 100 &\rightarrow 101\\
1 \times 010 + 1 \times 100 &\rightarrow 110\\
1 \times 001 + 1 \times 010 + 1 \times 100 &\rightarrow 111.
\end{aligned}
$$

Writing the generated vectors as the elements of a set, we get

$$\mathcal{V} = \{000, 001, 010, 100, 011, 101, 110, 111\}$$

which is a vector space.

To find a subspace of the vector space, we can first find a subset of the basis $\boldsymbol{B}$, and then, by taking the linear combinations of the vectors in this subset of the basis, we can generate the elements of the subspace.

We can choose a subset of $\boldsymbol{B}$ as

$$\boldsymbol{B}_s = \{001, 010\}.$$

By taking the linear combinations of the elements of $\boldsymbol{B}_s$, we can obtain the generated vectors, i.e., code-words, as

$$
\begin{aligned}
0 \times 001 &\rightarrow 000\\
1 \times 001 &\rightarrow 001\\
1 \times 010 &\rightarrow 010\\
1 \times 001 + 1 \times 010 &\rightarrow 011.
\end{aligned}
$$

Writing the generated vectors, i.e., code-words, as the elements of a set, we get

$$\boldsymbol{C} = \{000, 001, 010, 011\}$$

which is a linear block code of dimension 2.

Note that for a vector space there are more than a single basis, and considering the availability of many bases, we can construct a subspace in many different ways. For instance, assume that a vector space has 20 bases; then, the basis of a subspace can be formed considering many different subsets from these 20 bases.

**Example 2.2** The vector space of dimension 6 has the standard basis

$$\boldsymbol{B} = \{000001, 000010, 000100, 001000, 010000, 1000000\}.$$

Find the elements of vector space $\mathcal{V}$ generated by $\boldsymbol{B}$. How many elements do $\mathcal{V}$ have?

Find a subspace, i.e., code, using the generated vector space.

**Solution 2.2** The basis vectors span the vector space. In other words, the elements of the vector space can be obtained by taking the linear combinations of the basis vectors. Since there are 6 elements in the basis, it is possible to generate $2^6 = 64$ vectors by taking the linear combinations of the basis vectors. Thus, the vector space generated by $\boldsymbol{B}$ has 64 vectors. The generated vector space can be shown as

$$\mathcal{V} = \{000000, 000001, \ldots 011111, 111111\}.$$

To find a subspace of the vector space, we can first find a subset of the basis $\boldsymbol{B}$, and then, by taking the linear combinations of the vectors in this subset of the basis, we can generate the elements of the subspace.

We can choose a subset of $\boldsymbol{B}$ as

$$\boldsymbol{B}_s = \{100000, 010000, 000100\}.$$

By taking the linear combinations of the elements of $\boldsymbol{B}_s$, we can construct the linear code as in

$$\boldsymbol{C} = \{000000, 100000, 010000, 000100, 110000, 100100, 010100, 110100\}.$$

For the simplicity of illustration, we used the standard basis for the vector space of this example and used a subset of it to generate the elements of a subspace, i.e., code. However, note that there are many bases of the vector space under concern, and many subsets of them can be obtained for the basis of the subspace and many codes can be generated with the same dimension.

**Example 2.3** A linear code should always include the all-zero code-word. Is this statement correct or not? If it is correct, explain the reasoning behind it.

**Solution 2.3** A linear code is a vector subspace, and a vector subspace satisfies all the properties of a vector space. We know that a vector space is a vector set $\mathcal{V}$, and this vector set is a group under + operation, i.e., $(\mathcal{V}, +)$ is a group. This means that $\mathcal{V}$ has an identity element under + operation, and this identity element is the all-zero vector. Thus, a linear code always includes the all-zero vector as one of its code-words.

**Example 2.4** A linear code $\boldsymbol{C}$ is given as

$$\boldsymbol{C} = \{0000, 0001, 0010, 0011\}.$$

(a) What is the dimension of the vector space from which the linear code is generated?
(b) What is the dimension of the linear code?

**Solution 2.4**

(a) The number of elements, i.e., bits, in the code-words indicates the dimension of the vector space. Since there are four bits in the code-words, the dimension of the vector space is 4.
(b) If the dimension of the linear code is $k$, then there are $2^k$ code-words in the code. Since there are four code-words in the code, the dimension of the code can be found as

$$2^k = 4 \rightarrow k = 2.$$

## 2.2 Generator Matrix of a Linear Code

We can construct the generator matrix of a linear code using the basis vectors of the subspace from which the linear code is formed.

Let $\mathcal{V}$ be a vector space of dimension $n$ and $\boldsymbol{C}$ be a linear code, i.e., a subspace, whose basis is given as

$$\boldsymbol{B}_s = [\boldsymbol{b}_1, \boldsymbol{b}_2, \ldots, \boldsymbol{b}_k] \tag{2.1}$$

where row vectors $\boldsymbol{b}_i$, $i = 1 \ldots k$ are the basis vectors, and $k$ is the dimension of the subspace.

The span of the basis vectors is the linear code $\boldsymbol{C}$. Let's write the elements of $\boldsymbol{B}_s$ as the rows of a matrix, and let's denote this matrix by $\boldsymbol{G}$; then, $\boldsymbol{G}$ happens to be as

$$\boldsymbol{G} = \begin{bmatrix} \boldsymbol{b}_1 \\ \boldsymbol{b}_2 \\ \vdots \\ \boldsymbol{b}_k \end{bmatrix}_{k \times n} \tag{2.2}$$

where $k$ is the dimension of the linear code, and $n$ is the dimension of the vector space. The matrix $\boldsymbol{G}$ is called the generator matrix of the linear code. If we use the notation $\boldsymbol{g}_i$ for the place of $\boldsymbol{b}_i$, the generator matrix becomes as

$$\boldsymbol{G} = \begin{bmatrix} g_1 \\ g_2 \\ \vdots \\ g_k \end{bmatrix}_{k \times n}. \tag{2.3}$$

**Example 2.5** The dimension of a vector space is $n = 4$. Design a linear code with dimension $k = 2$, and find the generator matrix of the designed code.

**Solution 2.5** The standard basis of the vector space with dimension 4 can be written as

$$\boldsymbol{B} = \{0001, 0010, 0100, 1000\}.$$

The span of basis vectors in $\boldsymbol{B}$ gives us the vector space which can be written as

$$\mathcal{V} = \{0000, 0001, 0010, \ldots, 1111\}.$$

It is clear that $\mathcal{V}$ includes all the 4-tuples, i.e., all the binary vectors with 4 elements. To construct a linear code from the given vector space, first, we select a subset of the basis $\boldsymbol{B}$ as

$$\boldsymbol{B}_s = \{0010, 0100\}.$$

If we write the elements of $\boldsymbol{B}_s$ as the rows of a matrix, we obtain the generator matrix

$$\boldsymbol{G} = \begin{bmatrix} 0010 \\ 0100 \end{bmatrix}_{2 \times 4}$$

where the size of the matrix $2 \times 4$ gives information about the dimension of the code and vector space from which the code is formed.

Once we have the generator matrix of a linear code, we can generate the codewords by taking the span of the rows of the generator matrix. Considering this information, we can generate the linear code as in

$$\boldsymbol{C} = \{0000, 0010, 0100, 0110\}.$$

If we choose the basis other than the standard basis as

$$B = \{1001, 0110, 0101, 1000\},$$

select a subset of the basis $\boldsymbol{B}$

$$B_s = \{0110, 0101\},$$

and wring the elements of $\boldsymbol{B_s}$ as the rows of a matrix, we obtain the generator matrix

$$G = \begin{bmatrix} 0110 \\ 0101 \end{bmatrix}_{2\times 4}$$

**Exercise** Consider the vector space with the basis

$$B = \{10001, 00110, 00100, 11001, 10010\}.$$

Construct a linear code with dimension $k = 3$ from this vector space.

## 2.3 Hamming Weight of a Code-Word

Let $\boldsymbol{C}$ be a binary code and $\boldsymbol{c_i} \in \boldsymbol{C}$ be a code-word. If the code-word $\boldsymbol{c_i}$ is defined as

$$\boldsymbol{c_i} = \begin{bmatrix} c_{i1} & c_{i2} \ldots c_{in} \end{bmatrix} \text{ where } c_{im} \in F = \{0, 1\} \text{ and } m = 1, \ldots, n \tag{2.4}$$

then the Hamming weight of the code-word $\boldsymbol{c_i}$ is calculated using

$$d_H(\boldsymbol{c_i}) = \sum_{m=1}^{n} c_{im}. \tag{2.5}$$

That is, the number of ones in the code-word is called the Hamming weight of the code-word.

**Example 2.6** A binary code, i.e., a binary linear block code, is given as

$$C = \{00000, 10011, 01000, 11011\}.$$

Find the Hamming weight of each code-word for this code.

**Solution 2.6** Let's denote the code-words available in this code as

$$\boldsymbol{c_1} = [00000] \qquad \boldsymbol{c_2} = [10011] \qquad \boldsymbol{c_3} = [01000] \qquad \boldsymbol{c_4} = [11011].$$

According to the definition given in Eq. (2.5), we can calculate the Hamming weight of each code-word as

$$d_H(\boldsymbol{c}_1) = 0 \quad d_H(\boldsymbol{c}_2) = 3 \quad d_H(\boldsymbol{c}_3) = 1 \quad d_H(\boldsymbol{c}_4) = 4.$$

### 2.3.1 Hamming Distance

Let $\boldsymbol{C}$ be a binary code and $\boldsymbol{c}_i, \boldsymbol{c}_j \in \boldsymbol{C}$ be two code-words. If the code-words $\boldsymbol{c}_i, \boldsymbol{c}_j$ are defined as

$$\boldsymbol{c}_i = \left[c_{i1} \;\; c_{i2} \ldots c_{in}\right] \text{ where } c_{im} \in F = \{0, 1\} \text{ and } m = 1, \ldots, n$$
$$\boldsymbol{c}_j = \left[c_{j1} \;\; c_{j2} \ldots c_{jn}\right] \text{ where } c_{jm} \in F = \{0, 1\} \text{ and } m = 1, \ldots, n$$

then the Hamming distance between the code-words $\boldsymbol{c}_i$ and $\boldsymbol{c}_j$ is calculated using

$$d\left(\boldsymbol{c}_i, \boldsymbol{c}_j\right) = \sum_{m=1}^{n} \left(c_{im} + c_{jm}\right). \tag{2.6}$$

where + is the **mod-2** addition operation. Equation (2.6) can also be written as

$$d\left(\boldsymbol{c}_i, \boldsymbol{c}_j\right) = \text{sum}\left(\boldsymbol{c}_i + \boldsymbol{c}_j\right). \tag{2.7}$$

That is, Hamming distance is the number of positions in which $\boldsymbol{c}_i$ and $\boldsymbol{c}_j$ have different values.

Note that when the word "distance" appears, the readers usually think the "−" operation, and the distance expression can be considered as

$$d\left(\left|\boldsymbol{c}_i - \boldsymbol{c}_j\right|\right) = \sum_{m=1}^{n} \left|c_{im} - c_{jm}\right|. \tag{2.8}$$

However, our code-words are constructed using the binary field $F = \{0, 1\}$ for which mod-2 "+" and "·" operations are available. There is no "−" operation defined for the binary field elements. Thus, the formula in Eq. (2.8) has no meaning. The correct formula is the one in Eq. (2.6).

**Example 2.7** Code-words of a binary code are given as

$$\boldsymbol{c}_1 = [00000] \qquad \boldsymbol{c}_2 = [10011] \qquad \boldsymbol{c}_3 = [01000] \qquad \boldsymbol{c}_4 = [11011].$$

Find the Hamming distance between the code-words $\boldsymbol{c}_2$ and $\boldsymbol{c}_3$.

**Solution 2.7** According to the definition given in Eq. (2.6), we can calculate the Hamming distance between $\boldsymbol{c_3}$ and $\boldsymbol{c_4}$ as

$$d(\boldsymbol{c}_3, \boldsymbol{c}_4) = \sum_{m=1}^{5} \left( c_{im} + c_{jm} \right) \rightarrow$$

$$d(\boldsymbol{c}_3, \boldsymbol{c}_4) = (0+1) + (1+1) + (0+0) + (0+1) + (0+1) \rightarrow d(\boldsymbol{c}_3 + \boldsymbol{c}_4) = 3.$$

## 2.3.2 *Minimum Distance of a Linear Block Code*

Let $\boldsymbol{C}$ be a binary code and $\boldsymbol{c}_i$, $\boldsymbol{c}_j \in \boldsymbol{C}$ be two code-words. Considering all the code-word pairs $\boldsymbol{c}_i$, $\boldsymbol{c}_j$, the minimum distance of a linear block code is defined as

$$d_{\min} = \min \{ d_H(\boldsymbol{c}_i, \boldsymbol{c}_j) \; i,j = 1, \ldots, k \}, \quad i \neq j \tag{2.9}$$

which means that we consider the Hamming distance between all possible pairs $\boldsymbol{c}_i$, $\boldsymbol{c}_j$ and choose the smallest value as the minimum distance of the code.

Although Eq. (2.9) is the formal definition of the minimum distance, we can simplify it more. Using Eq. (2.7), we can write the expression in Eq. (2.9) as

$$d_{\min} = \min \{ \text{sum}(\boldsymbol{c}_i + \boldsymbol{c}_j) \; i,j = 1, \ldots, k \}, \quad i \neq j. \tag{2.10}$$

where + is the mod-2 addition operation. Since the summation of two code-words produces another code-word, the expression in Eq. (2.10) can be written as

$$d_{\min} = \min \{ \text{sum}(\boldsymbol{c}_l), l = 1, \ldots, k \}, \quad d_{\min} > 0 \tag{2.11}$$

where, substituting $d_H(\boldsymbol{c}_l)$ for sum($\boldsymbol{c}_l$), we obtain

$$d_{\min} = \min \{ d_H(\boldsymbol{c}_l), l = 1, \ldots, k \}, \quad d_{\min} > 0 \tag{2.12}$$

which means that the minimum distance of a linear block code is the minimum Hamming weight of all the code-words other than the zero code-word.

**Example 2.8** A binary code, i.e., a binary linear block code, is given as

$$\boldsymbol{C} = \{00000, 10011, 01000, 11011\}.$$

The code-words of this code are denoted as

$$\boldsymbol{c}_1 = [00000] \quad \boldsymbol{c}_2 = [10011] \quad \boldsymbol{c}_3 = [01000] \quad \boldsymbol{c}_4 = [11011].$$

The Hamming weight of each code-word can be calculated as

$$d_H(\boldsymbol{c}_1) = 0 \quad d_H(\boldsymbol{c}_2) = 3 \quad d_H(\boldsymbol{c}_3) = 1 \quad d_H(\boldsymbol{c}_4) = 4.$$

Find the minimum distance of $\boldsymbol{C}$.

**Solution 2.8** Minimum distance of a code is the minimum Hamming weight of all the code-words, i.e.,

$$d_{\min} = \min\{d_H(\boldsymbol{c}_l), l = 1, \ldots, k\}. \tag{2.13}$$

Using the given values in question, we can decide the minimum distance of the code as

$$d_{\min} = 1.$$

## 2.4 Performance Enhancement of Communication Systems and Encoding Operation

Let $\boldsymbol{C}$ be a binary code and $\boldsymbol{c}_i$, $\boldsymbol{c}_j \in \boldsymbol{C}$ be two code-words. Assume that we transmit $\boldsymbol{c}_i$, and at the receiver side, we get $\boldsymbol{c}_j$. The Hamming distance between $\boldsymbol{c}_i$ and $\boldsymbol{c}_j$ is $d_H(\boldsymbol{c}_i, \boldsymbol{c}_j)$ which indicates the number of positions by which two code-words differ.

It is clear that as the value of $d_H(\boldsymbol{c}_i, \boldsymbol{c}_j)$ increases, the probability of receiving $\boldsymbol{c}_j$ instead of $\boldsymbol{c}_i$ decreases, since more bits have to be flipped. If $d_H(\boldsymbol{c}_i, \boldsymbol{c}_j)$ has a small value, then the probability of receiving wrong code-word at the receiver side becomes larger.

Let's define a probability of error function $p_e(\boldsymbol{c}_i, \boldsymbol{c}_j)$ for receiving $\boldsymbol{c}_j$ instead of $\boldsymbol{c}_i$ at the receiver side as

$$p_e(\boldsymbol{c}_i, \boldsymbol{c}_j) = E(d(\boldsymbol{c}_i, \boldsymbol{c}_j)) \tag{2.14}$$

in which using

$$d(\boldsymbol{c}_i, \boldsymbol{c}_j) = \text{sum}(\boldsymbol{c}_i + \boldsymbol{c}_j)$$

we get

$$p_e(\boldsymbol{c}_i, \boldsymbol{c}_j) = E(\text{sum}(\boldsymbol{c}_i + \boldsymbol{c}_j)) \tag{2.15}$$

where using $\boldsymbol{c} = (\boldsymbol{c}_i + \boldsymbol{c}_j)$, we obtain

$$p_e(\boldsymbol{c}_i, \boldsymbol{c}_j) = E(\text{sum}(\boldsymbol{c})) \rightarrow p_e(\boldsymbol{c}_i, \boldsymbol{c}_j) = E(d_H(\boldsymbol{c})). \tag{2.16}$$

For the transmission through AWGN channel employing BPSK modulation, it is known that the function $E(\cdot)$ in Eq. (2.16) is the $Q(\cdot)$ function and the error expression in Eq. (2.16) is calculated using

$$p_e(\mathbf{c}_i, \mathbf{c}_j) = Q\left(\sqrt{\frac{d_H^2(c_l)}{N_0}}\right). \tag{2.17}$$

If we consider the transmission of all the code-words in our code, the code probability of error can be written as

$$P_e(\mathbf{C}) = \sum_{l=1}^{2^k} Q\left(\sqrt{\frac{d_H^2(c_l)}{N_0}}\right). \tag{2.18}$$

There may be many code-words with the same Hamming weight. Denoting $d_H^2(c_l)$ by $d_l^2$, we can write Eq. (2.18) as

$$P_e(\mathbf{C}) = \sum_l A_l Q\left(\sqrt{\frac{d_l^2}{N_0}}\right) \tag{2.19}$$

where $A_l$ indicates the number of code-words with Hamming weight $d_l$. When Eq. (2.19) is expanded, we obtain

$$P_e(\mathbf{C}) = A_1 Q\left(\sqrt{\frac{d_1^2}{N_0}}\right) + A_2 Q\left(\sqrt{\frac{d_2^2}{N_0}}\right) + \ldots \tag{2.20}$$

The function $Q(\cdot)$ is a nonlinear decreasing function, and the graph of this function is depicted in Fig. 2.1 where it is seen that as the input argument of the function gets large values, the function converges to zero.

In the summation expression of Eq. (2.20), the dominant term is the one in which $d_{\min}^2$ appears, i.e., the dominant term in the summation can be written as

$$A_{\min} Q\left(\sqrt{\frac{d_{\min}^2}{N_0}}\right). \tag{2.21}$$

We can approximate the summation expression in Eq. (2.20) considering the dominant terms as

$$P_e(\mathbf{C}) \approx A_{\min} Q\left(\sqrt{\frac{d_{\min}^2}{N_0}}\right). \tag{2.22}$$

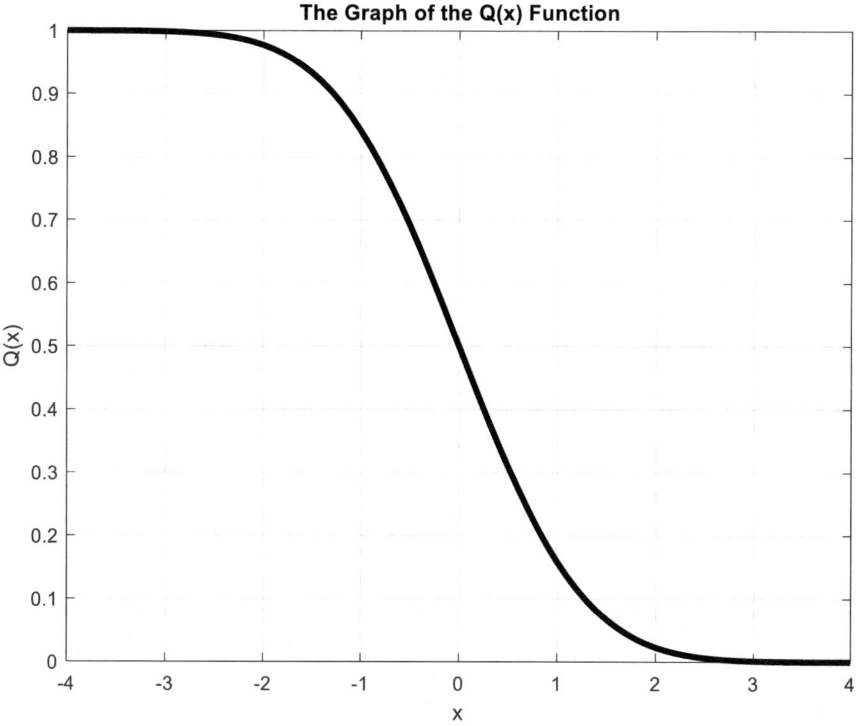


**Fig. 2.1** The graph of the $Q(\cdot)$ function

Equation (2.22) implies that if we increase the distance between vectors of a vector space, then the probability of the transmission error of the code decreases.

## 2.5 The Philosophy of Channel Encoding

Assume that we have $k$-dimensional data vector space, and we want to transmit the data available in the vector space via electronic communication devices. For this purpose, the data available in the binary vector space can be put into a matrix of size $m \times k$ where $k$ is the number of data bits in each row of the matrix, and there are $m = 2^k$ rows. The set of rows of the matrix is a vector space and the dimension of the data vector space is $k$. This vector space denoted by $\boldsymbol{W}$ can be written as

$$\boldsymbol{W} = [\boldsymbol{w}_1\, \boldsymbol{w}_2 \ldots \boldsymbol{w}_m] \quad m = 2^k \tag{2.23}$$

where

$$\boldsymbol{w_i} = \left[w_{i1}\ w_{i2} \ldots w_{ik}\right] \tag{2.24}$$

is a $k$-bit data vector to be transmitted. If we transmit the data vectors directly, the probability of the error at the receiver side approximately equals to

$$P_e(\boldsymbol{W}) \approx W_{\min} Q\left(\sqrt{\frac{w_{\min}^2}{N_0}}\right) \tag{2.25}$$

where $w_{\min}$ is the minimum Hamming weight of all the data-words, i.e., data vectors, and $W_{\min}$ is the number of data vectors with minimum Hamming weight $w_{\min}$.

Now, consider this question ("How can we decrease the transmission error?"), i.e., how can we decrease the value of $P_e(\boldsymbol{W})$ in Eq. (2.25)?"

***Answer:*** To decrease $P_e(\boldsymbol{W})$ in Eq. (2.25), we need to increase the value of $w_{\min}^2$. However, we cannot change the data to be transmitted. We should look for another solution.

Consider that there is another set of $m = 2^k$ vectors, but this vector set has a larger minimum Hamming weight. Instead of transmitting data vectors directly, we can make a mapping among the data vectors and the vectors available in other vector space and transmit the mapped vectors. And at the receiver side after demodulation operation, we can make de-mapping to recover the original data vectors.

So this solution seems to be feasible. Now we can state the encoding operation.

## 2.6 Encoding and Decoding Operations

Assume that we have a $k$-dimensional data vector space denoted by $\boldsymbol{W}$. We have $2^k$ data vectors in this vector space. Consider another vector space $\boldsymbol{\mathcal{V}}$ with dimension $n$ such that $n > k$. Obviously the vector space with dimension $n$ includes more vectors since $2^n > 2^k$.

Consider a $k$-dimensional subspace of $\boldsymbol{\mathcal{V}}$ denoted by $\boldsymbol{C}$ called code. There are $2^k$ code-words, i.e., vectors, in $\boldsymbol{C}$. However, the vectors in $\boldsymbol{C}$ include $n$-bits, whereas the vectors in $\boldsymbol{W}$ include $k$-bits although the dimensions of both $\boldsymbol{C}$ and $\boldsymbol{W}$ are equal to the same number $k$.

The mapping of the vectors from $\boldsymbol{W}$ to $\boldsymbol{C}$ is called encoding. Since the vectors in $\boldsymbol{C}$ include more bits than the vectors in $\boldsymbol{W}$, by transmitting the mapped vectors, i.e., code-words, instead of the data vectors, it may be possible to decrease the transmission error, since it is possible to have a larger minimum distance for $\boldsymbol{C}$ compared to $\boldsymbol{W}$.

Once transmission of the code-words is complete, at the receiver side, we perform de-mapping of the code-words to the data vectors. The de-mapping of the received code-words to the data vectors at the receiver side is called decoding. The graphical illustration of the encoding and decoding operations is depicted in Fig. 2.2.

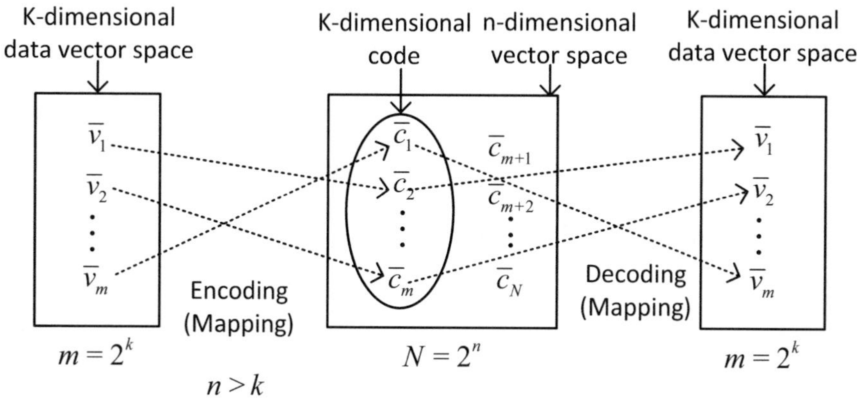


**Fig. 2.2** The graphical illustration of the encoding and decoding operations

**Example 2.9** The $k = 2$ dimensional data vector space is given as

$$\boldsymbol{W} = [00 \quad 01 \quad 10 \quad 11].$$

The data vectors of $\boldsymbol{W}$ can be denoted as

$$\boldsymbol{d}_1 = [00] \quad \boldsymbol{d}_2 = [01] \quad \boldsymbol{d}_3 = [10] \quad \boldsymbol{d}_4 = [11].$$

$n = 3$ dimensional vector space can be written as

$$\mathcal{V} = [000 \quad 001 \quad 010 \quad 011 \quad 100 \quad 101 \quad 110 \quad 111].$$

A $k = 2$ dimensional subspace of $\mathcal{V}$, i.e., a $k = 2$ dimensional code, is given as

$$\boldsymbol{C} = [000 \quad 001 \quad 100 \quad 101].$$

A mapping operation, i.e., an encoding operation, and de-mapping operation, i.e., decoding operation, among data-words and code-words are illustrated in Fig. 2.3.

**Example 2.10** The dimensions of a data vector space and a vector space are $k = 8$ and $n = 16$, respectively.

(a) How many data-words, i.e., vectors, are available in the data vector space?
(b) How many bits are available in the data vector space?
(c) How many vectors are available in the vector space?
(d) How many bits are available in the vector space?

**Solution 2.10**

(a) There are $2^k \to 2^8 = 256$ data-words available in the data vector space.
(b) There are $k \times 2^k \to 8 \times 2^8 = 2048$ bits available in the data vector space.

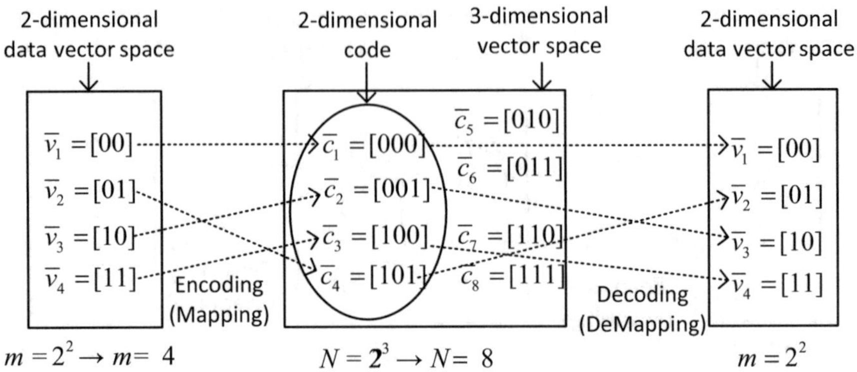


**Fig. 2.3** Example for encoding and decoding operations

(c) There are $2^n \rightarrow 2^{16} = 2^6 \times 2^{10} \rightarrow 2^{16} = 65{,}536$ vectors available in the vector space.

(d) There are $n \times 2^n \rightarrow 16 \times 2^{16} = 2^4 \times 2^{16} \rightarrow 1{,}048{,}576$ bits available in vector space.

**Example 2.11** We have two vector subspaces, i.e., codes, of dimension 1. The subspaces, i.e., codes, are given as

$$C_1 = [0000 \;\; 1010]$$

$$\boldsymbol{C}_2 = [00000000 \;\; 10101101].$$

Compare the transmission error probability of the code-words available in these two codes.

**Solution 2.11** Assume that the bits are transmitted through the binary symmetric channel whose graphical illustration is shown in Fig. 2.4. The crossover probability of the binary symmetric channel is $p = Prob(y|x) = 0.2$.

Assume that we use the first code $\boldsymbol{C}_1$ for transmission. Consider that we transmit the code-word 1010. If during the transmission two "1" flip, 000 will be available at the receiver. This is a transmission error that cannot be detected, and the probability of the transmission error can be calculated as

$$p_e = 0.2^2 \times 0.8^2 \rightarrow p_e = 0.0256. \tag{2.26}$$

On the other hand, if we use the code $\boldsymbol{C}_2$ for transmission and transmit the code-word 10101101, a transmission error occurs if five "1" flips. And this transmission error cannot be detected at the receiver side. The probability of this transmission error can be calculated as

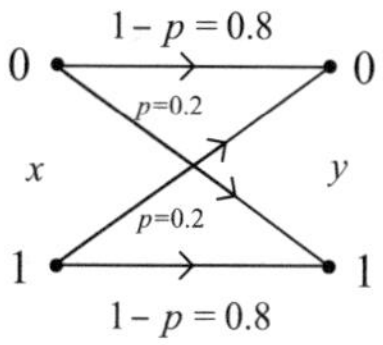


**Fig. 2.4** Binary symmetric channel with crossover probability $p = 0.2$

$$q_e = 0.2^5 \times 0.8^3 \rightarrow q_e \approx 0.000164. \tag{2.27}$$

If we compare Eqs. (2.26) and (2.27), we see that $q_e \ll p_e$.

### 2.6.1 Encoding Operation Using the Generator Matrix

Assume that we have $k$-dimensional data vector space, and a data vector in this vector space is given as

$$\boldsymbol{d}_i = [d_{i1}\ d_{i2} \ldots d_{ik}]\ \ i = 1, 2 \ldots, m \text{ where } m = 2^k.$$

We have a code of dimension $n$ such that $n > k$, and the generator matrix of the code is given as

$$\boldsymbol{G} = \begin{bmatrix} \boldsymbol{g}_1 \\ \boldsymbol{g}_2 \\ \vdots \\ \boldsymbol{g}_k \end{bmatrix}_{k \times n}$$

where $\boldsymbol{g}_i$, $i = 1, \ldots, k$ are the basis vectors of the code. The mapping of the data-words to code-words can be achieved using

$$\boldsymbol{c}_i = \boldsymbol{d}_i \times \boldsymbol{G} \rightarrow \boldsymbol{c}_i = [d_{i1}\ d_{i2} \ldots d_{ik}] \times \begin{bmatrix} \boldsymbol{g}_1 \\ \boldsymbol{g}_2 \\ \vdots \\ \boldsymbol{g}_k \end{bmatrix} \rightarrow \boldsymbol{c}_i = d_{i1}\boldsymbol{g}_1 + d_{i2}\boldsymbol{g}_2 + \ldots + d_{ik}\boldsymbol{g}_k \tag{2.28}$$

where the code-word $\boldsymbol{c}_i$ is a vector having $n$-bits, and it can be written as

$$\boldsymbol{c}_i = [c_{i1}\ c_{i2} \ldots c_{in}]\ \ i = 1, 2 \ldots, m \quad \text{where } m = 2^k. \tag{2.29}$$

When Eq. (2.28) is inspected, we see that the code-word is generated by taking the linear combination of the basis vectors considering the values of the bits in the data vector.

**Example 2.12** The generator matrix of a linear block code is given as

$$\boldsymbol{G} = \begin{bmatrix} 100011 \\ 010100 \\ 001010 \end{bmatrix}.$$

(a) What is the dimension of the data vector space, i.e., $k$=? How many data vectors are available?
(b) What is the dimension of the code, i.e., $n$=? How many code-words are available?
(c) Write the basis vectors of the code.
(d) Find the code-words after encoding the data-words

$$\boldsymbol{d}_1 = [101] \quad \boldsymbol{d}_2 = [111] \quad \boldsymbol{d}_3 = [011]$$

using the given generator matrix.

**Solution 2.12**
(a–b) The size of the generator matrix is $3 \times 6$ which means that $k = 3$ and $n = 6$, and this implies that the dimension of the data vector space is 3 and the dimension of the vector space is 6. There are $2^k = 2^3 \rightarrow 8$ data vectors and $2^n = 2^6 \rightarrow 64$ vectors from which 8 of them are selected for subspace and used for encoding, i.e., mapping operation.
(c) The basis vectors of the code are the rows of the generator matrix. The basis of the code can be written as

$$\boldsymbol{B} = \{100011, 010100, 001010\}.$$

(d) Employing

$$\boldsymbol{c}_i = \boldsymbol{d}_i \times \boldsymbol{G}$$

for the given data-vectors, we obtain the code-words as

$$\boldsymbol{c}_1 = \boldsymbol{d}_1 \times \boldsymbol{G} \rightarrow$$

$$\boldsymbol{c}_1 = [101] \times \begin{bmatrix} 100011 \\ 010100 \\ 001010 \end{bmatrix} \rightarrow$$

$$\boldsymbol{c}_1 = 1 \times [100011] + 0 \times [010100] + 1 \times [001010] \rightarrow$$

$$\boldsymbol{c}_1 = [101001]$$

$$\boldsymbol{c}_2 = \boldsymbol{d}_2 \times \boldsymbol{G} \rightarrow$$

$$\boldsymbol{c}_2 = [111] \times \begin{bmatrix} 100011 \\ 010100 \\ 001010 \end{bmatrix} \rightarrow$$

$$\boldsymbol{c}_2 = 1 \times [100011] + 1 \times [010100] + 1 \times [001010] \rightarrow$$

$$\boldsymbol{c}_2 = [111101]$$

$$\boldsymbol{c}_3 = \boldsymbol{d}_3 \times \boldsymbol{G} \rightarrow$$

$$\boldsymbol{c}_3 = [011] \times \begin{bmatrix} 100011 \\ 010100 \\ 001010 \end{bmatrix} \rightarrow$$

$$\boldsymbol{c}_3 = 0 \times [100011] + 1 \times [010100] + 1 \times [001010] \rightarrow$$

$$\boldsymbol{c}_3 = [011110].$$

## 2.7 Dual Code

We mentioned before that a code is a subspace of a vector space. A subspace has a dual subspace whose elements are orthogonal to the elements of the subspace. This means that a code has a dual code, and the dual code-words are orthogonal to the code-words.

If the dimensions of the vector space and subspace equal to $n$ and $k$, then the dimension of the dual subspace equals to $n - k$. This means that to construct a code, we construct $k$ basis vectors out of the $n$ basis vectors of the vector space, and the span of the $k$ basis vectors is the code, and some $n - k$ basis vectors can be used to construct the generator matrix of the dual code. Let's illustrate the concept by an example.

**Example 2.13** The standard basis of a vector space $\mathcal{V}$ is given as

$$\boldsymbol{B}_s = [0000001 \;\; 0000010 \;\; 0000100 \;\; 0001000 \;\; 0010000 \;\; 0100000 \;\; 1000000].$$

There are seven vectors in basis, and by taking the linear combinations of the basis vectors, we can generate $2^7 = 128$ vectors belonging to the vector space.

By taking the linear combinations of the basis vectors in $\boldsymbol{B}_s$, we can find another set of 7 linearly independent vectors which can be considered another basis for $\mathcal{V}$. Note that a basis for a vector space $\mathcal{V}$ containing $N$ elements is a set of $n$ linearly independent vectors included in $\mathcal{V}$ such that $N = 2^n$.

Considering the given information in the previous paragraph, another basis for $\mathcal{V}$ using the standard basis can be constructed as just changing the first element of $\boldsymbol{B}_s$ as

$$\boldsymbol{B}_{s1} = [0000011\ \ 0000010\ \ 0000100\ \ 0001000\ \ 0010000\ \ 0100000\ \ 1000000].$$

We can construct a code and its dual code as follows. A subset of $\boldsymbol{B}_{s1}$ can be selected as

$$\boldsymbol{B}_{s2} = [0000011\ \ 0100000\ \ 1000000].$$

The elements of $\boldsymbol{B}_{s2}$ can be used for the rows of a generator matrix of a block code as in

$$\boldsymbol{G} = \begin{bmatrix} 0000011 \\ 0100000 \\ 1000000 \end{bmatrix}.$$

By taking the linear combinations of the rows of $\boldsymbol{G}$, we obtain the code

$$\boldsymbol{C} = [0000000\ \ 0000011\ \ 0100000\ \ 1000000\ \ 0100011\ \ 1000011\ \ 1100000\ \ 1100011].$$

Now, let's construct the dual code. For this purpose, first, we write the basis vectors which appear in $\boldsymbol{B}_{s1}$ and do not appear in $\boldsymbol{B}_{s2}$ as

$$[0000010\ \ 0000100\ \ 0001000\ \ 0010000] \tag{2.30}$$

We can use all the vectors in Eq. (2.30) for the basis of the dual code. However, to make it a little bit different, let's take the first row of the generator matrix and the elements [ 0000100 0001000 0010000] from Eq. (2.30) and form the basis of the dual code as

$$\boldsymbol{B}_{s3} = [0000011\ \ 0000100\ \ 0001000\ \ 0010000]$$

for which the linear independence rule still holds.

By writing the elements of $\boldsymbol{B}_{s3}$ as rows, we obtain the generator matrix as

$$\boldsymbol{H} = \begin{bmatrix} 0000011 \\ 0000100 \\ 0001000 \\ 0010000 \end{bmatrix}.$$

By taking the linear combinations of the rows of $\boldsymbol{H}$, we obtain the dual code as

$$\begin{aligned} \boldsymbol{C_d} = [&0000000 \quad 0000011 \quad 0000100 \quad 0001000 \\ &0010000 \quad 0000111 \quad 0001011 \quad 0010011 \\ &0001100 \quad 0010100 \quad 0011000 \quad 0001111 \\ &0010111 \quad 0110111 \quad 0010100 \quad 0011111]. \end{aligned}$$

If the code-words in $\boldsymbol{C}$ and dual code-words in $\boldsymbol{C_d}$ are inspected, we see that, for $\forall \boldsymbol{c} \in \boldsymbol{C}$ and $\forall \boldsymbol{c_d} \in \boldsymbol{C_d}$, we have $\boldsymbol{c} \cdot \boldsymbol{c_d} = 0$. For instance, for $\boldsymbol{c} = [1000011]$ and $\boldsymbol{c_d} = [0110111]$, we have

$$\boldsymbol{c} \cdot \boldsymbol{c_d} = [1000011] \cdot [0110111] \rightarrow$$

$$\boldsymbol{c} \cdot \boldsymbol{c_d} = 1 \times 0 + 0 \times 1 + 0 \times 1 + 0 \times 0 + 0 \times 1 + 1 \times 1 + 1 \times 1 \rightarrow \boldsymbol{c} \cdot \boldsymbol{c_d} = \boldsymbol{0}.$$

Note that a code and its dual code may have common elements. In fact, even the dual of a code can be the code itself.

**Example 2.14** The dimensions of a vector space and the code obtained from the vector space are given as $n = 4$ and $k = 2$, respectively. Find a code such that its dual equals to the code itself.

**Solution 2.14** The required code can be generated trivially as

$$\boldsymbol{C} = [0000\ 0011\ 1100\ 1111].$$

**Property** Let $n$ be the dimension of a vector space. If the dimension of a code is $k$, then the dimension of the dual code is $n - k$.

## 2.8 Parity Check Matrix

Let $\boldsymbol{C}$ be a linear block code whose generator matrix is $\boldsymbol{G}$ with size $k \times n$, and the dual code of $\boldsymbol{C}$ be $\boldsymbol{C_d}$. The generator matrix of the dual code $\boldsymbol{C_d}$ denoted by $\boldsymbol{H}$ with size $(n - k) \times n$ is called the parity check matrix of the code $\boldsymbol{C}$. We have

$$\boldsymbol{G} \times \boldsymbol{H}^T = 0 \tag{2.31}$$

where $\boldsymbol{H}^T$ is the transpose of $\boldsymbol{H}$, or with the explicit size information, we can write

$$\boldsymbol{G_{k \times n}} \times \boldsymbol{H}^T_{n \times (n-k)} = 0 \tag{2.32}$$

**Example 2.15** The size of the parity check matrix of a linear block code is given as $4 \times 10$.

(a) Find the dimension of the code, and find the number of code-words in the code.
(b) How many vectors are available in the vector space?
(c) Find the dimension of the dual code, and find the number of code-words in the dual code.
(d) Write the size of the generator matrix of the code.

**Solution 2.15** Equating $(n - k) \times n$ to $4 \times 10$, we obtain $k = 6$, $n = 10$.

(a) The dimension of the code is $k = 6$, and the number of code-words can be calculated as

$$2^k \rightarrow 2^6 \rightarrow 64.$$

(b) The number of vectors in the vector space can be calculated as

$$2^n \rightarrow 2^{10} \rightarrow 1024.$$

(c) The dimension of the dual code is $n - k = 10 - 6 \rightarrow 4$. The number of dual code-words can be calculated as

$$2^{n-k} \rightarrow 2^4 \rightarrow 16.$$

(d) The size of the generator matrix is $k \times n = 6 \times 10$.

**Example 2.16** The standard basis of a vector space with dimension 5 is given as

$$\boldsymbol{B}_{\textbf{std}} = [00001 \;\; 00010 \;\; 00100 \;\; 01000 \;\; 10000].$$

Let's denote the basis vectors of the standard basis as

$$\boldsymbol{c}_1 = [00001] \quad \boldsymbol{c}_2 = [00010] \quad \boldsymbol{c}_3 = [00100] \quad \boldsymbol{c}_4 = [01000] \quad \boldsymbol{c}_5 = [10000].$$

Another basis for the vector space using the standard basis vectors can be formed as

$$\begin{gathered}[\boldsymbol{c}_1 \;\; \boldsymbol{c}_2 \;\; \boldsymbol{c}_3 \;\; \boldsymbol{c}_4 \;\; \boldsymbol{c}_5] \rightarrow [\boldsymbol{c}_1 \;\; (\boldsymbol{c}_2 + \boldsymbol{c}_1) \;\; (\boldsymbol{c}_3 + \boldsymbol{c}_1) \;\; (\boldsymbol{c}_4 + \boldsymbol{c}_1) \;\; (\boldsymbol{c}_5 + \boldsymbol{c}_1)] \rightarrow \\ [\boldsymbol{c}_1 \;\; (\boldsymbol{c}_2 + \boldsymbol{c}_1) \;\; (\boldsymbol{c}_3 + \boldsymbol{c}_1) \;\; (\boldsymbol{c}_4 + \boldsymbol{c}_1 + \boldsymbol{c}_2) \;\; (\boldsymbol{c}_5 + \boldsymbol{c}_1)] \rightarrow \\ [\boldsymbol{c}_1 \;\; (\boldsymbol{c}_2 + \boldsymbol{c}_1) \;\; (\boldsymbol{c}_3 + \boldsymbol{c}_1) \;\; (\boldsymbol{c}_4 + \boldsymbol{c}_1 + \boldsymbol{c}_2 + \boldsymbol{c}_3) \;\; (\boldsymbol{c}_5 + \boldsymbol{c}_1 + \boldsymbol{c}_3)]\end{gathered}$$

leading to

$$\boldsymbol{B}_s = [00001 \;\; 00011 \;\; 00101 \;\; 01111 \;\; 10101].$$

Note that in $\boldsymbol{B}_s$, there are no two identical basis vectors.
Let's selects a subset of $\boldsymbol{B}_s$ as

$$\boldsymbol{B}_{s1} = [00001 \quad 10101].$$

The elements of $\boldsymbol{B}_{s1}$ can be used to construct the generator matrix of a code as

$$\boldsymbol{G} = \begin{bmatrix} 00001 \\ 10101 \end{bmatrix}_{2\times 5}.$$

By taking all the possible linear combinations of the rows of the generator matrix $\boldsymbol{G}$, we obtain the linear code

$$\boldsymbol{C} = [00000 \;\; 00001 \;\; 10101 \;\; 10100].$$

Now, let's find the dual code. Let $\boldsymbol{c_d} = [a \; b \; c \; d \; e\;]$ be a dual code-word. As we stated before, for $\forall \boldsymbol{c} \in \boldsymbol{C}$ and $\forall \boldsymbol{c_d} \in \boldsymbol{C_d}$, we have $\boldsymbol{c} \cdot \boldsymbol{c_d} = 0$. Using the code-words in $\boldsymbol{C}$ for the equation $\boldsymbol{c} \cdot \boldsymbol{c_d}$=0, we obtain the equation set

$$\begin{gathered} d = 0 \qquad a + c + d = 0 \\ a + c = 0. \end{gathered} \tag{2.33}$$

The solution of Eq. (2.33) can be found as

$$\text{0X0X0} \qquad \text{1X1X0} \tag{2.34}$$

where $\text{X} \in F = \{0, 1\}$. Considering all the possible values of X in Eq. (2.34), we can get the dual code as

$$\boldsymbol{C_d} = [00000 \quad 10100 \quad 10110 \quad 11100 \quad 11110 \;\; 01000 \;\; 00010 \;\; 01010].$$

Since $\boldsymbol{C_d}$ has eight dual code-words, we can trivially choose three independent vectors from $\boldsymbol{C_d}$ as the basis vectors, and writing the chosen vectors as the row of a matrix, we can form the generator matrix of the dual code as

$$\boldsymbol{H} = \begin{bmatrix} 10100 \\ 10110 \\ 11100 \end{bmatrix}_{3\times 5}. \tag{2.35}$$

Notice that the rows of $\boldsymbol{H}$ are not available in $\boldsymbol{B}_s$.

**Example 2.17** Let $\boldsymbol{B}_s$ be a basis of a vector space. Let's select two subsets of $\boldsymbol{B}_s$ as $\boldsymbol{B}_{s1}$ and $\boldsymbol{B}_{s2}$ such that $\boldsymbol{B}_s = \boldsymbol{B}_{s1} \cup \boldsymbol{B}_{s2}$ and $\boldsymbol{B}_{s1} \cap \boldsymbol{B}_{s2} = \phi$. If I choose the elements of $\boldsymbol{B}_{s1}$ as the rows of a generator matrix of a code as, can I choose the elements of the $\boldsymbol{B}_{s2}$ as the rows of the generator matrix of the dual code?

**Solution 2.17** If we choose the elements of $\boldsymbol{B}_{s1}$ for the rows of a generator matrix, there is no guarantee that the elements of $\boldsymbol{B}_{s2}$ can be chosen for the rows of the generator matrix of the dual code. It may work, or it may not work. If the basis $\boldsymbol{B}_s$ is a standard basis, then it works; otherwise, there is no guarantee.

**Theorem** Let $\boldsymbol{C}$ be a linear block code whose generator and parity check matrices are $\boldsymbol{G}_{k \times n}$ and $\boldsymbol{H}_{(n-k) \times n}$. For any code-word $\boldsymbol{c}_{1 \times n} \in \boldsymbol{C}$, we have

$$\boldsymbol{c}_{1\times n} \times \boldsymbol{H}^T_{n\times(n-k)} = \boldsymbol{0}_{1\times(n-k)} \tag{2.36}$$

where $\boldsymbol{0}_{1 \times (n-k)}$ is all-zero vector with $(n - k)$ elements. Equation (2.36) can be written in a more compact form as

$$\boldsymbol{c}\boldsymbol{H}^T = \boldsymbol{0}. \tag{2.37}$$

***Proof*** The parity check matrix $\boldsymbol{H}$ is the generator matrix of the dual code, and the rows of $\boldsymbol{H}^T$ are a set of linearly independent dual code-words which can be used for the generation of all the other dual code-words via linear combination. Let's represent $\boldsymbol{H}$ by

$$\boldsymbol{H} = \begin{bmatrix} \boldsymbol{h}_1 \\ \boldsymbol{h}_2 \\ \vdots \\ \boldsymbol{h}_{n-k} \end{bmatrix}$$

where $\boldsymbol{h}_i = [h_{i1}\ h_{i2}\ldots h_{in}]$ are the row vectors with length $n$. For a given code-word $\boldsymbol{c}_j = [c_{j1}\ c_{j2}\ldots c_{jn}]$, the product $\boldsymbol{c}_j \times \boldsymbol{H}^T$ can be calculated as

$$\boldsymbol{c}_j \times \boldsymbol{H}^T = \begin{bmatrix} c_{j1}\ c_{j2} \ldots c_{jn} \end{bmatrix} \times \begin{bmatrix} h_{11} & h_{21} & & h_{(n-k)1} \\ h_{12} & h_{22} & \ldots & h_{(n-k)2} \\ \vdots & \vdots & & \vdots \\ h_{1n} & h_{2n} & & h_{(n-k)n} \end{bmatrix} \rightarrow$$

$$\boldsymbol{c}_j \times \boldsymbol{H}^T = \boldsymbol{c}_j \times \boldsymbol{h}_1^T + \boldsymbol{c}_j \times \boldsymbol{h}_2^T + \ldots + \boldsymbol{c}_j \times \boldsymbol{h}_{n-k}^T$$

where, employing the property $\boldsymbol{c} \cdot \boldsymbol{c}_d = 0$, we obtain $\boldsymbol{c}_j \times \boldsymbol{H}^T = 0$, or in a more general form, we can write

$$\boldsymbol{c} \times \boldsymbol{H}^T = \boldsymbol{0}.$$

**Exercise** Let $\boldsymbol{C}$ be a linear block code whose generator and parity check matrices are $\boldsymbol{G}$ and $\boldsymbol{H}$. Show that

$$\boldsymbol{G} \times \boldsymbol{H}^T = 0. \tag{2.38}$$

## 2.9 Systematic Form of a Generator Matrix

Using elementary row operations, the generator matrix of a linear block code can be put into the forms

$$\boldsymbol{G} = [\boldsymbol{I} \;\; \boldsymbol{P}] \text{ or } \boldsymbol{G} = [\boldsymbol{P} \;\; \boldsymbol{I}] \tag{2.39}$$

which are called the systematic forms of the generator matrix. The size of the matrices $\boldsymbol{I}$ and $\boldsymbol{P}$ is $k \times k$ and $k \times (n - k)$, respectively. Out of these two forms, the first one, i.e., $\boldsymbol{G} = [\boldsymbol{I} \;\; \boldsymbol{P}]$, is more widely used in the literature.

Note that every generator matrix may not be put into systematic form using only elementary row operations.

If the generator matrix is in systematic form, then using the formula

$$\boldsymbol{c} = \boldsymbol{d} \times \boldsymbol{G} \tag{2.40}$$

we obtain the code-word

$$\boldsymbol{c} = \boldsymbol{d} \times [\boldsymbol{I} \;\; \boldsymbol{P}] \rightarrow \boldsymbol{c} = [\boldsymbol{d} \times \boldsymbol{I} \;\; \boldsymbol{d} \times \boldsymbol{P}] \rightarrow \boldsymbol{c} = \left[\boldsymbol{d} \;\; \boldsymbol{d_p}\right] \tag{2.41}$$

where $\boldsymbol{d_p} = \boldsymbol{d} \times \boldsymbol{P}$. The first $k$-bits of the code-word are data bits, and the next $n - k$ bits are the parity bits.

**Example 2.18** The generator matrix of a linear block code is given as

$$\boldsymbol{G} = \begin{bmatrix} 10100 \\ 01001 \\ 01110 \end{bmatrix}.$$

Obtain the systematic form of this generator matrix.

**Solution 2.18** Using only elementary row operations, we can get the systematic form of the generator matrix as

$$\boldsymbol{G} = \begin{bmatrix} 10100 \\ 01001 \\ 01110 \end{bmatrix} : (R3 \leftarrow R2 + R3) :\rightarrow \boldsymbol{G}_1 = \begin{bmatrix} 10100 \\ 01001 \\ 00111 \end{bmatrix}$$

$$\boldsymbol{G}_1 = \begin{bmatrix} 10100 \\ 01001 \\ 00111 \end{bmatrix} : (R1 \leftarrow R1 + R3) : \rightarrow \boldsymbol{G}_s = \left[\begin{array}{c|c} 100 & 11 \\ 010 & 01 \\ \underbrace{001}_{I_{3\times 3}} & \underbrace{11}_{P_{3\times 2}} \end{array}\right].$$

We obtained the systematic form of the generator matrix in $\boldsymbol{G} = [\boldsymbol{I} \;\; \boldsymbol{P}]$ model.

**Example 2.19** The generator matrix of a linear block code is given as

$$\boldsymbol{G} = \begin{bmatrix} 10001 \\ 00100 \end{bmatrix}.$$

Obtain the systematic form of this generator matrix.

**Solution 2.19** Using only elementary row operations, we cannot put the generator matrix into systematic form. On the other hand, if we do column permutations, we can obtain the systematic form as

$$\boldsymbol{G} = \begin{bmatrix} 10001 \\ 00100 \end{bmatrix} : (C2 \leftrightarrow C3) : \boldsymbol{G}' = \begin{bmatrix} 10001 \\ 01000 \end{bmatrix}.$$

The obtained generator matrix $\boldsymbol{G}'$ corresponds to a different code which is an equivalent code to the one that can be constructed using the matrix $\boldsymbol{G}$.

**Example 2.20** The generator matrix of a linear block code is given as

$$\boldsymbol{G} = \begin{bmatrix} 10100 \\ 11011 \\ 01110 \end{bmatrix}.$$

Obtain the systematic form of this generator matrix.

**Solution 2.20** Using only elementary row operations, we can get the systematic form of the generator matrix as

$$\boldsymbol{G} = \begin{bmatrix} 10100 \\ 11011 \\ 01110 \end{bmatrix} : (R3 \leftarrow R1 + R3) :\rightarrow \boldsymbol{G}_1 = \begin{bmatrix} 10100 \\ 11011 \\ 11010 \end{bmatrix}$$

$$\boldsymbol{G}_1 = \begin{bmatrix} 10100 \\ 11011 \\ 11010 \end{bmatrix} : (R2 \leftarrow R2 + R3) :\rightarrow \boldsymbol{G}_2 = \begin{bmatrix} 10100 \\ 00001 \\ 11010 \end{bmatrix}$$

$$\boldsymbol{G}_2 = \begin{bmatrix} 10100 \\ 00001 \\ 11010 \end{bmatrix} : (R2 \leftrightarrow R3) : \rightarrow \boldsymbol{G}_s = \left[\begin{array}{c|c} 10 & 100 \\ 11 & 010 \\ \underbrace{00}_{\boldsymbol{P}} & \underbrace{001}_{\boldsymbol{I}} \end{array}\right].$$

We obtained the systematic form of the generator matrix in $\boldsymbol{G} = [\boldsymbol{P}\ \boldsymbol{I}]$ form.

Note that some generator matrices can be put into $\boldsymbol{G} = [\boldsymbol{I}\ \boldsymbol{P}]$, whereas some others can be put into $\boldsymbol{G} = [\boldsymbol{P}\ \boldsymbol{I}]$ form, and some may not have systematic form.

### 2.9.1 Construction of Parity Check Matrix from Systematic Generator Matrix

If the generator matrix of a linear block code is in systematic form

$$\boldsymbol{G}_{k\times n} = \left[\boldsymbol{I}_{k\times k}\ \ \boldsymbol{P}_{k\times(n-k)}\right] \tag{2.42}$$

then the parity check matrix of this code can be calculated as

$$\boldsymbol{H}_{(n-k)\times n} = \left[\boldsymbol{P}^T_{(n-k)\times k}\ \ \boldsymbol{I}_{(n-k)\times(n-k)}\right]. \tag{2.43}$$

**Example 2.21** The generator matrix of a linear block code is given as

$$\boldsymbol{G} = \begin{bmatrix} 10011 \\ 01001 \\ 00111 \end{bmatrix}.$$

Obtain the parity check matrix of this code.

**Solution 2.21** When the given generator matrix is inspected, we see that it is in the form

$$\boldsymbol{G} = [\boldsymbol{I}\ \ \boldsymbol{P}]$$

where

$$\boldsymbol{I} = \begin{bmatrix} 100 \\ 010 \\ 001 \end{bmatrix} \quad \boldsymbol{P} = \begin{bmatrix} 11 \\ 01 \\ 11 \end{bmatrix}.$$

The parity check matrix of the code can be obtained using the formula

$$\boldsymbol{H} = \left[\boldsymbol{P}^T\ \ \boldsymbol{I}\right]$$

as

$$\boldsymbol{H} = \begin{bmatrix} 10110 \\ 11101 \end{bmatrix}$$

where

$$\boldsymbol{P^T} = \begin{bmatrix} 101 \\ 111 \end{bmatrix} \ \boldsymbol{I} = \begin{bmatrix} 10 \\ 01 \end{bmatrix}.$$

As an extra information, let's calculate $\boldsymbol{G} \times \boldsymbol{H^T}$ as

$$\boldsymbol{G} \times \boldsymbol{H^T} = \begin{bmatrix} 10011 \\ 01001 \\ 00111 \end{bmatrix} \begin{bmatrix} 11 \\ 01 \\ 11 \\ 10 \\ 01 \end{bmatrix} \rightarrow \boldsymbol{G} \times \boldsymbol{H^T} = \begin{bmatrix} 00 \\ 00 \\ 00 \end{bmatrix}.$$

We see that

$$\boldsymbol{G} \times \boldsymbol{H^T} = \boldsymbol{0}.$$

## 2.10 Equal and Equivalent Codes

***Equal Codes***

Let $\boldsymbol{G_1}$ be the generator matrix of a linear block code. If we perform elementary row operations on $\boldsymbol{G_1}$ and obtain $\boldsymbol{G_2}$, the codes generated by $\boldsymbol{G_1}$ and $\boldsymbol{G_2}$ are said to be equal codes. This means that the same code-words are generated by both generator matrices, and data-words are mapped to the same code-word by both codes. Let's give an example to illustrate this concept.

**Example 2.22** The generator matrix of a linear block code is given as

$$\boldsymbol{G_a} = \begin{bmatrix} 10011 \\ 01001 \\ 00111 \end{bmatrix}.$$

Adding the first row to the second row, and exchanging the second and third rows, we obtain the matrix

$$G_b = \begin{bmatrix} 10011 \\ 00111 \\ 11010 \end{bmatrix}.$$

Using the formula

$$c = d \times G,$$

we can obtain the code-words for all possible data-words for the matrices $G_a$ and $G_b$ as indicated in Table 2.1. When Table 2.1 is inspected, we see that the same set of code-words are generated by both generator matrices; however, the mapping between data-words and code-words are different for both codes. For instance, if $d_a = [110]$ and $d_b = [001]$, the encoding operations $d_a \times G_a$ and $d_b \times G_b$ yield the same code-word $c = [11010]$.

***Equivalent Codes***

If we perform elementary row operations and *column permutations* on $G_1$ and obtain $G_2$, the code generated by $G_2$ is said to be equivalent to the code generated by $G_1$. Note that a code can have many equivalent codes, since many different column permutations are possible.

**Example 2.23** The generator matrix of a linear block code is given as

$$G_a = \begin{bmatrix} 10010 \\ 01001 \\ 00111 \end{bmatrix}.$$

Adding the first row to the third row, and exchanging the second and fourth columns, we obtain the matrix

$$G_b = \begin{bmatrix} 11000 \\ 00011 \\ 10101 \end{bmatrix}.$$

Using the formula

$$c = d \times G,$$

we can obtain the code-words for all possible data-words for the matrices $G_a$ and $G_b$ as indicated in Table 2.2. When Table 2.2 is inspected, we see that different sets of code-words are generated by both generator matrices.

**Table 2.1** Data-words and corresponding code-words

| $d$ | $c_a$ | $c_b$ |
|---|---|---|
| 000 | 00000 | 00000 |
| **001** | 00111 | **11010** |
| 010 | 01001 | 00111 |
| 011 | 01110 | 11101 |
| 100 | 10011 | 10011 |
| 101 | 10100 | 01001 |
| **110** | **11010** | 10100 |
| 111 | 11101 | 01110 |

**Table 2.2** Data-words and corresponding code-words

| $d$ | $c_a$ | $c_b$ |
|---|---|---|
| 000 | 00000 | 00000 |
| 001 | 00111 | 10101 |
| 010 | 01001 | 00011 |
| 011 | 01110 | 10110 |
| 100 | 10010 | 11000 |
| 101 | 10101 | 01101 |
| 110 | 11011 | 11011 |
| 111 | 11100 | 01110 |

## 2.11 Finding the Minimum Distance of a Linear Block Code Using Its Parity Check Matrix

The minimum distance of a linear block code can be determined using the parity check matrix of the code. We know that for a given code-word $\boldsymbol{c}$, we have

$$\boldsymbol{c} \times \boldsymbol{H}^T = 0$$

in which employing $\boldsymbol{c} = [c_1\ c_2 \ldots c_n]$ and

$$\boldsymbol{H} = [\boldsymbol{l}_1\ \boldsymbol{l}_2 \cdots \boldsymbol{l}_n] \tag{2.44}$$

where $\boldsymbol{l}_j$, $j = 1, \ldots, n$ are the column vectors consisting of $(n - k)$ bits, we obtain

$$c_1 \times \boldsymbol{l}_1^T + c_2 \times \boldsymbol{l}_2^T + \ldots c_n \times \boldsymbol{l}_n^T = 0. \tag{2.45}$$

When Eq. (2.45) is inspected, we see that the minimum distance of a linear block code equals to the minimum number of columns of the parity check matrix whose sum equals to zero.

***Note*** The linear block code obtained using the generator matrix $\boldsymbol{G}_{k \times n}$ will be denoted as $\boldsymbol{C}(\boldsymbol{n}, \boldsymbol{k})$.

**Example 2.24** The size of the generator matrix of a linear block code is $k \times n = 3 \times 6$. The parity check matrix of this code is in the form

$$\boldsymbol{H} = [\boldsymbol{l}_1 \ \boldsymbol{l}_2 \ \boldsymbol{l}_3 \ \boldsymbol{l}_4 \ \boldsymbol{l}_5 \ \boldsymbol{l}_6].$$

Let the minimum distance of the code be 3. Let $\boldsymbol{c} = [0\ 0\ 1\ 1\ 0\ 1]$ be a code-word with minimum distance. Using the formula

$$\boldsymbol{c} \times \boldsymbol{H}^T = 0$$

we obtain

$$0 \times \boldsymbol{l}_1^T + 0 \times \boldsymbol{l}_2^T + 1 \times \boldsymbol{l}_3^T + 1 \times \boldsymbol{l}_4^T + 0 \times \boldsymbol{l}_5^T + 1 \times \boldsymbol{l}_6^T = 0 \rightarrow \ \boldsymbol{l}_3^T + \boldsymbol{l}_4^T + \boldsymbol{l}_6^T = 0$$

where we see that the sum of the three, which is the minimum distance of the code, columns of the parity check matrix equals to zero.

**Example 2.25** The parity check matrix of a code is given as

$$\boldsymbol{H} = \begin{bmatrix} 1110100 \\ 1101010 \\ 1010001 \end{bmatrix}.$$

Find the minimum distance of this code.

**Solution 2.25** The sum of the columns $\boldsymbol{l}_1$, $\boldsymbol{l}_2$, and $\boldsymbol{l}_7$ equals to zero. We cannot find any two columns whose sum equals to zero. Then, the minimum distance of the code is 3.

**Properties of Linear Block Codes**

Let $\boldsymbol{C}(\boldsymbol{n}, \boldsymbol{k})$ be a linear block code with generator matrix $\boldsymbol{G}_{\boldsymbol{k} \times \boldsymbol{n}}$, and $\boldsymbol{c}_i, \boldsymbol{c}_j \in \boldsymbol{C}(\boldsymbol{n}, \boldsymbol{k})$ be two code-words. We have the properties

1. $\boldsymbol{c}_i + \boldsymbol{c}_j \in \boldsymbol{C}(\boldsymbol{n}, \boldsymbol{k})$.
2. All-zero code-word, i.e., $\boldsymbol{0}_{\boldsymbol{1} \times \boldsymbol{n}}$, is an element of $\boldsymbol{C}(\boldsymbol{n}, \boldsymbol{k})$.

## 2.12 Error Detection and Correction Capability of a Linear Block Code

Assume that a code-word is transmitted. At the receiver side, if the transmitted code-word is not received, we understand that some bit errors occurred during the transmission. This is called error detection. If we are able to identify the error locations, then it is possible to correct the bit errors. This is called error correction or forward error correction.

The error correction and detection capability of a linear block code is related to the minimum distance of the code. Depending on the minimum distance value, only error detection or both error detection and correction can be possible.

If the minimum distance of the linear block code $\boldsymbol{C}$ is $d_{\min}$, then this code can detect

$$t_d = d_{\min} - 1 \tag{2.46}$$

bit errors and can correct

$$t_c = \left\lfloor \frac{d_{\min} - 1}{2} \right\rfloor \tag{2.47}$$

bit errors, and $\lfloor \cdot \rfloor$ is the floor function.

Let $\overline{\boldsymbol{r}}$ be the received word. We decide on the transmitted code-word at the receiver side using the formula

$$\boldsymbol{c} = \{\boldsymbol{c}_i \mid d_H(\overline{\boldsymbol{r}}, \boldsymbol{c}_i) > d_H(\overline{\boldsymbol{r}}, \boldsymbol{c}_j)\; j = 1 \ldots 2^k, j \neq i\} \tag{2.48}$$

where $d_H(\overline{\boldsymbol{r}}, \boldsymbol{c}_i)$ and $d_H(\overline{\boldsymbol{r}}, \boldsymbol{c}_j)$ indicate the Hamming distances between $\overline{\boldsymbol{r}}$ and $\boldsymbol{c}_i$ and between $\overline{\boldsymbol{r}}$ and $\boldsymbol{c}_j$, respectively.

**Example 2.26** Consider the linear block code

$$\boldsymbol{C} = [0000 \;\; 1111].$$

The minimum distance of this code is $d_{\min} = 4$. The code-words are denoted as $\boldsymbol{c}_1 = [0000]$ and $\boldsymbol{c}_2 = [1111]$.

Assume that the code-word $\boldsymbol{c} = [1111]$ is transmitted. Then, we have the following scenarios:

Let's say that a single-bit error occurs, and the received word is $\overline{\boldsymbol{r}} = [1110]$. To find the transmitted code-word, we calculate the Hamming distance between the received word and the two candidate code-words in a sequential manner and choose the one which is closer to the received word in terms of Hamming distance. According to this information, we decide on the transmitted code-word as in

$$d_H(\overline{\boldsymbol{r}}, \boldsymbol{c}_1) = 3 \quad d_H(\overline{\boldsymbol{r}}, \boldsymbol{c}_2) = 1 \rightarrow \text{the transmitted code} - \text{word is } \boldsymbol{c}_2.$$

Assume that two-bit errors occur, and the received word is $\overline{\boldsymbol{r}} = [1100]$. The Hamming distances are

$$d_H(\overline{\boldsymbol{r}}, \boldsymbol{c}_1) = 2 \quad d_H(\overline{\boldsymbol{r}}, \boldsymbol{c}_2) = 2.$$

Since Hamming distances are equal to each other, we cannot decide on the transmitted code-word. But we are sure that some errors occurred during transmission, i.e., the availability of the errors is detected, since the received word is not a code-word.

Assume that three-bit errors occur, and the received word is $\bar{\boldsymbol{r}} = [1000]$. The Hamming distances are

$$d_H(\bar{\boldsymbol{r}}, \boldsymbol{c}_1) = 1 \quad d_H(\bar{\boldsymbol{r}}, \boldsymbol{c}_2) = 3.$$

We are sure that some errors occurred during transmission, i.e., the availability of the errors is detected. Since the received word is not a code-word, and considering the Hamming distances calculated, we choose the code-word $\boldsymbol{c}_1$ as the transmitted code-word. However, this decision is not correct, since $\boldsymbol{c}_2$ is transmitted. This means that if three-bit errors occur, we can detect the occurrence of the errors; however, we cannot decide on the transmitted code-word correctly, i.e., the erroneous bits cannot be corrected, and we make a wrong decision on the transmitted code-word.

Assume that four-bit errors occurred, and the received word is $\bar{\boldsymbol{r}} = [0000]$. Since $\bar{\boldsymbol{r}} = \boldsymbol{c}_1$, we accept that $\boldsymbol{c}_1$ is the transmitted code-word, and we assume that no errors occurred during the transmission. However, all our decisions are wrong. This means that if four error occurs during the transmission, we can neither detect nor correct the bit errors.

Now, using Eqs. (2.46) and (2.47), let's calculate the error detection and error correction capability of our code. Our code can detect

$$t_d = d_{\min} - 1 \rightarrow t_d = 4 - 1 \rightarrow t_d = 3$$

bit errors and can correct

$$t_c = \left\lfloor \frac{d_{\min} - 1}{2} \right\rfloor \rightarrow t_c = \left\lfloor \frac{4 - 1}{2} \right\rfloor \rightarrow t_c = \lfloor 1.5 \rfloor \rightarrow t_c = 1$$

bit error, and obtained numbers coincide with the discussion we made for possible transmission scenarios.

**Example 2.27** Consider the linear block code

$$\boldsymbol{C} = [00000 \;\; 11111].$$

The minimum distance of this code is $d_{\min} = 5$. The code-words are denoted as $\boldsymbol{c}_1 = [00000\ ]$ and $\boldsymbol{c}_2 = [11111\ ]$.

Assume that the code-word $\boldsymbol{c} = [11111\ ]$ is transmitted. Now, using Eqs. (2.46) and (2.47), let's calculate the error detection and error correction capability of our code. Our code can detect

$$t_d = d_{\min} - 1 \rightarrow t_d = 5 - 1 \rightarrow t_d = 4$$

bit errors and can correct

$$t_c = \left\lfloor \frac{d_{\min} - 1}{2} \right\rfloor \rightarrow t_c = \left\lfloor \frac{5-1}{2} \right\rfloor \rightarrow t_c = \lfloor 2 \rfloor \rightarrow t_c = 2$$

bit errors.

Assume that the code-word $\boldsymbol{c} = [11111]$ is transmitted. Let's say that a single-bit error occurs, and the received word is $\bar{\boldsymbol{r}} = [11110]$. Considering the Hamming distances, we decide on the transmitted code-word as

$$d_H(\bar{\boldsymbol{r}}, \boldsymbol{c}_1) = 4 \quad d_H(\bar{\boldsymbol{r}}, \boldsymbol{c}_2) = 1 \rightarrow \text{the transmitted code} - \text{word is } \boldsymbol{c}_2.$$

Assume that two-bit errors occurred, and the received word is $\bar{\boldsymbol{r}} = [11100]$. The Hamming distances are

$$d_H(\bar{\boldsymbol{r}}, \boldsymbol{c}_1) = 3 \quad d_H(\bar{\boldsymbol{r}}, \boldsymbol{c}_2) = 2 \rightarrow \text{the transmitted code} - \text{word is } \boldsymbol{c}_2.$$

Assume that three-bit errors occurred, and the received word is $\bar{\boldsymbol{r}} = [11000]$. The Hamming distances are

$$d_H(\bar{\boldsymbol{r}}, \boldsymbol{c}_1) = 2 \quad d_H(\bar{\boldsymbol{r}}, \boldsymbol{c}_2) = 3.$$

Considering the Hamming distances, we decide on $\boldsymbol{c}_1$ as the transmitted code-word. We are sure that some errors occurred during transmission, i.e., the availability of the errors is detected. Since the received word is not a code-word, and considering the Hamming distances calculated, we choose the code-word $\boldsymbol{c}_1$ as the transmitted code-word. However, this decision is not correct. Since $\boldsymbol{c}_2$ is transmitted. This means that if three-bit errors occur, we can detect the occurrence of the errors; however, we cannot decide on the transmitted code-word correctly, i.e., the erroneous bits cannot be corrected.

Assume that four-bit errors occurred, and the received word is $\bar{\boldsymbol{r}} = [10000]$. The Hamming distances are

$$d_H(\bar{\boldsymbol{r}}, \boldsymbol{c}_1) = 1 \quad d_H(\bar{\boldsymbol{r}}, \boldsymbol{c}_2) = 1.$$

Since Hamming distances are equal to each other, we cannot decide on the transmitted code-word. But we are sure that some errors occurred during transmission, i.e., the availability of the errors is detected, since the received word is not a code-word.

Assume that five-bit errors occurred, and the received word is $\bar{\boldsymbol{r}} = [00000]$. Since $\bar{\boldsymbol{r}} = \boldsymbol{c}_1$, we accept that $\boldsymbol{c}_1$ is the transmitted code-word and no errors occurred during the transmission. However, all our decisions are wrong. This means that if five errors occur during the transmission, we can neither detect nor correct the bit errors.

We said that the error detection and correction capability of a linear block code can be calculated using the formulas

$$t_d = d_{\min} - 1 \qquad t_c = \left\lfloor \frac{d_{\min} - 1}{2} \right\rfloor. \tag{2.49}$$

However, this does not mean that more errors cannot be detected or corrected. The numbers obtained from these formulas are the guaranteed numbers. That means that for every case these numbers are valid. In fact, some code-words with minimum distances can correct or detect more errors than the ones indicated in these formulas. Let's show it with an example.

**Example 2.28** Consider the linear block code given as

$$\boldsymbol{C} = [000000 \quad 111000 \quad 000111 \quad 111111].$$

The minimum distance of this code is $d_{\min} = 3$. This code can detect

$$t_d = d_{\min} - 1 \rightarrow t_d = 3 - 1 \rightarrow t_d = 2$$

bit errors, and can correct

$$t_c = \left\lfloor \frac{d_{\min} - 1}{2} \right\rfloor \rightarrow t_c = \left\lfloor \frac{3 - 1}{2} \right\rfloor \rightarrow t_c = \lfloor 1 \rfloor \rightarrow t_c = 1$$

bit error.

This means that two-bit errors can be detected and single-bit error can be corrected for every code-word, but this does not mean that some code-words cannot detect or correct more errors.

For example, consider the transmission of the code-word $\boldsymbol{c} = [111111]$, and assume that three-bit errors occurred and the received word is $\bar{\boldsymbol{r}} = [101010]$. It is clear that $\bar{\boldsymbol{r}} \notin \boldsymbol{C}$, and this implies that some errors occurred during the communication. As we see, more errors than $d_{\min} - 1$ can be detected for some code-words.

**Example 2.29** The generator matrix of a linear block code is given as

$$\boldsymbol{G} = \begin{bmatrix} 1001100 \\ 0100011 \\ 0010111 \end{bmatrix}.$$

The code generated by this matrix can be obtained as

$$\boldsymbol{C} = [0000000 \ 1001100 \ 0100011 \ 0010111 \ 1101111 \ 1011011 \ 0110100 \ 1111000].$$

The code-words are denoted as

$$\boldsymbol{c}_1 = [0000000] \ \boldsymbol{c}_2 = [1001100] \ \boldsymbol{c}_3 = [0100011] \ \boldsymbol{c}_4 = [0010111]$$
$$\boldsymbol{c}_5 = [1101111] \ \boldsymbol{c}_6 = [1011011] \ \boldsymbol{c}_7 = [0110100] \ \boldsymbol{c}_8 = [1111000].$$

The minimum distance of this code is $d_{\min} = 3$. This code can detect

$$t_d = d_{\min} - 1 \rightarrow t_d = 3 - 1 \rightarrow t_d = 2$$

bit errors, and can correct

$$t_c = \left\lfloor \frac{d_{\min} - 1}{2} \right\rfloor \rightarrow t_c = \left\lfloor \frac{3 - 1}{2} \right\rfloor \rightarrow t_c = \lfloor 1 \rfloor \rightarrow t_c = 1$$

bit error.

This means that 2-bit errors can be detected and 1-bit error can be corrected for every code-word, but this does not mean that more code-words cannot detect or correct more errors.

For example, consider the transmission of the code-word $\boldsymbol{c} = [0000000]$, and assume that two-bit errors occurred and the received word is $\bar{\boldsymbol{r}} = [1000001]$. It is clear that $\bar{\boldsymbol{r}} \notin \boldsymbol{C}$, and this implies that some errors occurred during the communication. The Hamming distances between the received word and code-words can be calculated as

$$d_H(\bar{\boldsymbol{r}}, \boldsymbol{c}_1) = 2 \quad d_H(\bar{\boldsymbol{r}}, \boldsymbol{c}_2) = 3 \quad d_H(\bar{\boldsymbol{r}}, \boldsymbol{c}_3) = 3 \quad d_H(\bar{\boldsymbol{r}}, \boldsymbol{c}_4) = 4$$
$$d_H(\bar{\boldsymbol{r}}, \boldsymbol{c}_5) = 4 \quad d_H(\bar{\boldsymbol{r}}, \boldsymbol{c}_6) = 3 \quad d_H(\bar{\boldsymbol{r}}, \boldsymbol{c}_7) = 5 \quad d_H(\bar{\boldsymbol{r}}, \boldsymbol{c}_8) = 4.$$

Considering the calculated Hamming distances, we can decide on the transmitted code-word as $\boldsymbol{c}_1$, i.e., all-zero code-word. This result shows that more than

$$\left\lfloor \frac{d_{\min} - 1}{2} \right\rfloor \tag{2.50}$$

bit errors may be corrected; however, corrections are not guaranteed for every case. On the other hand, we are sure that every code-word can correct

$$\left\lfloor \frac{d_{\min} - 1}{2} \right\rfloor$$

bit errors for sure.

**Example 2.30** The error correction capability of a linear block code is given as $t_c$. What can be the minimum distance of this linear block code?

**Solution 2.30** If $d_{\min}$ is an odd number, i.e., $d_{\min} = 2m + 1$, then using the equation

$$t_c = \left\lfloor \frac{d_{\min} - 1}{2} \right\rfloor$$

we obtain

$$t_c = m$$

which implies that

$$d_{\min} = 2t_c + 1.$$

On the other hand, if $d_{\min}$ is an even number, i.e., $d_{\min} = 2m$, then using the equation

$$t_c = \left\lfloor \frac{d_{\min} - 1}{2} \right\rfloor$$

we obtain

$$t_c = \left\lfloor m - \frac{1}{2} \right\rfloor \rightarrow t_c = m - 1$$

which implies that

$$d_{\min} = 2t_c + 2.$$

In summary, for a given $t_c$, the minimum distance of the code can be either equal to

$$d_{\min} = 2t_c + 1$$

or be equal to

$$d_{\min} = 2t_c + 2.$$

**Example 2.31** If a linear block code can correct $t_c = 5$-bit errors, what can be the minimum distance of this linear block code?

**Solution 2.31** The minimum distance of the code can either be calculated as

$$d_{\min} = 2t_c + 1 \rightarrow d_{\min} = 11$$

or it can be calculated as

$$d_{\min} = 2t_c + 2 \rightarrow d_{\min} = 12.$$

## 2.13 Hamming Spheres

Let $\boldsymbol{v}$ be a $n$-dimensional vector, i.e., word with $n$-bits. The sphere whose center is indicated by the vector $\boldsymbol{v}$ and its radius is determined by the Hamming distance $r$ is called the Hamming sphere.

A Hamming sphere has $r$ orbits, and the vectors in each orbit have equal distance to the center.

**Example 2.32** Assume that we have a vector space $\mathcal{V}$ with dimension $n = 4$.

(a) Draw a Hamming sphere with radius $r = 1$, and at the center of the sphere, all-zero vector resides.
(b) Draw a Hamming sphere with radius $r = 2$, and at the center of the sphere, all-zero vector resides.

**Solution 2.32**

(a) For the Hamming sphere with radius $r = 1$, there is a single orbit, and the words at this orbit have equal Hamming distance to the center, and the value of the Hamming distance is 1. The required Hamming sphere is drawn in Fig. 2.5.
(b) In this case, there are two orbits with radiuses $r_1 = 1$ and $r_2 = 2$. The words at orbit-1 have equal distances $r_1$ to the center, and the words at orbit-2 have equal distances $r_2$ to the center. In the orbit-1, we have

$$\binom{4}{1} = 4$$

words, i.e., vectors, and in orbit-2, we have

$$\binom{4}{2} = 6$$

words. The required Hamming sphere is drawn in Fig. 2.6.

**Example 2.33** Assume that we have a vector space $\mathcal{V}$ with dimension $n = 4$. Consider the Hamming sphere with radius $r = 4$ around the all-zero vector. We can consider four different Hamming spheres with radiuses

$$r_1 = 1 \quad r_2 = 2 \quad r_3 = 3 \quad r_4 = 4.$$

The Hamming sphere with radius $r_1 = 1$ has a single orbit, and the number of vectors in this orbit can be calculated using

$$\binom{4}{1}.$$

The Hamming sphere with radius $r_2 = 2$ has two orbits, and the number of vectors in these orbits are

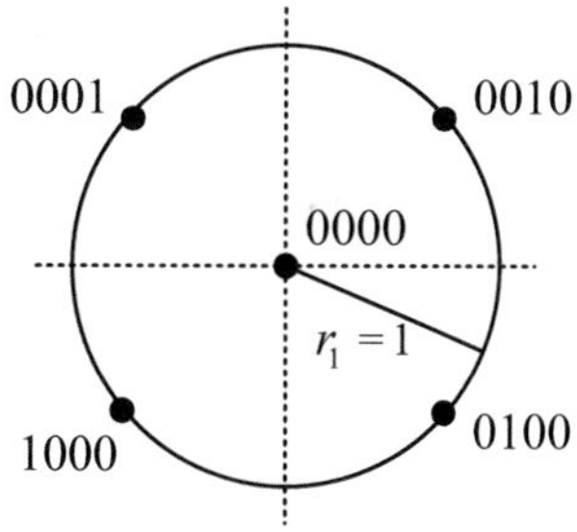


**Fig. 2.5** Hamming sphere with radius $r = 1$

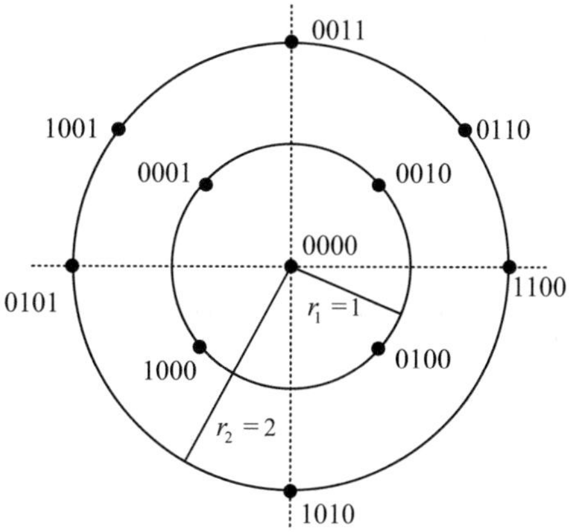


**Fig. 2.6** Hamming spheres with radiuses $r_1 = 1$ and $r_2 = 2$

$$\binom{4}{1} \text{ and } \binom{4}{2}.$$

The total number of vectors in Hamming sphere with radius $r_2 = 2$ including the vector at the center can be calculated as

$$1 + \binom{4}{1} + \binom{4}{2}.$$

The Hamming sphere with radius $r_3 = 3$ has three orbits, and the number of vectors in these orbits are

$$\binom{4}{1} \text{ and } \binom{4}{2} \text{ and } \binom{4}{3}.$$

The total number of vectors in Hamming sphere with radius $r_3 = 3$ is

$$1 + \binom{4}{1} + \binom{4}{2} + \binom{4}{3}.$$

The Hamming sphere with radius $r_4 = 4$ has four orbits, and the number of vectors in these orbits are

$$\binom{4}{1} \text{ and } \binom{4}{2} \text{ and } \binom{4}{3} \text{ and } \binom{4}{4}.$$

The total number of vectors in Hamming sphere with radius $r_4 = 4$ is

$$1 + \binom{4}{1} + \binom{4}{2} + \binom{4}{3} + \binom{4}{4}$$

which is equal to $2^n = 2^4$, and this number is the total number of vectors in vector space $\mathcal{V}$ whose dimension is $n = 4$.

**Example 2.34** The generator matrix of a linear block code is given as

$$\boldsymbol{G} = \begin{bmatrix} 1001100 \\ 0100011 \\ 0010111 \end{bmatrix}.$$

Write all code-words. How many bit errors can this code correct, i.e., $t_c = ?$ Draw Hamming spheres with radiuses $t_c$ around each code-word, and determine all the words inside each Hamming sphere.

**Solution 2.34** By taking all the possible linear combinations of the rows of the generator matrix, we can obtain the code

$$\boldsymbol{C} = [0000000\ \ 1001100\ \ 0100011\ \ 0010111\ \ 1101111\ \ 1011011\ \ 0110100\ \ 1111000]$$

from which we can determine the minimum distance of the code as

$$d_{\min} = 3.$$

The designed code can correct

$$t_c = \left\lfloor \frac{d_{\min} - 1}{2} \right\rfloor \rightarrow t_c = 1$$

bit error and can detect

$$t_d = d_{\min} - 1 \rightarrow t_d = 2$$

bit errors. The Hamming spheres with radiuses $r = 1$ around each code-word are depicted in Fig. 2.7.

There are eight words inside each Hamming sphere including the code-word at the center. Since there are eight Hamming spheres, the total number of words inside the spheres equals to $8 \times 8 = 64$. On the other hand, the number of words in the vector space equals to $2^7 = 128$. This means that $128 - 64 = 64$ words are available outside the spheres.

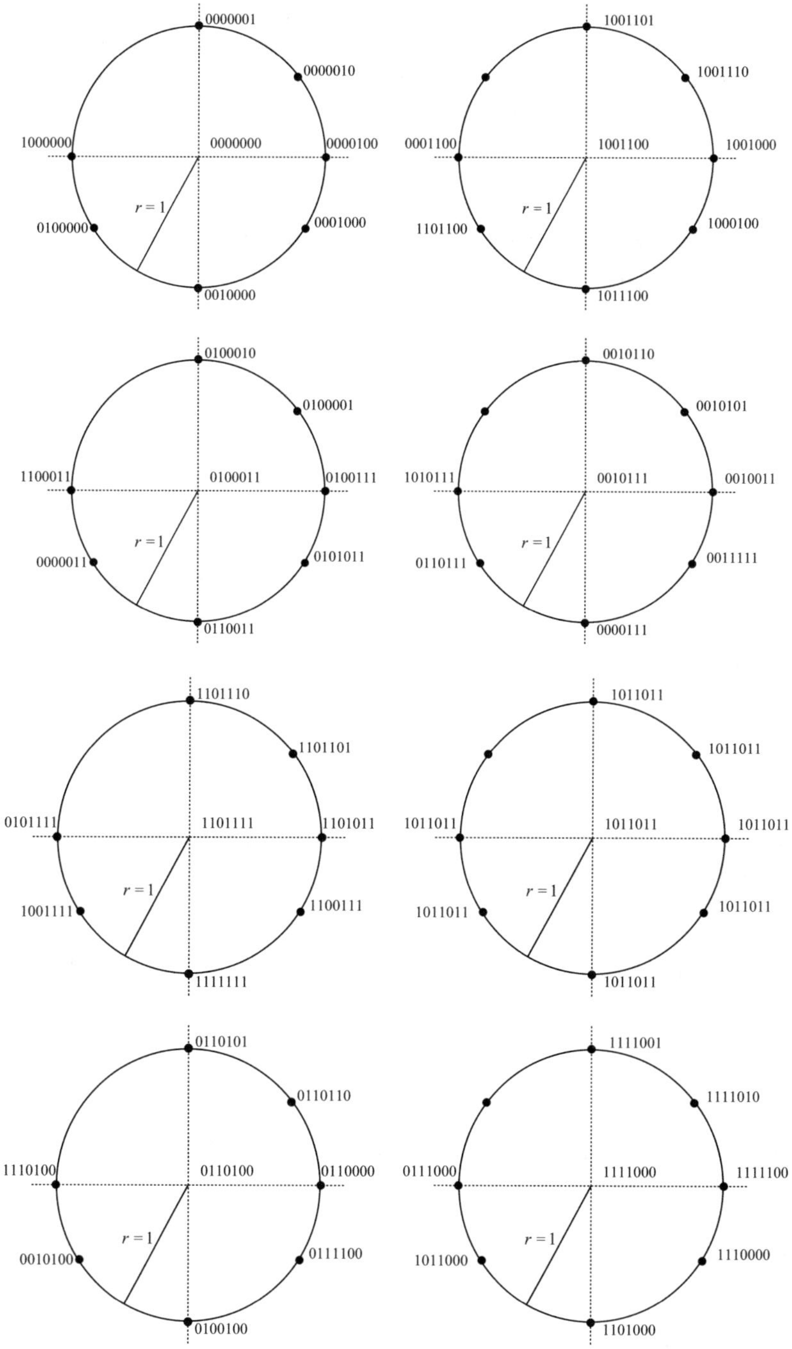


**Fig. 2.7** Hamming spheres with radius $r = 1$ around code-words

Assume that we design a code with 16 code-words, i.e., we have 16 Hamming spheres, then the total number of words inside the spheres would be $16 \times 8 = 128$, and this means that all the words of the vector space are inside the Hamming spheres, i.e., no word is left outside the spheres.

### *2.13.1 The Total Number of Words Inside a Hamming Sphere with Radius* r

Given a vector space of dimension $n$, the number of words, i.e., vectors, inside a Hamming sphere with radius $r$ can be calculated using

$$N_H = \sum_{j=0}^{r} \binom{n}{j}. \tag{2.51}$$

**Example 2.35** Considering a vector space of dimension $n = 8$, calculate the total number of words inside a Hamming sphere with radius $r = 4$.

**Solution 2.35** Using the formula

$$N_H = \sum_{j=0}^{r} \binom{n}{j}$$

for the given values, we calculate the total number of words inside a Hamming sphere with radius $r = 4$ as

$$N_H = \sum_{j=0}^{4} \binom{8}{j} \rightarrow N_H = \binom{8}{0} + \binom{8}{1} + \binom{8}{2} + \binom{8}{3} + \binom{8}{4} \rightarrow N_H = 163.$$

## 2.14 Some Simple Bounds for the Minimum Distances of Linear Block Codes

In this subsection, we will provide some bound expressions for $d_{\min}$ of linear block codes.

***Singleton Bound***
For a $\boldsymbol{C}(n, k)$ linear block code, the minimum distance is bounded by

$$d_{\min} \leq n - k + 1 \tag{2.52}$$

which is called singleton bound.

***Proof*** Consider a data-word of $k$-bits such that the Hamming weight of the data-word is 1, i.e., there is only a single one and all the other bits are zero. The systematic code-word obtained from this data-word contains $n - k$ parity bits. If all the parity bits are equal to 1, the Hamming weight of the code becomes equal to $n - k + 1$, and this implies that

$$d_{\min} \leq n - k + 1.$$

***Maximum Distance Separable Codes***
A linear block code $\boldsymbol{C}(n,k)$ is a maximum distance separable code if its minimum distance satisfies the equality

$$d_{\min} = n - k + 1. \tag{2.53}$$

For a maximum distance separable code, the Hamming spheres with code-words at their centers contain all the words of the vector space. No words are left outside the Hamming spheres.

**Example 2.36** The minimum distance of a linear block code $\boldsymbol{C}(n = 10, k = 5)$ is given as

$$d_{\min} = 8.$$

Comment on the given minimum distance.

**Solution 2.36** The minimum distance of the code should satisfy the bound

$$d_{\min} \leq n - k + 1.$$

However, for the given parameters, we have

$$d_{\min} \leq n - k + 1 \rightarrow 8 \leq 10 - 5 + 1$$

which is not correct. This means that for the given $n$ and $k$ values, it is not possible to design a linear block code with minimum distance $d_{\min} = 8$.

***Hamming Bound***
Assume that the linear block code $\boldsymbol{C}(n,k)$ can correct $t_c$ number of errors. The Hamming bound for this code is defined as

$$\sum_{j=0}^{t_c} \binom{n}{j} \leq 2^{n-k}. \tag{2.54}$$

***Proof*** The total number of words in the vector space, from which the linear block code is constructed, can be calculated as $2^n$. There are $2^k$ code-words, and using these code-words, we can construct $2^k$ Hamming spheres centered at the code-words. A Hamming sphere includes

$$\sum_{j=0}^{t_c} \binom{n}{j} \tag{2.55}$$

number of words. The total number of words inside all the Hamming spheres can be calculated as

$$2^k \sum_{j=0}^{t_c} \binom{n}{j} \tag{2.56}$$

where $2^k$ indicates the number of Hamming spheres. Since the total number of words inside the Hamming spheres is smaller than or equal to the total number of words in the vector space, we can write the inequality

$$2^k \sum_{j=0}^{t_c} \binom{n}{j} \leq 2^n \tag{2.57}$$

from which we obtain the inequality

$$\sum_{j=0}^{t_c} \binom{n}{j} \leq 2^{n-k} \tag{2.58}$$

which is called the Hamming bound.

**Example 2.37** Using the Hamming bound for the code $\boldsymbol{C}(n = 10, k)$ for which the minimum distance is given as $d_{\min} = 5$, determine the largest possible value of $k$.

**Solution 2.37** The error correction capability of the code can be calculated as

$$t_c = \left\lfloor \frac{d_{\min} - 1}{2} \right\rfloor \rightarrow t_c = 2.$$

Using the given parameters, we can write the Hamming bound as

$$2^k \sum_{j=0}^{t_c} \binom{n}{j} \leq 2^n \rightarrow 2^k \sum_{j=0}^{2} \binom{10}{j} \leq 2^{10}. \quad (2.27) \tag{2.59}$$

When Eq. (2.59) is expanded, we obtain

$$2^k \leq \frac{1024}{\binom{10}{0}+\binom{10}{1}+\binom{10}{2}} \rightarrow 2^k \leq \frac{1024}{56} \rightarrow 2^k \leq 18.2 \rightarrow k \leq 4$$

from which it is clear that the largest value of $k$ can be 4.

***Gilbert-Varshamov Bound***

There exists a linear block code $\boldsymbol{C}(n,k)$ with minimum distance $d_{\min}$, if the bound

$$\binom{n-1}{0}+\binom{n-1}{1}+\ldots\binom{n-1}{d_{\min}-1} < 2^{n-k} \tag{2.60}$$

is satisfied for the given parameters.

Using Eq. (2.60), we can obtain the bound

$$2^k \geq \frac{2^{n-1}}{\binom{n-1}{0}+\binom{n-1}{1}+\ldots\binom{n-1}{d_{\min}-1}}. \tag{2.61}$$

**Example 2.38** Does there exist a linear block code $\boldsymbol{C}(n=5,k=2)$ with minimum distance $d_{\min}=4$?

**Solution 2.38** Employing Eq. (2.61) for the given parameters, we obtain

$$2^2 \geq \frac{2^4}{\binom{4}{0}+\binom{4}{1}+\binom{4}{2}+\binom{4}{3}} \rightarrow 4 > \frac{16}{15}$$

which is correct, and this means that a code exists with the given parameters.

***Plotkin Bound***

For a linear block code $\boldsymbol{C}(n,k)$, the minimum distance satisfies the bound

$$d_{\min} \leq n\frac{2^{k-1}}{2^k-1} \tag{2.62}$$

which is called the Plotkin bound. For large $k$ values, the Plotkin bound can be approximated as

$$d_{\min} \approx \frac{n}{2}. \tag{2.63}$$

## Problems

1. What is a vector space? State the properties of a vector space.
2. What is a subspace? How do we determine whether a vector space is a subspace or not?
3. What is a linear block code? Is it a subspace or an ordinary vector set?
4. The prime field $F_3$ is defined as

$$F_3 = \{0, 1, 2\}$$

$$+ \rightarrow \text{Mod} - 3 \text{ addition}$$

$$* \rightarrow \text{Mod} - 3 \text{ multiplication.}$$

Assume that, using the field elements, we construct a vector space with dimension 4, and $\boldsymbol{v}_1$ and $\boldsymbol{v}_2$ are two vectors chosen from this vector space. We calculate the subtraction of these two vectors as

$$\boldsymbol{v}_1 - \boldsymbol{v}_3. \tag{P2.1}$$

Is Eq. (P2.1) a meaningful operation?

5. A linear block code should always include all-zero code-word as one of its code-words. Is this statement correct or not? If it is correct, then why it is correct, and explain it. If it is not correct, then explain why it is not correct.
6. Using the binary field elements, construct a vector space with dimension 4, and from this vector space, find a subspace with dimension 3.
7. Write the standard basis elements of a vector space with dimension 5, and using the standard basis, generate three different bases of this vector space.
8. What is encoding? Explain it very briefly without mentioning any formula.
9. What is decoding? Explain it very briefly without mentioning any formula.
10. Why do we do encoding and decoding operations? Explain the reasoning behind them.
11. Assume that we have two binary linear block codes with the same dimension and length, i.e., code-words include the same number of bits and the number of code-words in both codes is the same. Which one do you choose? And which criteria is important in your choice?
12. The generator matrix of a binary linear block code is given as

$$\boldsymbol{G} = \begin{bmatrix} 101001 \\ 110110 \\ 101101 \end{bmatrix}.$$

Determine all the code-words for this code, and calculate the Hamming weight of each code-word. What is the minimum distance of this code?

13. The generator matrix of a binary linear block code is given as

$$G = \begin{bmatrix} 1001001 \\ 1110010 \\ 1010100 \\ 1001001 \end{bmatrix}.$$

Find the systematic form of this matrix and determine the parity check matrix of this code.

14. A binary subspace with dimension 2 is given as

$$B = \{10010, 01010, 00110, 10001\}.$$

Find a dual space of this subspace.

15. The parity check matrix of a code is the generator matrix of its dual code. Is this statement correct or not?
16. The parity check matrix of a code is given as

$$H = \begin{bmatrix} 1001100 \\ 0110100 \\ 1011010 \\ 1011001 \end{bmatrix}.$$

Find the generator matrix of this code.

17. The generator matrix of a code is given as

$$G = \begin{bmatrix} 1001001 \\ 1110010 \\ 1010110 \end{bmatrix}.$$

   (a) How many data-words can be encoded using this generator matrix?
   (b) What is the dimension of the vector space from which this code is constructed?
   (c) What is the dimension of code?

(d) How many vectors are available in the vector space from which this code is generated?
(e) How many code-words are available in this code?

18. The parity check matrix of a binary linear block code is given as

$$\boldsymbol{H} = \begin{bmatrix} 1001100 \\ 1100111 \\ 1011100 \end{bmatrix}.$$

Find the minimum distance of this code.

19. If I sum two code-words belonging to a linear block code, I obtain another code-word. Explain the reasoning behind this statement.
20. The minimum distance of a linear block code is given as

$$d_{\min} = 7.$$

(a) How many errors can this code detect?
(b) How many errors can this code correct?

21. The error correction capability of a linear block code is $t_c = 5$. What can be the minimum distance of this code?
22. We are informed that a linear block code can correct $t_c = 3$ bit errors. Does this mean that four errors can never be corrected? Explain your reasoning behind your answer.
23. Consider a vector space with dimension 5. Write all the vectors of this vector space. Locate all the vectors of this vector space on Hamming spheres with different radiuses, and at the center of the innermost Hamming sphere, we have the all-zero vector.
24. The minimum distance of a linear block code $\boldsymbol{C}(n = 15, k = 7)$ is given as $d_{\min} = 10$. Comment on the correctness of this minimum distance using the singleton bound.
25. What is maximum distance separable code? Explain it.
26. Write the Hamming bound formula.
27. Using the Hamming bound formula for the code $\boldsymbol{C}(n = 16, k)$ for which the minimum distance is given as $d_{\min} = 6$, determine the largest possible value of $k$.

# Chapter 3
# Syndrome Decoding and Some Important Linear Block Codes

## 3.1 Syndrome Decoding of Linear Block Codes

In this section, we will explain the decoding of linear block codes using syndromes.

***Cosets of a Linear Clock Code***
The binary field $F$ is given as

$$F = \{0, 1\}$$
$$\oplus \rightarrow \text{Mod} - 2 \text{ addition operation}$$
$$\otimes \rightarrow \text{Mod} - 2 \text{ multiplication operation.}$$

Let $\mathcal{V}$ be an $n$-dimensional vector space constructed using the elements of binary field and $\boldsymbol{C}$ be $k$-dimensional linear block code which is a subspace of $\mathcal{V}$. We can write the code $\boldsymbol{C}$ as

$$\boldsymbol{C} = [\boldsymbol{c}_1 \, \boldsymbol{c}_2 \ldots \boldsymbol{c}_M] \tag{3.1}$$

where $\boldsymbol{c}_i$, $i = 1, 2, \ldots, M$ are the code-words consisting of $n$-bits and $M = 2^k$. The number of words in the vector space $\mathcal{V}$ can be calculated as $N = 2^n$.

Let $\boldsymbol{w}$ be a word which appears in $\mathcal{V}$, and it does not appear in $\boldsymbol{C}$ and in the previously formed cosets. A coset of $\boldsymbol{C}$ is obtained as

$$\boldsymbol{C}_s = \boldsymbol{w} \oplus \boldsymbol{C} \rightarrow \boldsymbol{C}_s = [(\boldsymbol{w} \oplus \boldsymbol{c}_1) \ \ (\boldsymbol{w} \oplus \boldsymbol{c}_2) \ldots (\boldsymbol{w} \oplus \boldsymbol{c}_M)]. \tag{3.2}$$

The cosets of $\boldsymbol{C}$ contain the same number of elements as $\boldsymbol{C}$. The total number of cosets of $\boldsymbol{C}$ can be calculated as



O. Gazi, *Forward Error Correction via Channel Coding*,
https://doi.org/10.1007/978-3-030-33380-5_3

$$\frac{2^n}{2^k} - 1 \rightarrow 2^{n-k} - 1. \tag{3.3}$$

While determining $\boldsymbol{w}$, we pay attention to choose it among the words having smallest Hamming weights, and the first element of a coset is called coset leader.

**Example 3.1** A linear block code $\boldsymbol{C}(n = 4, k = 3)$ is given as

$$\boldsymbol{C} = [0000 \ \ 1001 \ \ 1101 \ \ 0111 \ \ 0100 \ \ 1110 \ \ 1010 \ \ 1011].$$

Find all the cosets of $\boldsymbol{C}$.

**Solution 3.1** There is

$$2^{n-k} - 1 = 2^{4-3} - 1 \rightarrow 1$$

coset, and we can find the single coset of $\boldsymbol{C}$ as

$$\text{let } \boldsymbol{w} = [1000] \text{ and } \boldsymbol{w} \in \boldsymbol{\mathcal{V}}, \boldsymbol{w} \notin \boldsymbol{C} \rightarrow \boldsymbol{C}_s = \boldsymbol{w} \oplus \boldsymbol{C} \rightarrow$$

$$\boldsymbol{C}_s = [1000 \oplus 0000 \ \ 1000 \oplus 1001 \ \ 1000 \oplus 1101 \ \ 1000 \oplus 0111 \ \ 1000 \oplus 0100 \\ 1000 \oplus 1110 \ \ 1000 \oplus 1010 \ \ 1000 \oplus 1011] \rightarrow$$

$$\boldsymbol{C}_s = [1000 \ \ 0001 \ \ 0101 \ \ 1111 \ \ 1100 \ \ 0110 \ 0010 \ 0011].$$

It is clear that the vector space can be written as

$$\boldsymbol{\mathcal{V}} = \boldsymbol{C} \cup \boldsymbol{C}_s. \tag{3.4}$$

**Example 3.2** A linear block code $\boldsymbol{C}(n = 4, k = 2)$ is given as

$$\boldsymbol{C} = [0000 \ \ 1001 \ \ 0101 \ \ 1100 \,].$$

Find all the cosets of $\boldsymbol{C}$.

**Solution 3.2** There are

$$2^{n-k} - 1 = 2^{4-2} - 1 \rightarrow 3$$

cosets, and we can find the cosets of $\boldsymbol{C}$ as follows:

The first coset can be found as

$$\text{let } \boldsymbol{w} = [1000] \text{ and } \boldsymbol{w} \in \boldsymbol{\mathcal{V}}, \boldsymbol{w} \notin \boldsymbol{C} \rightarrow \boldsymbol{C}_{s1} = \boldsymbol{w} \oplus \boldsymbol{C} \rightarrow$$

$$\boldsymbol{C}_{s1} = [1000 \oplus 0000 \;\; 1000 \oplus 1001 \; 1000 \oplus 0101 \;\; 1000 \oplus 1100\,] \rightarrow$$

$$\boldsymbol{C}_{s1} = [1000 \;\; 0001 \;\; 1101 \;\; 0100\,].$$

The second coset can be found as

$$\text{let } \boldsymbol{w} = [0010] \text{ and } \boldsymbol{w} \in \boldsymbol{\mathcal{V}}, \boldsymbol{w} \notin \boldsymbol{C}, \boldsymbol{w} \notin \boldsymbol{C}_{s1} \rightarrow \boldsymbol{C}_{s2} = \boldsymbol{w} \oplus \boldsymbol{C} \rightarrow$$

$$\boldsymbol{C}_{s2} = [0010 \oplus 0000 \;\; 0010 \oplus 1001 \;\; 0010 \oplus 0101 \;\; 0010 \oplus 1100\,] \rightarrow$$

$$\boldsymbol{C}_{s2} = [0010 \;\; 1011 \;\; 0111 \;\; 1110\,].$$

The third coset can be found as

$$\text{let } \boldsymbol{w} = [1010] \text{ and } \boldsymbol{w} \in \boldsymbol{\mathcal{V}}, \boldsymbol{w} \notin \boldsymbol{C} \rightarrow \boldsymbol{C}_{s3} = \boldsymbol{w} \oplus \boldsymbol{C} \rightarrow$$

$$\boldsymbol{C}_{s3} = [1010 \oplus 0000 \;\; 1010 \oplus 1001 \; 1010 \oplus 0101 \;\; 1010 \oplus 1100\,] \rightarrow$$

$$\boldsymbol{C}_{s3} = [1010 \;\; 0011 \;\; 1111 \;\; 0110\,].$$

It is clear that the vector space can be written as

$$\boldsymbol{\mathcal{V}} = \boldsymbol{C} \cup \boldsymbol{C}_{s1} \cup \boldsymbol{C}_{s2} \cup \boldsymbol{C}_{s3}. \tag{3.5}$$

## 3.2 Standard Array

Using the code and its cosets, we can form a matrix such that the first row of the matrix is the code and the other rows are cosets. This matrix is also called standard array. For instance, using the code and cosets of the previous example, we can form a standard array as

$$\overline{S}_A = \begin{bmatrix} \boldsymbol{C} & : 0000\ \ 1001\ \ 0101\ \ 1100 \\ \boldsymbol{C}_{s1} & : 1000\ \ 0001\ \ 1101\ \ 0100 \\ \boldsymbol{C}_{s2} & : 0010\ \ 1011\ \ 0111\ \ 1110 \\ \boldsymbol{C}_{s3} & : 1010\ \ 0011\ \ 1111\ \ 0110 \end{bmatrix}. \tag{3.6}$$

The standard array contains $2^{n-k}$ rows and $2^k$ columns, and the total number of elements in the standard array equals to $2^{n-k} \times 2^k = 2^n$ which is the total number of elements in the vector space.

Note that as we have mentioned before, the first element in each coset is called the coset leader.

For a $t_c$ error-correcting linear block code, we draw a horizontal line in the standard array to separate the rows whose coset leaders have Hamming weights less than or equal to $t_c$ from the rows whose coset leaders have Hamming weight greater than $t_c$.

**Example 3.3** The generator matrix of a linear block code is given as

$$\boldsymbol{G} = \begin{bmatrix} 10111 \\ 01101 \end{bmatrix}_{2\times 5}.$$

Find the code generated by this generator matrix, and find the cosets of the code and construct the standard array.

**Solution 3.3** The code generated by $\boldsymbol{G}$ can be obtained as

$$\boldsymbol{C} = [00000\ \ 10111\ \ 01101\ \ 11010].$$

The minimum distance of this code is

$$d_{\min} = 3.$$

This code can detect

$$t_d = d_{\min} - 1 \rightarrow t_d = 3 - 1 \rightarrow t_d = 2$$

bit errors and can correct

$$t_c = \left\lfloor \frac{d_{\min} - 1}{2} \right\rfloor \rightarrow t_c = \left\lfloor \frac{3 - 1}{2} \right\rfloor \rightarrow t_c = \lfloor 1 \rfloor \rightarrow t_c = 1$$

bit error.

**Table 3.1** Standard array for Example 3.3

| $C$ | 00000 | 10111 | 01101 | 11010 |
|---|---|---|---|---|
| | | | | |
| | | | | |
| | | | | |
| | | | | |
| | | | | |
| | | | | |
| | | | | |

For the obtained code, we have $2^{n-k} - 1 = 2^{5-2} - 1 \rightarrow 7$ cosets. The size of the standard array, i.e., matrix, is $2^{n-k} \times 2^k = 8 \times 4$, and the total number of elements in the standard array is 32.

Now, let's form the standard array as follows. First, we write the code as the first row of the array as in Table 3.1.

To determine the first coset, we first determine $\boldsymbol{w}$ which has the smallest Hamming weight. In fact, while determining the words $\boldsymbol{w}$, we pay attention to the error-correcting capability of the code and select the words $\boldsymbol{w}$ such that, if possible, their Hamming weights are smaller than or equal to $t_c$. Considering this, we can choose $\boldsymbol{w} = [10000]$, and determine the first coset as

$$\boldsymbol{w} = [10000] \text{ and } \boldsymbol{w} \in \mathcal{V}, \boldsymbol{w} \notin \boldsymbol{C} \rightarrow \boldsymbol{C}_{s1} = \boldsymbol{w} \oplus \boldsymbol{C} \rightarrow$$

$$\boldsymbol{C}_{s1} = [10000 \oplus 00000 \;\; 10000 \oplus 10111 \;\; 10000 \oplus 01101 \;\; 10000 \oplus 11010] \rightarrow$$

$$\boldsymbol{C}_{s1} = [10000 \;\; 00111 \;\; 11101 \;\; 01010].$$

Placing the first coset into the second row, we update the standard array as in Table 3.2.

The second coset can be found as

$$\text{let } \boldsymbol{w} = [01000] \text{ and } \boldsymbol{w} \in \mathcal{V}, \boldsymbol{w} \notin \boldsymbol{C}, \boldsymbol{w} \notin \boldsymbol{C}_{s1} \rightarrow \boldsymbol{C}_{s2} = \boldsymbol{w} \oplus \boldsymbol{C} \rightarrow$$

$$\boldsymbol{C}_{s2} = [01000 \oplus 00000 \;\; 01000 \oplus 10111 \;\; 01000 \oplus 01101 \;\; 01000 \oplus 11010] \rightarrow$$

$$\boldsymbol{C}_{s2} = [01000 \;\; 11111 \;\; 00101 \;\; 10010].$$

Placing the second coset into the third row, we get the standard array as in Table 3.3.

**Table 3.2** Standard array formation for Example 3.3

| | | | | |
|---|---|---|---|---|
| $\boldsymbol{C}$ | 00000 | 10111 | 01101 | 11010 |
| $\boldsymbol{C}_{s1}$ | 10000 | 00111 | 11101 | 01010 |
| | | | | |
| | | | | |
| | | | | |
| | | | | |
| | | | | |
| | | | | |

**Table 3.3** Standard array formation for Example 3.3

| | | | | |
|---|---|---|---|---|
| $\boldsymbol{C}$ | 00000 | 10111 | 01101 | 11010 |
| $\boldsymbol{C}_{s1}$ | 10000 | 00111 | 11101 | 01010 |
| $\boldsymbol{C}_{s2}$ | 01000 | 11111 | 00101 | 10010 |
| | | | | |
| | | | | |
| | | | | |
| | | | | |
| | | | | |

The third coset can be found as

$$\text{let } \boldsymbol{w} = [00100] \text{ and } \boldsymbol{w} \in \boldsymbol{\mathcal{V}}, \boldsymbol{w} \notin \boldsymbol{C}, \boldsymbol{w} \notin \boldsymbol{C}_{s1}, \boldsymbol{w} \notin \boldsymbol{C}_{s2} \rightarrow \boldsymbol{C}_{s3} = \boldsymbol{w} \oplus \boldsymbol{C} \rightarrow$$

$$\boldsymbol{C}_{s3} = [00100 \oplus 00000 \;\; 00100 \oplus 10111 \;\; 00100 \oplus 01101 \;\; 00100 \oplus 11010\,] \rightarrow$$

$$\boldsymbol{C}_{s3} = [00100 \;\; 10011 \;\; 01001 \;\; 11110].$$

Placing the third coset into the fourth row, we get the standard array as in Table 3.4.

The fourth coset can be found as

$$\text{let } \boldsymbol{w} = [00010] \text{ and } \boldsymbol{w} \in \boldsymbol{\mathcal{V}}, \boldsymbol{w} \notin \boldsymbol{C}, \boldsymbol{w} \notin \boldsymbol{C}_{si}, i = 1, \ldots, 3 \rightarrow \boldsymbol{C}_{s4} = \boldsymbol{w} \oplus \boldsymbol{C} \rightarrow$$

$$\boldsymbol{C}_{s4} = [00010 \oplus 00000 \;\; 00010 \oplus 10111 \;\; 00010 \oplus 01101 \;\; 00010 \oplus 11010\,] \rightarrow$$

**Table 3.4** Standard array formation for Example 3.3

| | | | | |
|---|---|---|---|---|
| $C$ | 00000 | 10111 | 01101 | 11010 |
| $C_{s1}$ | 10000 | 00111 | 11101 | 01010 |
| $C_{s2}$ | 01000 | 11111 | 00101 | 10010 |
| $C_{s3}$ | 00100 | 10011 | 01001 | 11110 |
| | | | | |
| | | | | |
| | | | | |
| | | | | |

**Table 3.5** Standard array formation for Example 3.3

| | | | | |
|---|---|---|---|---|
| $C$ | 00000 | 10111 | 01101 | 11010 |
| $C_{s1}$ | 10000 | 00111 | 11101 | 01010 |
| $C_{s2}$ | 01000 | 11111 | 00101 | 10010 |
| $C_{s3}$ | 00100 | 10011 | 01001 | 11110 |
| $C_{s4}$ | 00010 | 10101 | 01111 | 11000 |
| | | | | |
| | | | | |
| | | | | |

$$\boldsymbol{C}_{s4} = [00010\ \ 10101\ \ 01111\ \ 11000].$$

Placing the fourth coset into the fifth row, we get the standard array as in Table 3.5.

The fifth coset can be found as

$$\text{let } \boldsymbol{w} = [00001] \text{ and } \boldsymbol{w} \in \boldsymbol{\mathcal{V}}, \boldsymbol{w} \notin \boldsymbol{C}, \boldsymbol{w} \notin \boldsymbol{C}_{si}, i = 1, \ldots, 4 \rightarrow \boldsymbol{C}_{s5} = \boldsymbol{w} \oplus \boldsymbol{C} \rightarrow$$

$$\boldsymbol{C}_{s5} = [00001 \oplus 00000\ \ 00001 \oplus 10111\ \ 00001 \oplus 01101\ \ 00001 \oplus 11010\,] \rightarrow$$

$$\boldsymbol{C}_{s5} = [00001\ \ 10110\ \ 01100\ \ 11011].$$

Placing the fifth coset into the sixth row, we get the standard array as in Table 3.6.

For the sixth coset, there is no word $\boldsymbol{w}$ left with the Hamming weight "1." Then, we need to search a word $\boldsymbol{w}$ with Hamming weight "2" such that it should not be available in Table 3.6. Considering this, we can choose $\boldsymbol{w} = [00011]$ and determine the sixth coset as

**Table 3.6** Standard array formation for Example 3.3

| $C$ | 00000 | 10111 | 01101 | 11010 |
|---|---|---|---|---|
| $C_{s1}$ | 10000 | 00111 | 11101 | 01010 |
| $C_{s2}$ | 01000 | 11111 | 00101 | 10010 |
| $C_{s3}$ | 00100 | 10011 | 01001 | 11110 |
| $C_{s4}$ | 00010 | 10101 | 01111 | 11000 |
| $C_{s5}$ | 00001 | 10110 | 01100 | 11011 |
| | | | | |
| | | | | |

$$w = [00011] \text{ and } w \in \mathcal{V}, w \notin C, w \notin C_{si}, i = 1, \ldots, 5 \rightarrow C_{s6} = w \oplus C \rightarrow$$

$$C_{s6} = [00011 \oplus 00000 \;\; 00011 \oplus 10111 \;\; 00011 \oplus 01101 \;\; 00011 \oplus 11010\,] \rightarrow$$

$$C_{s6} = [00011 \;\; 10100 \;\; 01110 \;\; 11001].$$

Placing the sixth coset into the seventh row, we get the standard array as in Table 3.7.

For the seventh coset, we need to search a word $w$ with Hamming weight 2 such that it should not be available in Table 3.7. Considering this, we can choose $w = [00110]$ and determine the sixth coset as

$$w = [00110] \text{ and } w \in \mathcal{V}, w \notin C, w \notin C_{si}, i = 1, \ldots, 6 \rightarrow C_{s7} = w \oplus C \rightarrow$$

$$C_{s7} = [00110 \oplus 00000 \;\; 00110 \oplus 10111 \;\; 00110 \oplus 01101 \;\; 00110 \oplus 11010\,] \rightarrow$$

$$C_{s7} = [00110 \;\; 10001 \;\; 01011 \;\; 11100].$$

Placing the seventh coset into the eighth row, we get the standard array as in Table 3.8.

If we omit the leftmost column of Table 3.8, and draw a horizontal line below to the coset leaders with Hamming weight less than or equal to $t_c$, we get the final form of standard array as in Table 3.9.

**Table 3.7** Standard array formation for Example 3.3

| $C$ | 00000 | 10111 | 01101 | 11010 |
|---|---|---|---|---|
| $C_{s1}$ | 10000 | 00111 | 11101 | 01010 |
| $C_{s2}$ | 01000 | 11111 | 00101 | 10010 |
| $C_{s3}$ | 00100 | 10011 | 01001 | 11110 |
| $C_{s4}$ | 00010 | 10101 | 01111 | 11000 |
| $C_{s5}$ | 00001 | 10110 | 01100 | 11011 |
| $C_{s6}$ | 00011 | 10100 | 01110 | 11001 |
| | | | | |

**Table 3.8** Standard array formation for Example 3.3

| $C$ | 00000 | 10111 | 01101 | 11010 |
|---|---|---|---|---|
| $C_{s1}$ | 10000 | 00111 | 11101 | 01010 |
| $C_{s2}$ | 01000 | 11111 | 00101 | 10010 |
| $C_{s3}$ | 00100 | 10011 | 01001 | 11110 |
| $C_{s4}$ | 00010 | 10101 | 01111 | 11000 |
| $C_{s5}$ | 00001 | 10110 | 01100 | 11011 |
| $C_{s6}$ | 00011 | 10100 | 01110 | 11001 |
| $C_{s7}$ | 00110 | 10001 | 01011 | 11100 |

**Table 3.9** Standard array formation for Example 3.3

| | | | |
|---|---|---|---|
| 00000 | 10111 | 01101 | 11010 |
| 10000 | 00111 | 11101 | 01010 |
| 01000 | 11111 | 00101 | 10010 |
| 00100 | 10011 | 01001 | 11110 |
| 00010 | 10101 | 01111 | 11000 |
| 00001 | 10110 | 01100 | 11011 |
| 00011 | 10100 | 01110 | 11001 |
| 00110 | 10001 | 01011 | 11100 |

## 3.3 Error Correction Using Standard Array

The code whose standard array is given in Table 3.9 can correct single errors, and the first row of the table is the code; the other rows are the cosets of this code. We can use the standard array for error correction operations. Let's illustrate the concept by an example.

**Table 3.10** Transmitted and received words

| | | | |
|---|---|---|---|
| 00000 | **10111** | 01101 | 11010 |
| 10000 | 00111 | 11101 | 01010 |
| 01000 | 11111 | 00101 | 10010 |
| 00100 | 10011 | 01001 | 11110 |
| 00010 | 10101 | 01111 | 11000 |
| 00001 | **10110** | 01100 | 11011 |
| 00011 | 10100 | 01110 | 11001 |
| 00110 | 10001 | 01011 | 11100 |

**Example 3.4** Assume that the code-word $\boldsymbol{c_1} = [10111]$ is transmitted, and during transmission, a single-bit error occurs. Assume that the error word is $\boldsymbol{e} = [00001]$. Then, the received word happens to be $\boldsymbol{r} = \boldsymbol{c_1} + \boldsymbol{e} \rightarrow \boldsymbol{r} = [10110^*]$ where the erroneous bit is indicated by a $*$. The transmitted code-word and the received word are indicated in bold-black and bold-red colors in Table 3.10.

The received word is an element of a coset that is above the bold line, and looking at the coset leader, we can determine the error word as $\boldsymbol{e} = [00001]$. This is illustrated in Table 3.11.

Adding the error word to the received word, we find the transmitted code-word. That is

$$\boldsymbol{c} = \boldsymbol{r} + \boldsymbol{e} \rightarrow \boldsymbol{c} = [10110] + [00001] \rightarrow \boldsymbol{c} = [10111].$$

Now, assume that $\boldsymbol{c_1} = [10111]$ is transmitted, and during transmission, two-bit errors occur. Assume that the error word is $\boldsymbol{e} = [00011]$. Then, the received word happens to be

$$\boldsymbol{r} = \boldsymbol{c}_1 + \boldsymbol{e} \rightarrow \boldsymbol{r} = [10100].$$

The transmitted code-word and the received word are indicated in bold-black and bold-red colors in Table 3.12. The received word is an element of a coset, and looking at the coset leader, we can determine the error word as $\boldsymbol{e} = [00011]$. This is illustrated in Table 3.12.

Adding the error word to the received word, we find the transmitted code-word. That is

$$\boldsymbol{c} = \boldsymbol{r} + \boldsymbol{e} \rightarrow \boldsymbol{c} = [10100] + [00011] \rightarrow \boldsymbol{c} = [10111].$$

Although our code is a single-error-correcting code, we were able to correct two-bit errors using the standard array.

**Table 3.11** Determination of coset leader

| | | | |
|---|---|---|---|
| 00000 | **10111** | 01101 | 11010 |
| 10000 | 00111 | 11101 | 01010 |
| 01000 | 11111 | 00101 | 10010 |
| 00100 | 10011 | 01001 | 11110 |
| 00010 | 10101 | 01111 | 11000 |
| **00001** ← | **10110** | 01100 | 11011 |
| 00011 | 10100 | 01110 | 11001 |
| 00110 | 10001 | 01011 | 11100 |

**Table 3.12** Case of two-bit errors

| | | | |
|---|---|---|---|
| 00000 | **10111** | 01101 | 11010 |
| 10000 | 00111 | 11101 | 01010 |
| 01000 | 11111 | 00101 | 10010 |
| 00100 | 10011 | 01001 | 11110 |
| 00010 | 10101 | 01111 | 11000 |
| 00001 | 10110 | 01100 | 11011 |
| **00011** ← | **10100** | 01110 | 11001 |
| 00110 | 10001 | 01011 | 11100 |

Now, assume that $\boldsymbol{c}_1 = [10111]$ is transmitted, and during transmission, two-bit errors occur. Assume that the error word is $\boldsymbol{e} = [01001]$. Then, the received word happens to be

$$\boldsymbol{r} = \boldsymbol{c}_1 + \boldsymbol{e} \rightarrow \boldsymbol{r} = [11110].$$

The transmitted code-word and the received word are indicated in bold-black and bold-red colors in Table 3.13. The received word is an element of a coset, and looking at the coset leader, we can determine the error word as $\boldsymbol{e} = [00100]$. This is illustrated in Table 3.13.

Adding the error word to the received word, we find the transmitted code-word as shown in Table 3.13. That is

$$\boldsymbol{c} = \boldsymbol{r} + \boldsymbol{e} \rightarrow \boldsymbol{c} = [11110] + [00100] \rightarrow \boldsymbol{c} = [11010].$$

However, this result is wrong. As we see, our code couldn't correct two-bit errors.

Our code can correct single-error bits for sure. However, more errors can sometimes be corrected also, but this can happen only for some specific cases.

**Table 3.13** Case of two-bit errors

| | | | |
|---|---|---|---|
| 00000 | **10111** | 01101 | **11010** |
| 10000 | 00111 | 11101 | 01010 |
| 01000 | 11111 | 00101 | 10010 |
| **00100** | 10011 | 01001 | **11110** |
| 00010 | 10101 | 01111 | 11000 |
| 00001 | 10110 | 01100 | 11011 |
| 00011 | 10100 | 01110 | 11001 |
| 00110 | 10001 | 01011 | 11100 |

## 3.4 Syndrome

Let $\boldsymbol{c}$ be a code-word transmitted, and $\boldsymbol{r}$ is the received word such that $\boldsymbol{r} = \boldsymbol{c} + \boldsymbol{e}$ where $\boldsymbol{e}$ is the error word. We know that

$$\boldsymbol{c} \times \boldsymbol{H}^T = \boldsymbol{0}. \tag{3.7}$$

The Eq. (3.7) implies that if

$$\boldsymbol{r} \times \boldsymbol{H}^T = \boldsymbol{0} \tag{3.8}$$

then we accept that no error did occur during the transmission, and $\boldsymbol{r} = \boldsymbol{c}$. On the other hand, if

$$\boldsymbol{r} \times \boldsymbol{H}^T \neq \boldsymbol{0} \tag{3.9}$$

then we understand that some bit errors occurred during the transmission.

**Definition** The syndrome for the received word $\boldsymbol{r}$ is defined as

$$\boldsymbol{s} = \boldsymbol{r} \times \boldsymbol{H}^T. \tag{3.10}$$

If $\boldsymbol{s} = \boldsymbol{0}$, then we accept that no error occurred during the transmission; otherwise, we accept that some errors occurred during the transmission.

If we substitute $\boldsymbol{r} = \boldsymbol{c} + \boldsymbol{e}$ in Eq. (3.10), we get

$$\boldsymbol{s} = (\boldsymbol{c} + \boldsymbol{e}) \times \boldsymbol{H}^T \rightarrow \boldsymbol{s} = \underbrace{\boldsymbol{c} \times \boldsymbol{H}^T}_{=\boldsymbol{0}} + \boldsymbol{e} \times \boldsymbol{H}^T \rightarrow \boldsymbol{s} = \boldsymbol{e} \times \boldsymbol{H}^T.$$

That is, we can define the syndrome for the received word $\boldsymbol{r}$ as

$$s = e \times H^T. \tag{3.11}$$

**Example 3.5** The dimension of a vector space is given as $n = 6$. Using this vector space, a linear block code is designed, and the parity check matrix of this linear block code is provided as

$$H = \left[h_1^T \; h_2^T \; h_3^T \; h_4^T \; h_5^T \; h_6^T\right]$$

where $h_i^T, i = 1, \ldots, n$ are the column vectors of length $n - k$ where $k = 3$. Assume that the designed code is a single-error-correcting code. Using single-error pattern

$$e_1 = [100000]$$

a syndrome of this code is calculated as

$$s_1 = e_1 \times H^T \rightarrow s_1 = [100000] \times \begin{bmatrix} h_1 \\ h_2 \\ h_3 \\ h_4 \\ h_5 \\ h_6 \end{bmatrix} \rightarrow s_1 = h_1. \tag{3.12}$$

For the other single-error patterns

$$e_2 = [010000] \;\; e_3 = [001000] \;\; e_4 = [000100] \;\; e_5 = [000010] \; e_6 = [000001]$$

using $s_i = e_i \times H^T$, we can calculate the syndromes as

$$s_2 = h_2 \quad s_3 = h_3 \quad s_4 = h_4 \quad s_5 = h_5 \quad s_6 = h_6. \tag{3.13}$$

From Eqs. (3.12) and (3.13),we see that for single-error-correcting codes, the syndromes are columns of parity check matrix transposed.

Assume that the designed code is a double-error-correcting code. For single-error patterns, the syndromes happen to be the columns of the parity check matrix. On the other hand, for double-error patterns, the syndromes are obtained by summing any two columns of the parity check matrix considering the error pattern. For instance, if the error pattern is

$$e_1 = [100001],$$

the syndrome is calculated as

$$s_1 = e_1 \times H^T \rightarrow s_1 = [100001] \times \begin{bmatrix} h_1 \\ h_2 \\ h_3 \\ h_4 \\ h_5 \\ h_6 \end{bmatrix} \rightarrow s_1 = h_1 + h_6. \tag{3.14}$$

If the designed code can correct $t_c$ errors, then a syndrome equals to the sum of $t_c$ columns of parity check matrix. For instance, for three-error-correcting code, the syndrome for the error pattern

$$e_1 = [011001]$$

is calculated as

$$s_1 = e_1 \times H^T \rightarrow s_1 = [011001] \times \begin{bmatrix} h_1 \\ h_2 \\ h_3 \\ h_4 \\ h_5 \\ h_6 \end{bmatrix} \rightarrow s_1 = h_2 + h_3 + h_6 \tag{3.15}$$

which is the sum of the transposed three columns of parity check matrix.

### 3.4.1 Syndrome Table

Standard array may be difficult to construct for some codes. Instead syndrome table is used for its easiness. Syndrome table can be accepted as a concise representation of the standard array.

For a $t_c$-bit error-correcting code, we can construct a syndrome table including all the error patterns with Hamming weights $\leq t_c$, and using the syndrome table, we can correct the transmission bit errors.

We can construct a table involving $2^{n-k} - 1$ rows just as we construct the standard array. However, for those error patterns involving more than $t_c$-bit errors, correction is not guaranteed. For this reason, we usually construct the syndrome table for those error patterns with Hamming weights less than or equal to $t_c$.

**Example 3.6** Let's construct the syndrome table of a single-error-correcting code whose parity check matrix is given as

$$\boldsymbol{H} = \begin{bmatrix} 011100 \\ 101010 \\ 110001 \end{bmatrix}.$$

Although our code is a single-error-correcting code, let's construct the syndrome table such that it contains $2^{n-k} - 1$ rows.

First, considering all the single-error patterns, and using the equation $\boldsymbol{s} = \boldsymbol{e} \times \boldsymbol{H}^T$, we can construct part of the table as in Table 3.14.

Now, let's choose the double-error pattern as

$$\boldsymbol{e} = [000011].$$

If we calculate the syndrome using $\boldsymbol{s} = \boldsymbol{e} \times \boldsymbol{H}^T$, we obtain

$$\boldsymbol{s} = [011].$$

However, this syndrome appears in the last row of the table. For this reason, we should change the error pattern. In fact, we should continue trying the error patterns until we obtain a syndrome not appearing in the table.

If we choose the double-error pattern as

$$\boldsymbol{e} = [001001]$$

and calculate the syndrome using $\boldsymbol{s} = \boldsymbol{e} \times \boldsymbol{H}^T$, we obtain

$$\boldsymbol{s} = [111].$$

Thus, our complete syndrome table happens to be as in Table 3.15.

If we choose other error patterns involving two or more "1's," the generated syndromes will be a replica of those available in the second column of Table 3.15.

For instance, if we choose the error pattern as $\boldsymbol{e} = [110001]$, we obtain the syndrome $\overline{s} = [111]$. Interestingly, for some error patterns involving more than one "1," we can calculate the syndrome as $\boldsymbol{s} = [000]$ which means error-free transmission. However, this is not correct. The reason for obtaining zero syndromes for nonzero error patterns is that the error patterns are beyond the error correction capability of our code. For instance, for error pattern $\boldsymbol{e} = [100011]$, the syndrome happens to be $\boldsymbol{s} = [000]$.

**Table 3.14** Syndrome table for Example 3.6

| $e$ | $s = e \times H^T$ |
|---|---|
| 000001 | 001 |
| 000010 | 010 |
| 000100 | 100 |
| 001000 | 110 |
| 010000 | 101 |
| 100000 | 011 |
| | |

**Table 3.15** Syndrome table for Example 3.6

| $e$ | $s = e \times H^T$ |
|---|---|
| 000001 | 001 |
| 000010 | 010 |
| 000100 | 100 |
| 001000 | 110 |
| 010000 | 101 |
| 100000 | 011 |
| 001001 | 111 |

## 3.5 Syndrome Decoding

Assume that we transmit the code-word $\boldsymbol{c}$, and $\boldsymbol{r} = \boldsymbol{c} + \boldsymbol{e}$ is the received word. To identify the error pattern $\boldsymbol{e}$ at the receiver side, we can employ syndrome decoding which needs the syndrome table of the linear block code. The syndrome decoding operation is outlined as follows:

*Syndrome Decoding*

1. Using the received word and parity check matrix of the linear block code, calculate the syndrome

$$s = \boldsymbol{r} \times \boldsymbol{H}^T. \tag{3.16}$$

2. Locate the calculated syndrome in the syndrome table and find the corresponding error pattern $\boldsymbol{e}$.
3. Perform the error correction operation using

$$c = r + e. \tag{3.17}$$

**Example 3.7** The parity check matrix of a single-error-correcting linear block code is given as

$$H = \begin{bmatrix} 001011 \\ 101100 \\ 110001 \end{bmatrix}.$$

(a) Obtain the syndrome table of this code.
(b) Assume that the code-word $c = [100111]$ is transmitted. Determine the decoder's estimate about the transmitted code-word when $c$ incurs the error patterns:

1. $e_1 = [010000]$ 2) $e_2 = [110000]$ 3) $e_3 = [011010]$ 4) $e_4 = [011101]$.

**Solution 3.7** (a) Our code is a single-error-correcting code, and for this reason, it is sufficient to use only single-error patterns while constructing the syndrome table. First, we place the single-error patterns to the left column as shown in Table 3.16.

In the sequel, we calculate the syndromes using $s = e \times H^T$ and place them into the right column. Note that for single-error patterns, syndromes are the transposed columns of the parity check matrix. For instance, for the error pattern $e = [100000]$, the syndrome is the transpose of the first column of the parity check matrix, i.e., the syndrome is

$$s = [100000] \times \begin{bmatrix} 001011 \\ 101100 \\ 110001 \end{bmatrix}^T \rightarrow s = \begin{bmatrix} 0 \\ 1 \\ 1 \end{bmatrix}^T \rightarrow s = [011].$$

When all the syndromes are calculated and placed to the right column, we get the syndrome Table 3.17.

(b) Using the error pattern $e_1 = [010000]$, the received word is calculated as

$$r = c + e_1 \rightarrow r = [100111] + [010000] \rightarrow r = [110111].$$

At the receiver side, we first calculate the syndrome using the received word as

**Table 3.16** Syndrome table formation for Example 3.7

| $e$ | $s = e \times H^T$ |
|---|---|
| 000001 | |
| 000010 | |
| 000100 | |
| 001000 | |
| 010000 | |
| 100000 | |

**Table 3.17** Syndrome table for Example 3.7

| $e$ | $s = e \times H^T$ |
|---|---|
| 000001 | 101 |
| 000010 | 100 |
| 000100 | 010 |
| 001000 | 110 |
| 010000 | 001 |
| 100000 | 011 |

$$s = r \times H^T \rightarrow s = [110111] \times \begin{bmatrix} 001011 \\ 101100 \\ 110001 \end{bmatrix}^T \rightarrow$$

$$s = \begin{bmatrix} 0 \\ 1 \\ 1 \end{bmatrix}^T + \begin{bmatrix} 0 \\ 0 \\ 1 \end{bmatrix}^T + \begin{bmatrix} 0 \\ 1 \\ 0 \end{bmatrix}^T + \begin{bmatrix} 1 \\ 0 \\ 0 \end{bmatrix}^T + \begin{bmatrix} 1 \\ 0 \\ 1 \end{bmatrix}^T \rightarrow$$

$$s = \begin{bmatrix} 0 \\ 0 \\ 1 \end{bmatrix}^T \rightarrow s = [001].$$

Since syndrome is not equal to zero vector, we understand that bit errors did occur during the transmission. That is, errors are detected. We search the syndrome in the syndrome table and find the corresponding error pattern. This is illustrated in Table 3.18.

In the last stage, we add the error pattern found from the syndrome table to the received word to correct the transmission errors as

$$c = r + e \rightarrow c = [110111] + [010000] \rightarrow c = [100111]. \quad (3.18)$$

**Table 3.18** Syndrome table

| $e$ | $s = e \times H^T$ |
|---|---|
| 000001 | 101 |
| 000010 | 100 |
| 000100 | 010 |
| 001000 | 110 |
| **010000** ← | **001** |
| 100000 | 011 |

When we inspect the code-word in Eq. (3.18), we see that it is the correct transmitted code-word. Since a single-bit error did occur during the transmission, our code was able to correct it.

2. Using the error pattern $\boldsymbol{e_2} = [110000]$, the received word is calculated as

$$\boldsymbol{r} = \boldsymbol{c} + \boldsymbol{e}_2 \rightarrow \boldsymbol{r} = [100111] + [110000] \rightarrow \boldsymbol{r} = [010111].$$

At the receiver side, we first calculate the syndrome using the received word as

$$\boldsymbol{s} = \boldsymbol{r} \times \boldsymbol{H}^T \rightarrow \boldsymbol{s} = [010111] \times \begin{bmatrix} 001011 \\ 101100 \\ 110001 \end{bmatrix}^T \rightarrow \boldsymbol{s}$$

$$= \begin{bmatrix} 0 \\ 0 \\ 1 \end{bmatrix}^T + \begin{bmatrix} 0 \\ 1 \\ 0 \end{bmatrix}^T + \begin{bmatrix} 1 \\ 0 \\ 0 \end{bmatrix}^T + \begin{bmatrix} 1 \\ 0 \\ 1 \end{bmatrix}^T \rightarrow$$

$$\boldsymbol{s} = \begin{bmatrix} 0 \\ 1 \\ 0 \end{bmatrix}^T \rightarrow \boldsymbol{s} = [010].$$

Since syndrome is not equal to zero vector, we understand that bit errors did occur during the transmission. That is, errors are detected. We search the syndrome in the syndrome table and find the corresponding error patter. This is illustrated in Table 3.19.

In the last stage, we add the error pattern found from the syndrome table to the received word to correct the transmission errors as

**Table 3.19** Determination of error pattern using syndrome table

| $e$ | $s = e \times H^T$ |
|---|---|
| 000001 | 101 |
| 000010 | 100 |
| 000100← | 010 |
| 001000 | 110 |
| 010000 | 001 |
| 100000 | 011 |

$$c = r + e \rightarrow c = [010111] + [000100] \rightarrow c = [010011]. \quad (3.19)$$

For the word obtained in Eq. (3.19), we check $c \times H^T = \overline{0}$ condition to determine whether the obtained word is a code-word or not. The obtained word satisfies the condition $c \times H^T = \overline{0}$. That means that it is a code-word. However, we see that it is NOT the correct transmitted code-word. Since two-bit errors did occur during the transmission, our code was NOT able to correct it.

3. Using the error pattern $e_3 = [011010]$, the received word is calculated as

$$r = c + e_3 \rightarrow r = [100111] + [011010] \rightarrow r = [111101].$$

At the receiver side, we first calculate the syndrome using the received word as

$$s = r \times H^T \rightarrow s = [111101] \times \begin{bmatrix} 001011 \\ 101100 \\ 110001 \end{bmatrix}^T \rightarrow$$

$$s = \begin{bmatrix} 0 \\ 1 \\ 1 \end{bmatrix}^T + \begin{bmatrix} 0 \\ 0 \\ 1 \end{bmatrix}^T + \begin{bmatrix} 1 \\ 1 \\ 0 \end{bmatrix}^T + \begin{bmatrix} 0 \\ 1 \\ 0 \end{bmatrix}^T + \begin{bmatrix} 1 \\ 0 \\ 1 \end{bmatrix}^T \rightarrow$$

$$s = \begin{bmatrix} 0 \\ 1 \\ 1 \end{bmatrix}^T \rightarrow s = [011].$$

Since syndrome is not equal to zero vector, we understand that bit errors did occur during the transmission. That is, errors are detected. We search the syndrome in the syndrome table and find the corresponding error pattern. This is illustrated in Table 3.20.

In the last stage, we add the error pattern found from the syndrome table to the received word to correct the transmission errors as follows:

$$\boldsymbol{c} = \boldsymbol{r} + \boldsymbol{e} \rightarrow \boldsymbol{c} = [111101] + [100000] \rightarrow \boldsymbol{c} = [011101]. \tag{3.20}$$

For the word obtained in Eq. (3.20), we check $\boldsymbol{c} \times \boldsymbol{H}^T = \boldsymbol{0}$ condition to determine whether the obtained word is a code-word or not. The obtained word satisfies the condition $\boldsymbol{c} \times \boldsymbol{H}^T = \boldsymbol{0}$. That means it is a code-word. However, we see that it is NOT the correct transmitted code-word. Since three-bit errors did occur during the transmission, our code was NOT able to correct it.

4. Using the error pattern $\boldsymbol{e_4} = [011101]$, the received word is calculated as

$$\boldsymbol{r} = \boldsymbol{c} + \boldsymbol{e_4} \rightarrow \boldsymbol{r} = [100111] + [011101] \rightarrow \boldsymbol{r} = [111010].$$

At the receiver side, we first calculate the syndrome using the received word as

$$\boldsymbol{s} = \boldsymbol{r} \times \boldsymbol{H}^T \rightarrow \boldsymbol{s} = [011010] \times \begin{bmatrix} 001011 \\ 101100 \\ 110001 \end{bmatrix}^T \rightarrow \boldsymbol{s} = [000].$$

We see that $\boldsymbol{s} = [000]$, and this means that no error did occur during the transmission. However, this is wrong result. Since four-bit errors occurred during the transmission, our code was not able to detect or correct the errors.

**Table 3.20** Case of three errors

| $\boldsymbol{e}$ | $\boldsymbol{s} = \boldsymbol{e} \times \boldsymbol{H}^T$ |
|---|---|
| 000001 | 101 |
| 000010 | 100 |
| 000100 | 010 |
| 001000 | 110 |
| 010000 | 001 |
| **100000** ← | **011** |

**Example 3.8** The parity check matrix of a linear block code is given as

$$\boldsymbol{H} = \begin{bmatrix} 0010110 \\ 1011000 \\ 1100010 \\ 1111111 \end{bmatrix}.$$

Find the syndrome corresponding to the error pattern

$$\boldsymbol{e} = [0100101].$$

**Solution 3.8** When the syndrome equation

$$\boldsymbol{s} = \boldsymbol{e} \times \boldsymbol{H}^T$$

is used for the given error pattern and parity check matrix, we obtain

$$\boldsymbol{s} = \begin{bmatrix} 0 \\ 0 \\ 1 \\ 1 \end{bmatrix}^T + \begin{bmatrix} 1 \\ 0 \\ 0 \\ 1 \end{bmatrix}^T + \begin{bmatrix} 0 \\ 0 \\ 0 \\ 1 \end{bmatrix}^T \rightarrow \boldsymbol{s} = [1011]$$

where the it is seen that the transposed column vectors are chosen according to the indexes of the "1's" appearing in the error pattern, i.e., "1's" appear at the 2. , 5. ,and 7. positions in the error vector, and we choose the columns at the 2. , 5. ,and 7. positions in the parity check matrix.

## 3.6 Some Well-Known Linear Block Codes

In this section, we will provide a brief information about some linear block codes well-known in the literature.

### *3.6.1 Single Parity Check Codes*

The generator matrix of a single parity check code is given as

$$G = \left[\begin{array}{c|c} 100\cdots00 & 1 \\ 010\cdots00 & 1 \\ \vdots & \vdots \\ 000\cdots01 & 1 \end{array}\right]_{k\times n}. \tag{3.21}$$

When the data-word

$$\boldsymbol{d} = [d_1\ d_2 \ldots d_k] \tag{3.22}$$

is encoded using the generator matrix in Eq. (3.21), we obtain

$$\boldsymbol{c} = \boldsymbol{d} \times \boldsymbol{G} \rightarrow \boldsymbol{c} = [d_1\ d_2 \ldots d_k\ p] \text{ where } p = d_1 + d_2 + \ldots d_k. \tag{3.23}$$

**Example 3.9** The generator matrix of a single parity check code is given as

$$G = \left[\begin{array}{c|c} 1000 & 1 \\ 0100 & 1 \\ 0010 & 1 \\ 0001 & 1 \end{array}\right]_{4\times 5}.$$

Find the parity check matrix of this code.

**Solution 3.9** If $\boldsymbol{G}$ is in the form $\boldsymbol{G} = [\boldsymbol{I}|\boldsymbol{P}]$, then the parity check matrix is obtained as

$$\boldsymbol{H} = \left[\boldsymbol{P}^{\boldsymbol{T}}|\boldsymbol{I}\right].$$

Accordingly, we get the parity check matrix as

$$\boldsymbol{H} = [1111|1]_{1\times 5}.$$

### 3.6.2 Repetition Codes

The generator matrix of a repetition code is given as

$$\boldsymbol{G} = [11 \ldots 1]_{1\times n}. \tag{3.24}$$

When the data bit

$$\boldsymbol{d} = [d_1] \tag{3.25}$$

is encoded using the generator matrix in Eq. (3.24), we obtain

$$\boldsymbol{c} = \boldsymbol{d} \times \boldsymbol{G} \rightarrow \boldsymbol{c} = [d_1\ d_1 \ldots d_1]. \tag{3.26}$$

**Example 3.10** The generator matrix of a repetition code is given as

$$\boldsymbol{G} = [11111]$$

Find the parity check matrix of this code.

**Solution 3.10** If $\boldsymbol{G}$ is in the form $\boldsymbol{G} = [\boldsymbol{I} | \boldsymbol{P}]$, then the parity check matrix is obtained as

$$\boldsymbol{H} = \left[\boldsymbol{P}^T | \boldsymbol{I}\right].$$

Accordingly, we get the parity check matrix as

$$\boldsymbol{H} = \left[\begin{array}{c|c} 1 & 1000 \\ 1 & 0100 \\ 1 & 0010 \\ 1 & 0001 \end{array}\right]_{4\times5}.$$

### 3.6.3 Golay Code

The Golay code is discovered in the late of 1940s. The generator matrix of the Golay code is in the form

$$\boldsymbol{G} = [\boldsymbol{I}|\boldsymbol{P}]_{12\times24} \tag{3.27}$$

where $\boldsymbol{I}$ is the identity matrix with size $12 \times 12$, $\boldsymbol{P}$ is the parity matrix with size $12 \times 12$, and it is defined as

$$P = \begin{bmatrix} 011111111111 \\ 111011100010 \\ 110111000101 \\ 101110001011 \\ 111100010110 \\ 111000101101 \\ 110001011011 \\ 100010110111 \\ 100101101110 \\ 101011011100 \\ 110110111000 \\ 101101110001 \end{bmatrix}. \tag{3.28}$$

The Golay code is self-dual code, that is, $\boldsymbol{G}\boldsymbol{G}^T = \boldsymbol{0}$.

The Hamming weight of any code-word is a multiple of 4, i.e., the Hamming weights of the code-words can be 0, 8, 12, 16, or 24. The minimum distance of the Golay code is 8, and this means that the Golay code is a three-error-correcting linear block code.

Golay codes were used for the encoding and decoding of the science and engineering data in Voyager 1 and Voyager 2 spacecrafts which were launched toward Jupiter and Saturn in 1977.

**Exercise** Find the number of rows and columns for the standard array of the Golay code.

### 3.6.4 Reed-Muller Codes

Reed-Muller code is a block code $\boldsymbol{C}(n, k)$ whose code parameters $k$ and $n$ are calculated according to

$$k = m + 1 \quad n = 2^m \tag{3.29}$$

where $m$ is an integer such that $m \geq 1$. A Reed-Muller code is indicated by

$$RM(m).$$

The generator matrix of the Reed-Muller code $RM(1)$ is defined as

$$\boldsymbol{G}_1 = \begin{bmatrix} 11 \\ 01 \end{bmatrix}.$$

The generator matrix of the Reed-Muller code *RM*(*m*) is calculated in a recursive manner as

$$\boldsymbol{G}_m = \begin{bmatrix} \boldsymbol{G}_{m-1} & \boldsymbol{G}_{m-1} \\ 0\ldots0 & 1\ldots1 \end{bmatrix} \rightarrow \boldsymbol{G}_m = \begin{bmatrix} \boldsymbol{G}_{m-1} & \boldsymbol{G}_{m-1} \\ \boldsymbol{0} & \overline{\boldsymbol{1}} \end{bmatrix}. \tag{3.30}$$

The minimum distance of the Reed-Muller code *RM*(*m*) is

$$d_{\min} = 2^{m-1} \tag{3.31}$$

which implies that the error detection and correction capability of the Reed-Muller code *RM*(*m*) can be calculated using

$$t_d = 2^{m-1} - 1 \tag{3.32}$$

and

$$t_c = \left\lfloor \frac{2^{m-1} - 1}{2} \right\rfloor. \tag{3.33}$$

**Example 3.11** Find the generator matrices of the Reed-Muller codes *RM*(2), *RM*(3), and determine minimum distances of the corresponding codes.

**Solution 3.11** Using the $\boldsymbol{G}_1$ matrix which is given as

$$\boldsymbol{G}_1 = \begin{bmatrix} 11 \\ 01 \end{bmatrix}$$

and employing the recursion

$$\boldsymbol{G}_m = \begin{bmatrix} \boldsymbol{G}_{m-1} & \boldsymbol{G}_{m-1} \\ \boldsymbol{0} & \overline{\boldsymbol{1}} \end{bmatrix}$$

for $m = 2$ and $m = 3$, we obtain the generator matrices of the Reed-Muller codes *RM* (2), *RM*(3) as

$$G_2 = \begin{bmatrix} G_1 \, G_1 \\ \mathbf{0} \;\; \bar{\mathbf{1}} \end{bmatrix} \rightarrow G_2 = \begin{bmatrix} 11 & 11 \\ 01 & 01 \\ 00 & 11 \end{bmatrix} \rightarrow G_2 = \begin{bmatrix} 1111 \\ 0101 \\ 0011 \end{bmatrix}$$

and

$$G_3 = \begin{bmatrix} G_2 \, G_2 \\ \mathbf{0} \;\; \bar{\mathbf{1}} \end{bmatrix} \rightarrow G_3 = \begin{bmatrix} 1111 & 1111 \\ 0101 & 0101 \\ 0011 & 0011 \\ 0000 & 1111 \end{bmatrix} \rightarrow G_3 = \begin{bmatrix} 11111111 \\ 01010101 \\ 00110011 \\ 00001111 \end{bmatrix}.$$

The minimum distances of the Reed-Muller codes *RM*(2), *RM*(3) can be calculated using the formula

$$d_{\min} = 2^{m-1}$$

as

$$d_{\min} = 2^{2-1} \rightarrow d_{\min} = 2$$

and

$$d_{\min} = 2^{3-1} \rightarrow d_{\min} = 4.$$

**Exercise** Find the generator matrix of *RM*(4) and find its minimum distance.

### 3.6.5 Hamming Codes

Hamming codes are single-error-correcting linear block codes. For a Hamming code $\boldsymbol{C}(n, k)$, the code parameters $k$ and $n$ are calculated according to

$$n = 2^m - 1 \quad k = n - m \tag{3.34}$$

where $m$ is an integer such that $m \geq 2$. In Eq. (3.34), using the second equation, i.e., using $k = n - m$, we can also write that

$$m = n - k \tag{3.35}$$

which indicates the number of parity bits used for the transmission of data-word consisting of $k$-bits.

The parity check matrix of the Hamming code $\boldsymbol{C}(n, k)$ can be constructed by placing all the $m$-bit words as the columns of $\boldsymbol{H}$.

**Example 3.12** Obtain the parity check and generator matrices of the Hamming code for which $m = 3$.

**Solution 3.12** Using $m = 3$ bits, we can write 7 nonzero bit vectors, and these vectors and their integer equivalents can be listed as

$$001 \rightarrow 1$$

$$010 \rightarrow 2$$

$$011 \rightarrow 3$$

$$100 \rightarrow 4$$

$$101 \rightarrow 5$$

$$110 \rightarrow 6$$

$$111 \rightarrow 7.$$

We construct the parity check matrix of the Hamming code by placing the transpose of the binary equivalents of the digits listed in the previous paragraph as columns of the matrix, that is, we obtain the parity check matrix as

$$\boldsymbol{H} = \begin{bmatrix} \mathbf{1} & \mathbf{2} & \mathbf{3} & \mathbf{4} & \mathbf{5} & \mathbf{6} & \mathbf{7} \\ \downarrow & \downarrow & \downarrow & \downarrow & \downarrow & \downarrow & \downarrow \\ 0 & 0 & 0 & 1 & 1 & 1 & 1 \\ 0 & 1 & 1 & 0 & 0 & 1 & 1 \\ 1 & 0 & 1 & 0 & 1 & 0 & 1 \end{bmatrix} \rightarrow \boldsymbol{H} = \begin{bmatrix} 0001111 \\ 0110011 \\ 1010101 \end{bmatrix}. \tag{3.36}$$

The order of the columns of the parity check matrix is not an important issue. For this reason, while we construct the parity check matrix of the Hamming code, we prefer to construct it in the form

$$\boldsymbol{H} = \left[\boldsymbol{P}^T | \boldsymbol{I}\right] \tag{3.37}$$

which can be used to find the generator matrix of the code in the form $\boldsymbol{G} = [\boldsymbol{I}|\boldsymbol{P}^T]$. Considering this issue, we can construct the parity check matrix of the Hamming code with parameter $m = 3$ as in

$$\boldsymbol{H} = \left[\begin{array}{cccc|ccc} \mathbf{3} & \mathbf{5} & \mathbf{6} & \mathbf{7} & \mathbf{4} & \mathbf{2} & \mathbf{1} \\ \downarrow & \downarrow & \downarrow & \downarrow & \downarrow & \downarrow & \downarrow \\ 0 & 1 & 1 & 1 & 1 & 0 & 0 \\ 1 & 0 & 1 & 1 & 0 & 1 & 0 \\ \underbrace{1 \quad 1 \quad 0 \quad 1}_{\boldsymbol{P}^T} & & & & \underbrace{0 \quad 0 \quad 1}_{\boldsymbol{I}} & & \end{array}\right] \rightarrow \boldsymbol{H} = \begin{bmatrix} 0111100 \\ 1011010 \\ 1101001 \end{bmatrix}. \tag{3.38}$$

Permuting the columns of $\boldsymbol{P}^T$ in Eq. (3.38), it is possible to write $4\,! = 24$ different parity check matrices in systematic form. However, all the generated matrices have the same error-correcting capability. Hence, it is sufficient to construct one of them.

The generator matrix of the Hamming code corresponding to the parity check matrix in Eq. (3.38) can be obtained using the equation $\boldsymbol{G} = [\boldsymbol{I}|\boldsymbol{P}]$ as

$$\boldsymbol{G} = \left[\begin{array}{c|c} 1000 & 011 \\ 0100 & 101 \\ 0010 & 110 \\ 0001 & 111 \end{array}\right].$$

It can also be verified that

$$\boldsymbol{G} \times \boldsymbol{H}^T = \mathbf{0}. \tag{3.39}$$

**Example 3.13** Find the parity check matrix and generator matrix of the Hamming code in systematic form for $m = 4$.

**Solution 3.13** The parity check matrix of the required Hamming code can be constructed using the bit vectors containing $m = 4$ bits as the columns of the matrix. Note that the order of the column vectors is not an important issue. We can form the systematic form of the parity check matrix as

$$H = \begin{bmatrix} \begin{matrix} 3 & 5 & 6 & 7 & 9 & 10 & 11 & 12 & 13 & 14 & 15 \\ \downarrow & \downarrow & \downarrow & \downarrow & \downarrow & \downarrow & \downarrow & \downarrow & \downarrow & \downarrow & \downarrow \\ 0 & 0 & 0 & 0 & 1 & 1 & 1 & 1 & 1 & 1 & 1 \\ 0 & 1 & 1 & 1 & 0 & 0 & 0 & 1 & 1 & 1 & 1 \\ 1 & 0 & 1 & 1 & 0 & 1 & 1 & 0 & 0 & 1 & 1 \\ 1 & 1 & 0 & 1 & 1 & 0 & 1 & 0 & 1 & 0 & 1 \end{matrix} & \left| \begin{matrix} 8 & 4 & 2 & 1 \\ \downarrow & \downarrow & \downarrow & \downarrow \\ 1 & 0 & 0 & 0 \\ 0 & 1 & 0 & 0 \\ 0 & 0 & 1 & 0 \\ 0 & 0 & 0 & 1 \end{matrix} \right. \end{bmatrix} \rightarrow$$

$$\underbrace{\phantom{0\;0\;0\;0\;1\;1\;1\;1\;1\;1\;1}}_{P^T} \qquad \underbrace{\phantom{1\;0\;0\;0}}_{I}$$

$$H = \begin{bmatrix} 00001111111 & 1000 \\ 01110001111 & 0100 \\ 10110110011 & 0010 \\ 11011010101 & 0001 \end{bmatrix}. \tag{3.40}$$

## 3.7   Non-Systematic Form of Generator and Parity Check Matrices of the Hamming Codes

Let $G$ and $H$ be the generator and parity check matrices of a Hamming code. The matrices $G, H$ are denoted as

$$G = [g_1\, g_2 \cdots g_n]_{k \times n} \quad H = \left[ h_1^T\, h_2^T \cdots h_n^T \right]_{(n-k) \times n} \tag{3.41}$$

where $g_i = [g_{i1}\, g_{i2} \cdots g_{ik}]^T$, $i = 1 \ldots n$ and $h_i^T = \left[ h_{i1}\, h_{i2} \ldots h_{i(n-k)} \right]^T$ are column vectors.

Employing

$$G \times H^T = 0$$

in Eq. (3.41), we obtain

$$[g_1\, g_2 \ldots g_n] \times \begin{bmatrix} \boldsymbol{h}_1 \\ \boldsymbol{h}_2 \\ \vdots \\ \boldsymbol{h}_n \end{bmatrix} = \mathbf{0} \rightarrow g_1 \times \boldsymbol{h}_1 + g_2 \times \boldsymbol{h}_2 + \ldots g_n \times \boldsymbol{h}_n = \mathbf{0}. \quad (3.42)$$

Assume that the parity check matrix of a Hamming code is in non-systematic form. To obtain the generator matrix corresponding to this parity check matrix, we benefit from Eq. (3.42). That is, to find the generator matrix corresponding to a non-systematic parity check matrix, first, we obtain the systematic form of $\boldsymbol{H}$ using column permutations only and note the permutation information. Next, using the systematic form of $\boldsymbol{H}$, we can obtain the systematic form of the generator matrix $\boldsymbol{G}$. In the last step, we apply the inverse permutation information on the columns of $\boldsymbol{G}$ and obtain the generator matrix in non-systematic form.

**Example 3.14** The non-systematic parity check matrix of a Hamming code is given in Eq. (3.43). Obtain the generator matrix corresponding to the parity check matrix in Eq. (3.43)

$$\boldsymbol{H} = \begin{bmatrix} \mathbf{1} & \mathbf{2} & \mathbf{3} & \mathbf{4} & \mathbf{5} & \mathbf{6} & \mathbf{7} \\ \downarrow & \downarrow & \downarrow & \downarrow & \downarrow & \downarrow & \downarrow \\ 0 & 0 & 0 & 1 & 1 & 1 & 1 \\ 0 & 1 & 1 & 0 & 0 & 1 & 1 \\ 1 & 0 & 1 & 0 & 1 & 0 & 1 \end{bmatrix} \rightarrow \boldsymbol{H} = \begin{bmatrix} 0001111 \\ 0110011 \\ 1010101 \end{bmatrix}. \quad (3.43)$$

**Solution 3.14** Using column permutations, we put the parity check matrix into the systematic form as in Eq. (3.44)

$$\boldsymbol{H}_s = \left[\begin{array}{cccc|ccc} \mathbf{3} & \mathbf{5} & \mathbf{6} & \mathbf{7} & \mathbf{4} & \mathbf{2} & \mathbf{1} \\ \downarrow & \downarrow & \downarrow & \downarrow & \downarrow & \downarrow & \downarrow \\ 0 & 1 & 1 & 1 & 1 & 0 & 0 \\ 1 & 0 & 1 & 1 & 0 & 1 & 0 \\ \underbrace{1 \quad 1 \quad 0 \quad 1}_{\boldsymbol{P}^T} & & & & \underbrace{0 \quad 0 \quad 1}_{\boldsymbol{I}} & & \end{array}\right] \rightarrow \boldsymbol{H}_s = \begin{bmatrix} 0111100 \\ 1011010 \\ 1101001 \end{bmatrix}. \quad (3.44)$$

The permutation information is noted as in Fig. 3.1.

Using Eq. (3.44), we obtain the generator matrix in systematic form as

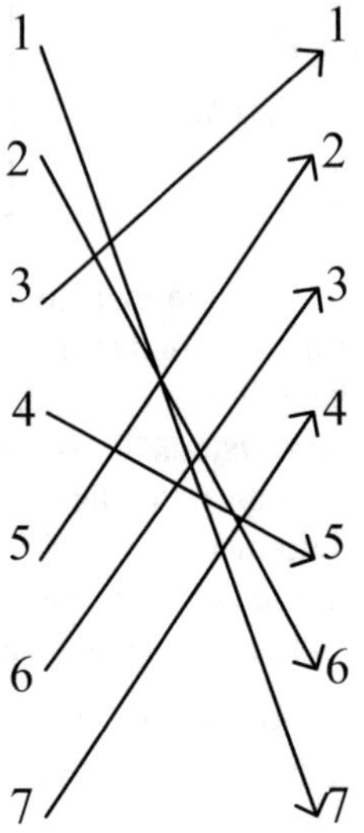


**Fig. 3.1** Permutation information

$$G_s = \begin{bmatrix} 1000 & 011 \\ 0100 & 101 \\ 0010 & 110 \\ 0001 & 111 \end{bmatrix}.$$

In the last step, we perform reverse permutation operation on the columns of $G_s$. We consider a new matrix $G_{ns}$ of the same size as $G_s$, and column 7 of $G_s$ is used as column 1 of $G_{ns}$, column 6 of $G_s$ is used as column 2 of $G_{ns}$, column 2 of $G_s$ is used as column 3 of $G_{ns}$, and so on resulting in

$$G_s = \begin{bmatrix} 1000 & 011 \\ 0100 & 101 \\ 0010 & 110 \\ 0001 & 111 \end{bmatrix} \to G_{ns} = \begin{bmatrix} 1110 & 000 \\ 1001 & 100 \\ 0101 & 010 \\ 1101 & 001 \end{bmatrix}.$$

## 3.8 Extended Hamming Codes

Let $H$ be the parity check matrix of a Hamming code. The parity check matrix of the extended linear block code is obtained adding a row vector consisting of all ones and adding a column vector consisting of all zeros as shown in

$$H_e = \left[\begin{array}{c|c} H & \begin{array}{c} 0 \\ 0 \\ \vdots \\ 0 \end{array} \\ \hline \multicolumn{2}{c}{11\ldots1} \end{array}\right].$$

If $\boldsymbol{G}$ is the generator matrix corresponding to $\boldsymbol{H}$, the generator matrix corresponding to $\boldsymbol{H_e}$ is obtained by concatenating a column vector to $\boldsymbol{G}$, and the bits of the column are obtained by summing the bits in each row of $\boldsymbol{G}$.

**Example 3.15** The parity check matrix of a Hamming code is given as

$$\boldsymbol{H} = \begin{bmatrix} 0111100 \\ 1011010 \\ 1101001 \end{bmatrix}.$$

Extend the given Hamming code and obtain the parity check and generator matrices of the extended Hamming code.

**Solution 3.15** First, we concatenate all zero vectors as the column of the parity check matrix as shown in

$$\boldsymbol{H} = \begin{bmatrix} 0111100 \\ 1011010 \\ 1101001 \end{bmatrix} \rightarrow \boldsymbol{H}_{e1} = \begin{bmatrix} 0111100\mathbf{0} \\ 1011010\mathbf{0} \\ 1101001\mathbf{0} \end{bmatrix}.$$

Next, we concatenate all ones vectors as the last row of the parity check matrix as in

$$\boldsymbol{H}_{e1} = \begin{bmatrix} 01111000 \\ 10110100 \\ 11010010 \end{bmatrix} \rightarrow \boldsymbol{H}_e = \begin{bmatrix} 0111100\mathbf{0} \\ 1011010\mathbf{0} \\ 1101001\mathbf{0} \\ \mathbf{11111111} \end{bmatrix}.$$

The generator matrix corresponding to the parity check matrix $\boldsymbol{H}$ can be obtained as

$$\boldsymbol{G} = \begin{bmatrix} 1000011 \\ 0100101 \\ 0010110 \\ 0001111 \end{bmatrix}.$$

Summing the rows of $\boldsymbol{G}$ and concatenating it as the last column of the generator matrix, we can get the generator matrix of the extended Hamming code as in

$$\boldsymbol{G_e} = \begin{bmatrix} 1000011 \rightarrow 1+0+0+0+0+1+1 \rightarrow \mathbf{1} \\ 0100101 \rightarrow 0+1+0+0+1+0+1 \rightarrow \mathbf{1} \\ 0010110 \rightarrow 0+0+1+0+1+1+0 \rightarrow \mathbf{1} \\ 0001111 \rightarrow 0+0+0+1+1+1+1 \rightarrow \mathbf{0} \end{bmatrix} \rightarrow \boldsymbol{G_e} = \begin{bmatrix} 1000011\mathbf{1} \\ 0100101\mathbf{1} \\ 0010110\mathbf{1} \\ 0001111\mathbf{0} \end{bmatrix}.$$

If we check the equation

$$\boldsymbol{G_e} \times \boldsymbol{H_e^T} = \mathbf{0}$$

we see that it is satisfied, i.e.,

$$\begin{bmatrix} 1000011\mathbf{1} \\ 0100101\mathbf{1} \\ 0010110\mathbf{1} \\ 0001111\mathbf{0} \end{bmatrix} \times \begin{bmatrix} 011\mathbf{1} \\ 101\mathbf{1} \\ 110\mathbf{1} \\ 111\mathbf{1} \\ 100\mathbf{1} \\ 010\mathbf{1} \\ 001\mathbf{1} \\ \mathbf{0001} \end{bmatrix} = \begin{bmatrix} 0000 \\ 0000 \\ 0000 \\ 0000 \end{bmatrix}.$$

## 3.9 Syndrome Decoding of Hamming Codes

Hamming codes are single-error-correcting codes. Due to this reason, for the construction of syndrome tables, it is sufficient to consider only single-error patterns, and for the single-error patterns, the syndromes correspond to the columns of the parity check matrix.

**Example 3.16** The parity check matrix of a Hamming code is given as

$$H = \begin{bmatrix} 0111100 \\ 1011010 \\ 1101001 \end{bmatrix}.$$

Obtain the syndrome table of this Hamming code.

**Solution 3.16** Considering only single-error patterns, we can construct the syndrome table as in Table 3.21 where it is clear that the syndromes are the columns of the parity check matrix.

## 3.10 Shortened and Extended Linear Codes

***Code Shortening***

Using a designed linear block code $\boldsymbol{C}(n, k)$, it is possible to obtain a shortened code

$$\boldsymbol{C}(n-i, k-i). \tag{3.45}$$

If $\boldsymbol{G}$ and $\boldsymbol{H}$ are the generator and parity check matrices of the linear block code $\boldsymbol{C}(n, k)$, the generator matrix of the shortened code $\boldsymbol{C}(n-i, k-i)$ can be obtained by omitting the first $i$ rows and columns of $\boldsymbol{G}$, and the parity check matrix of the shortened code is obtained by omitting the first $i$ columns of $\boldsymbol{H}$.

**Example 3.17** The generator and parity check matrices of $\boldsymbol{C}(9,5)$ linear block code are given as

$$\boldsymbol{G} = \begin{bmatrix} 100000111 \\ 010001101 \\ 001001011 \\ 000101110 \\ 000010011 \end{bmatrix} \quad \boldsymbol{H} = \begin{bmatrix} 011101000 \\ 110100100 \\ 101110010 \\ 111010001 \end{bmatrix}.$$

Using code shortening, obtain the generator and parity check matrices of the linear block code $\boldsymbol{C}(7,3)$.

**Solution 3.17** Omitting the first two rows and columns of $\boldsymbol{G}$, we get the generator matrix of the shortened code as

**Table 3.21** Syndrome table for Example 3.16

| $e$ | $s = e \times H^T$ |
|---|---|
| 0000001 | 001 |
| 0000010 | 010 |
| 0000100 | 100 |
| 0001000 | 111 |
| 0010000 | 110 |
| 0100000 | 101 |
| 1000000 | 011 |

$$\boldsymbol{G} = \begin{bmatrix} 100000111 \\ 010001101 \\ 001001011 \\ 000101110 \\ 000010011 \end{bmatrix} \rightarrow \boldsymbol{G}_r = \begin{bmatrix} 1001011 \\ 0101110 \\ 0010011 \end{bmatrix}.$$

Omitting the first two columns of $\boldsymbol{H}$, we get the parity check matrix of the shortened code as

$$\boldsymbol{H} = \begin{bmatrix} 011101000 \\ 110100100 \\ 101110010 \\ 111010001 \end{bmatrix} \rightarrow \boldsymbol{H}_r = \begin{bmatrix} 1101000 \\ 0100100 \\ 1110010 \\ 1010001 \end{bmatrix}.$$

We can verify that

$$\boldsymbol{G}_r \times \boldsymbol{H}_r^T = \boldsymbol{0}.$$

***Code Extension***

A linear block code $\boldsymbol{C}(n, k)$ can be extended to another linear block code

$$\boldsymbol{C}(n+1, k). \tag{3.46}$$

If $\boldsymbol{G}$ and $\boldsymbol{H}$ are the generator and parity check matrices of the linear block code $\boldsymbol{C}(n, k)$, the generator matrix of the extended code $\boldsymbol{C}(n + 1, k)$ can be obtained by concatenating a column to the last column of $\boldsymbol{G}$, and the concatenated column is obtained by summing all the other columns, i.e., by summing the bits in each row separately.

**Example 3.18** The generator matrix of $\boldsymbol{C}(9,5)$ linear block code is given as

$$\boldsymbol{G} = \begin{bmatrix} 100000111 \\ 010000101 \\ 001001011 \\ 000101011 \\ 000011010 \end{bmatrix}$$

Using code extension, obtain the generator matrix of the linear block code $\boldsymbol{C}(10,5)$.

**Solution 3.18** We can obtain the generator matrix of the extended code $\boldsymbol{C}(10,5)$ as

$$\boldsymbol{G} = \begin{bmatrix} 100000111 \rightarrow 1+0+0+0+0+0+1+1+1 \rightarrow \mathbf{0} \\ 010000101 \rightarrow 0+1+0+0+0+0+1+0+1 \rightarrow \mathbf{1} \\ 001001011 \rightarrow 0+0+1+0+0+1+0+1+1 \rightarrow \mathbf{0} \\ 000101011 \rightarrow 0+0+0+1+0+1+0+1+1 \rightarrow \mathbf{1} \\ 000011010 \rightarrow 0+0+0+1+0+1+0+1+1 \rightarrow \mathbf{0} \end{bmatrix} \rightarrow$$

$$\boldsymbol{G_e} = \begin{bmatrix} 100000111\mathbf{0} \\ 010000101\mathbf{1} \\ 001001011\mathbf{0} \\ 000101011\mathbf{1} \\ 000011010\mathbf{0} \end{bmatrix}. \tag{3.47}$$

Since the generator matrix of the extended code in Eq. (3.47) has systematic form, the parity check matrix can easily be found using the equation $\boldsymbol{H} = [\boldsymbol{P}^{\boldsymbol{T}} | \boldsymbol{I}]$ as

$$\boldsymbol{H_e} = \begin{bmatrix} 0011110000 \\ 1100001000 \\ 1011100100 \\ 1111000010 \\ \mathbf{0101}000001 \end{bmatrix}. \tag{3.48}$$

## Problems

1. The generator matrix of a binary linear block code is given as

$$G = \begin{bmatrix} 100110 \\ 110011 \\ 101101 \end{bmatrix}.$$

Find all the code-words, and construct the standard table for this code.

2. The generator matrix of a binary linear block code is given as

$$G = \begin{bmatrix} 1001100 \\ 1100111 \\ 1111011 \end{bmatrix}.$$

Determine the standard array of the dual code.

3. The generator matrix of a binary linear block code is given as

$$G = \begin{bmatrix} 11010101 \\ 01001010 \\ 10111010 \end{bmatrix}.$$

Obtain the syndrome table of this code.

4. The parity check matrix of a single-error-correcting binary linear block code is given as

$$H = \begin{bmatrix} 0010111 \\ 0101011 \\ 1001101 \end{bmatrix}.$$

(a) Obtain the syndrome table of this code.

(b) Find the generator matrix of this code, and encode the data-word $\boldsymbol{d} = [1011]$.
(c) For part (b), assume that during the transmission, a single-bit error occurs, and the error pattern is given as $\boldsymbol{e} = [0010]$. Determine the decoder's estimate using the syndrome table.

5. Find the generator and parity check matrices of the Reed-Muller code for $m = 4$. Determine the minimum distance of this code.
6. Determine the generator and parity check matrices of the Hamming code for $r = 3$, and obtain the parity check matrix of the extended Hamming code.
7. Syndromes are noting but the linear combinations of the transposed columns of the parity check matrix according to the error patterns. Determine whether this statement is correct or not.
8. What is the minimum distance of a Golay code?
9. The generator matrix of a linear block code $\boldsymbol{C}(7, 3)$ is given as

$$\boldsymbol{G} = \begin{bmatrix} 11100100 \\ 01011010 \\ 11101101 \end{bmatrix}.$$

(a) Determine the systematic form of the given generator matrix.
(b) Find the parity check matrix.
(c) Find the minimum distance of the code.
(d) Decide on the error detection and correction capability of this code.
(e) Construct the syndrome table of this code.
(f) Extend this code to a $\boldsymbol{C}(8, 3)$ linear block code. Find the generator and parity check matrices of the extended code.

10. The parity check matrix of a linear block code $\boldsymbol{C}(7, 4)$ is given as

$$\boldsymbol{H} = \begin{bmatrix} 1110100 \\ 0111010 \\ 1101001 \end{bmatrix}.$$

(a) Find the generator matrix in systematic form.
(b) Find the minimum distance of the code using the parity check matrix.
(c) Decide on the error detection and correction capability of this code.
(d) Let $t_c$ be the number of bit errors that the code can correct. Determine the number of words inside a Hamming sphere of radius $t_c$; at the center of the sphere, a code-word exists.
(e) Construct the syndrome table of this code.

(f) Extend this code to a $\boldsymbol{C}(8,4)$ linear block code. Find the generator and parity check matrices of the extended code, and besides, find the minimum distance of the extended code.
(g) Decide on the error detection and correction capability of the extended code.
(h) Verify the singleton and Hamming bounds for this code and its extended version.

# Chapter 4
# Cyclic Codes

## 4.1 Cyclic Codes

Before explaining the cyclic codes, let's briefly explain the cyclic shift operation.

***Cyclic Shift***
Let $\boldsymbol{C}(n, k)$ be a linear block code, and $\boldsymbol{c} = [c_{n-1} \ldots c_1 \; c_0]$ be a code-word. A cyclic leftward shift of $\boldsymbol{c}$ is defined as

$$\mathrm{CLS}(\boldsymbol{c}) = [c_{n-2} \ldots c_0 \; c_{n-1}]. \tag{4.1}$$

Similarly, a cyclic rightward shift of $\boldsymbol{c}$ is defined as

$$\mathrm{CRS}(\boldsymbol{c}) = [c_0 \; c_{n-1} \ldots c_2 \; c_1]. \tag{4.2}$$

***Cyclic Code***
A linear block code $\boldsymbol{C}(n, k)$ is a cyclic code if a cyclic shift of any code-word produces another code-word of $\boldsymbol{C}(n, k)$.

**Example 4.1** The linear block code given as

$$\boldsymbol{C} = [000 \;\; 110 \;\; 101 \;\; 011]$$

is a cyclic code. We can verify that $\forall \boldsymbol{c} \in \boldsymbol{C}$, $\mathrm{CLS}(\boldsymbol{c}) \in \boldsymbol{C}$ and $\mathrm{CRS}(\boldsymbol{c}) \in \boldsymbol{C}$; for instance,

$$\mathrm{CLS}(\,110) \rightarrow 101 \in \boldsymbol{C}$$



O. Gazi, *Forward Error Correction via Channel Coding*,
https://doi.org/10.1007/978-3-030-33380-5_4

$$\text{CRS}(\ 011) \rightarrow 101 \in \boldsymbol{C}.$$

Note that for a linear block code, the sum of any two code-words produces another code-word, and this property is also valid for cyclic codes, since cyclic codes are a class of linear block codes.

**Example 4.2** The linear block code

$$\boldsymbol{C} = [0000000\ \ 1011100\ \ 0101110\ \ 0010111\ \ 1110010\ \ 0111001\ \ 1001011\ \ 1100101]$$

is a cyclic code. We can verify that $\forall \boldsymbol{c} \in \boldsymbol{C}$, CLS($\boldsymbol{c}$) $\in \boldsymbol{C}$ and CRS($\boldsymbol{c}$) $\in \boldsymbol{C}$; for instance,

$$\text{CLS}(\ 1011100) \rightarrow 0111001 \in \boldsymbol{C}$$

$$\text{CRS}(\ 1001011) \rightarrow 1100101 \in \boldsymbol{C}.$$

## 4.2 Polynomials and Cyclic Codes

An $n$-bit word $\boldsymbol{w} = [a_{n\ -\ 1} \ldots a_1\ a_0]$ can be represented in polynomial form as

$$w(x) = a_{n-1}x^{n-1} + a_{n-2}x^{n-2} + \ldots a_0 x^0. \tag{4.3}$$

**Example 4.3** The word $\boldsymbol{w_1} = [0110010]$ can be represented in polynomial form as

$$w_1(x) = 0 \times x^6 + 1 \times x^5 + 1 \times x^4 + 0 \times x^3 + 0 \times x^2 + 1 \times x^1 + 0 \times x^0 \rightarrow$$
$$w_1(x) = x^5 + x^4 + x.$$

**Example 4.4** The word $\boldsymbol{w_2} = [001011]$ can be represented in polynomial form as

$$w_2(x) = 0 \times x^5 + 0 \times x^4 + 1 \times x^3 + 0 \times x^2 + 1 \times x^1 + 1 \times x^0 \rightarrow$$
$$w_2(x) = x^3 + x + 1.$$

In a polynomial $p(x)$, the largest power of $x$ is called the degree of $p(x)$.

Note that, in our book, we are only interested in the words whose elements are bits chosen from the binary field $F = \{0, 1\}$ for which mod-2 addition and multiplication operations are used.

Polynomials can be added, multiplied, and divided. Since polynomial coefficients are field elements, while summing two polynomials, we pay attention to sum the coefficients of the same $x^p$ terms, and while summing the coefficients, we use the operation defined for the field elements. In our case, that is the mod-2 addition operation.

Note that we did not mention the subtraction of two polynomials, since there is no subtraction operation defined for the elements of binary field; then, it becomes unlogical to consider the subtraction of two polynomials.

**Exercise** Divide the polynomial $w_1(x) = x^{12} + x^7 + x^4 + x^3 + 1$ by $w_2(x) = x^3 + x^2 + 1$, and find the dividend and remainder polynomials.

**Exercise** Divide the polynomial $w_1(x) = x^5 + x^3 + x^2 + 1$ by $w_2(x) = x^3 + 1$, and find the dividend and remainder polynomials.

**Definition** Let $w_1(x)$ and $w(x)$ be polynomials with binary coefficients. The remainder polynomial which is obtained after the division of $w_1(x)$ by $w(x)$ is denoted as

$$r(x) = R_{w(x)}[w_1(x)] \tag{4.4}$$

where $R_{w(x)}[\cdot]$ is the remainder function.

***Note*** To calculate the remainder polynomial $r(x)$ obtained after the division of $w_1(x)$ by $w(x)$, we equate $w(x)$ to zero, write the highest-order term in terms of the lower-degree ones, put this expression in $w_1(x)$, and continue like this until we get a polynomial with order less than the order of $w(x)$.

**Example 4.5** Let $w_1(x) = x^4 + x + 1$ and $w(x) = x^2 + x + 1$. Find $r(x) = R_{w(x)}[w_1(x)]$.

**Solution 4.5** From $w(x) = 0$, we obtain

$$x^2 = x + 1. \tag{4.5}$$

When Eq. (4.5) is used in $x^4 + x + 1$ repeatedly, we get

$$x^4 + x + 1 \rightarrow (x^2)^2 + x + 1 \rightarrow (x + 1)^2 + x + 1 \rightarrow x^2 + x \rightarrow x + 1 + x \rightarrow 1.$$

Hence, we have $r(x) = 1$.

***Cyclic Shift Operation by Polynomials***

The cyclic shift of an $n$-bit word can be expressed using the polynomials. The polynomial representation of the word $\boldsymbol{w} = [a_{n-1} \ldots a_1 \, a_0]$ can be formed as

$$w(x) = a_{n-1}x^{n-1} + a_{n-2}x^{n-2} + \ldots + a_1x^1 + a_0x^0. \tag{4.6}$$

When the word $\boldsymbol{w} = [a_{n-1} \ldots a_1 \, a_0]$ is cyclically shifted toward left, we obtain

$$\boldsymbol{w}_1 = [a_{n-2} \ldots a_1 \, a_0 \, a_{n-1}]$$

which can be represented in polynomial form as

$$w_1(x) = a_{n-2}x^{n-1} + a_{n-3}x^{n-2} + \ldots + a_0x^1 + a_{n-1}x^0. \tag{4.7}$$

The relation between the polynomials $w(x)$ and $w_1(x)$ can be written as

$$w_1(x) = R_{x^n+1}[x \times w(x)]. \tag{4.8}$$

In fact, the polynomial $x \times w(x)$ can be written as

$$x \times w(x) = a_{n-1}x^n + a_{n-2}x^{n-1} + \ldots + a_1x^2 + a_0x^1.$$

To find the remainder polynomial $w_1(x)$, we substitute $x^n + 1 = 0 \to x^n = 1$ in Eq. (4.8), and we obtain

$$a_{n-2}x^{n-1} + \ldots + a_1x^2 + a_0x^1 + a_{n-1} \tag{4.9}$$

which is the polynomial in Eq. (4.7).

**Example 4.6** The word $\boldsymbol{w} = [101100]$ can be represented in polynomial form as

$$w(x) = 1 \times x^5 + 0 \times x^4 + 1 \times x^3 + 1 \times x^2 + 0 \times x^1 + 0 \times x^0 \to$$
$$w(x) = x^5 + x^3 + x^2.$$

When the word $\boldsymbol{w} = [101100]$ is cyclically shifted toward left, we obtain

$$\boldsymbol{w_1} = CSL(\boldsymbol{w}) \to \boldsymbol{w_1} = [011001]$$

which can be represented in polynomial form as

$$w_1(x) = x^4 + x^3 + x^0. \tag{4.10}$$

We can show that the polynomial $w_1(x)$ can be obtained as

$$w_1(x) = R_{x^6+1}[x \times w(x)]. \tag{4.11}$$

Indeed, the polynomial $x \times w(x)$ can be formed as

$$x \times w(x) = x^6 + x^4 + x^3. \tag{4.12}$$

Substituting $x^6 = 1$ in Eq. (4.12), we get the remainder polynomial which is obtained after the division of Eq. (4.12) by $x^6 + 1$, and the obtained remainder polynomial is the same as Eq. (4.10).

***Note*** The degree of $w_1(x)$ obtained as

$$w_1(x) = R_{x^n+1}[x \times w(x)] \tag{4.13}$$

does not exceed $n - 1$.

## 4.3 The Generator Polynomial for Cyclic Codes

The generator polynomial of the cyclic code $\boldsymbol{C}(n, k)$ has the form

$$g(x) = a^{n-k}x^{n-k} + a_{n-k-1}x^{n-k-1} + \ldots + a_2x^2 + a_1x^1 + a_0x^0 \tag{4.14}$$

where $a^{n-k} = a_0 = 1$ and $a_i \in F = \{0, 1\}$, $i = 1, \ldots, n - k - 1$. The degree of the generator polynomial is $n - k$ which is equal to the number of parity bits in code-words, i.e., number of redundant bits available in code-words.

The generator polynomial is selected from the smallest-degree code-word polynomials, and all the other code-word polynomials have degree greater than or equal to $n - k$.

Once we have the generator polynomial of a cyclic code, we can get a code-word for an information word just by multiplying the information word polynomial by the generator polynomial. The data-word polynomials have degree less than or equal to $k - 1$.

## 4.4 Non-Systematic Encoding of Cyclic Codes

Assume that we have a cyclic code with generator polynomial $g(x)$. For the data-word $\boldsymbol{d} = [d_{k-1} \ldots d_1\, d_0]$, let $d(x)$ be the polynomial representation of this data-word. The encoding operation for the data-word $\boldsymbol{d} = [d_{k-1} \ldots d_1\, d_0]$ can be achieved using their polynomial representations as

$$c(x) = d(x)g(x). \tag{4.15}$$

**Example 4.7** Hamming codes are cyclic codes. The generator polynomial of the cyclic code

$$\boldsymbol{C}(n=7,k=4)$$

is given as

$$g(x)=x^3+x+1.$$

Encode the data-word $\boldsymbol{d}=[0101]$ using polynomial multiplication.

**Solution 4.7** The polynomial representation of the data-word can be obtained as

$$d(x)=x^2+1.$$

The code-word polynomial for the given data-word polynomial can be calculated as

$$c(x)=d(x)g(x)\rightarrow c(x)=\left(x^2+1\right)\left(x^3+x+1\right)\rightarrow c(x)=x^5+x^2+x+1. \tag{4.16}$$

The code-word polynomial obtained in Eq. (4.16) can be written in bit vector form as

$$\boldsymbol{c}=[0100111]. \tag{4.17}$$

Note that the code-word length is $n=7$. For this reason, when code-word polynomials are represented by bits, we pay attention to the length of the bit vector and pad the most significant bits by zeros when the polynomial representation involves less number of bits than the code-word length.

## 4.5 Systematic Encoding of Cyclic Codes

In systematic code-words, data-bits and parity bits are placed into two concatenated bit vectors, and a systematic code-word has the form of either

$$\boldsymbol{c}=[\boldsymbol{d}\,\boldsymbol{p}]\rightarrow\boldsymbol{c}=[d_{k-1}\,d_{k-2}\ldots d_0\,p_{n-k-1}\,p_{n-k-2}\cdots p_0]$$

or

$$\boldsymbol{c}=[\boldsymbol{p}\,\boldsymbol{d}]\rightarrow\boldsymbol{c}=[p_{n-k-1}\,p_{n-k-2}\cdots p_0\,d_{k-1}\,d_{k-2}\ldots d_0].$$

Before explaining the systematic encoding of cyclic codes, let's prepare ourselves considering some fundamental concepts.

***Zero Padding of Bit Vectors***

Let the data vector be represented as

$$\boldsymbol{d} = [d_{k-1} \ldots d_1\ d_0]. \tag{4.18}$$

If we concatenate $(n - k)$ zeros to the end of $\boldsymbol{d}$, we obtain

$$\boldsymbol{d}_1 = \left[ d_{k-1} \ldots d_1\ d_0\ \underbrace{0\,0 \ldots 0}_{n-k\ zeros} \right]. \tag{4.19}$$

The polynomial form of $\boldsymbol{d}$ is

$$d(x) = d_{k-1}x^{k-1} + d_{k-2}x^{k-2} + \ldots + d_1x^1 + d_0x^0 \tag{4.20}$$

and the polynomial form of $\boldsymbol{d}_1$ is

$$d_1(x) = d_{k-1}x^{n-1} + d_{k-2}x^{n-2} + \ldots + d_1x^{n-k+1} + d_0x^{n-k}. \tag{4.21}$$

If we compare Eq. (4.20) to (4.21), we see that Eq. (4.21) can be written in terms of Eq. (4.20) as

$$d_1(x) = x^{n-k}d(x) \tag{4.22}$$

which means that multiplying a polynomial by $x^{n-k}$ equals to zero padding its bit vector representation by $(n - k)$ zeros.

In short, multiplying a data vector $\boldsymbol{d}$ by $x^{n-k}$, we obtain the zero padded vector

$$\left[ \boldsymbol{d}\ \ \boldsymbol{0}_{1\times(n-k)} \right]. \tag{4.23}$$

**Example 4.8** A data vector and its polynomial representation are given as

$$\boldsymbol{d} = [0101] \rightarrow d(x) = x^2 + 1.$$

If we multiply the polynomial $d(x)$ by $x^3$, we obtain

$$d_1(x) = x^3d(x) \rightarrow d_1(x) = x^5 + x^3$$

and the bit vector representation of $d_1(x)$ can be formed as

$$\boldsymbol{d}_1 = [0101\mathbf{000}].$$

**Example 4.9** Consider the polynomials

$$g(x) = x^3 + x + 1 \quad d_1(x) = x^5 + x^4 + x^3.$$

(a) Find the remainder polynomial $p(x) = R_{g(x)}[d_1(x)]$.
(b) Using the result of part (a), find the remainder polynomial $p_1(x) = R_{g(x)}[d_1(x) + p(x)]$.

**Solution 4.9** (a) Equating the polynomial $g(x) = x^3 + x + 1$ to zero, we obtain

$$x^3 = x + 1. \tag{4.24}$$

When Eq. (4.24) is used in $d_1(x) = x^5 + x^4 + x^3$, we obtain

$$x^5 + x^4 + x^3 \rightarrow x^3x^2 + x^3x + x^3 \rightarrow (x+1)x^2 + (x+1)x + x + 1$$

$$\rightarrow x^3 + x^2 + x^2 + x + x + 1 \rightarrow x^3 + 1 \rightarrow x + 1 + 1 \rightarrow x.$$

Hence, $p(x) = x$.

(b) We can write

$$p_1(x) = R_{g(x)}[d_1(x) + p(x)]$$

as

$$p_1(x) = R_{g(x)}[d_1(x)] + R_{g(x)}[p(x)]$$

which can be calculated as

$$p_1(x) = x + x \rightarrow p_1(x) = 0.$$

**Example 4.10** Let $g(x)$ be a polynomial in the form

$$g(x) = a^{n-k}x^{n-k} + a_{n-k-1}x^{n-k-1} + \ldots + a_2x^2 + a_1x^1 + a_0x^0.$$

Let $p(x) = R_{g(x)}[d_1(x)]$ be a remainder polynomial where

$$d_1(x) = d_{k-1}x^{n-1} + d_{k-2}x^{n-2} + \ldots + d_1x^{n-k+1} + d_0x^{n-k}.$$

Find the remainder polynomial

$$p_1(x) = R_{g(x)}[d_1(x) + p(x)].$$

**Answer 4.10**

$$p_1(x) = 0.$$

### 4.5.1 Code-Word in Systematic Form

The code-word in systematic form

$$\boldsymbol{c}_{1\times n} = \left[\boldsymbol{d}_{1\times k} \ \boldsymbol{p}_{1\times(n-k)}\right] \tag{4.25}$$

can be written as

$$\boldsymbol{c}_{1\times n} = \left[\boldsymbol{d}_{1\times k} \ \boldsymbol{p}_{1\times(n-k)}\right] \rightarrow \boldsymbol{c}_{1\times n} = \left[\boldsymbol{d}_{1\times k} \ \boldsymbol{0}_{1\times(n-k)}\right] + \left[\boldsymbol{0} \ \boldsymbol{p}_{1\times(n-k)}\right] \tag{4.26}$$

which can be represented in polynomial form as

$$c(x) = x^{n-k}d(x) + p(x) \tag{4.27}$$

where $p(x)$ has the form

$$p(x) = p_{n-k-1}x^{n-k-1} + p_{n-k-2}x^{n-k-2} + \ldots + p_1x^1 + p_0x^0. \tag{4.28}$$

The degree of $p(x)$ is $n - k - 1$ which is smaller than the degree of the generator polynomial

$$g(x) = a^{n-k}x^{n-k} + a_{n-k-1}x^{n-k-1} + \ldots + a_2x^2 + a_1x^1 + a_0x^0 \tag{4.29}$$

whose degree is $n - k$. We conclude that

$$R_{g(x)}[p(x)] = p(x).$$

Since $c(x)$ is a code-word polynomial, it should be in the form

$$c(x) = d(x)g(x) \tag{4.30}$$

where $d(x)$ and $g(x)$ are data-word and generator polynomials. Equation (4.30) implies that

$$R_{g(x)}[c(x)] = 0. \tag{4.31}$$

When Eq. (4.27) is used in Eq. (4.31), we obtain

$$R_{g(x)}\left[x^{n-k}d(x) + p(x)\right] = 0 \rightarrow R_{g(x)}\left[x^{n-k}d(x)\right] + \underbrace{R_{g(x)}[p(x)]}_{=p(x)} = 0 \rightarrow$$

$$R_{g(x)}\left[x^{n-k}d(x)\right] = p(x). \tag{4.32}$$

Now let's gather the results we obtained in formulas (4.27) and (4.32) and summarize the systematic encoding algorithm of the cyclic codes as follows.

*Systematic Encoding of Cyclic Codes*

1. Obtain the polynomial form $d(x)$ of the data-word $\boldsymbol{d}$ and calculate $x^{n-k}d(x)$.
2. Find the remainder parity polynomial $p(x) = R_{g(x)}[x^{n-k}d(x)]$.
3. Form the code-word in systematic form as

$$c(x) = x^{n-k}d(x) + p(x).$$

**Example 4.11** The generator polynomial of $\boldsymbol{C}(n = 7, k = 4)$ cyclic code is given as

$$g(x) = x^3 + x + 1.$$

Systematically encode the data-word $\boldsymbol{d} = [0101]$ and obtain the code-word in systematic form.

**Solution 4.11** The parameters of the code $\boldsymbol{C}(n = 7, k = 4)$ are $n = 7$, $k = 4$. The polynomial representation of the data-word $\boldsymbol{d} = [0101]$ can be written as

$$d(x) = x^2 + 1.$$

We can obtain the systematic code-word for the given data-word as follows.

1. First, we multiply $d(x)$ by $x^{n-k}$ as in

$$x^{n-k}d(x) \rightarrow x^3 \times (x^2 + 1) \rightarrow x^{n-k}d(x) = x^5 + x^3.$$

2. In the second step, we calculate the remainder polynomial $p(x) = R_{g(x)}[x^{n-k}d(x)]$. For this purpose, we write $g(x) = 0$ from which we get $x^3 = x + 1$, and using the equality $x^3 = x + 1$ in $x^{n-k}d(x)$ repeatedly, we obtain the remainder polynomial as in

$$x^5 + x^3 \to x^3x^2 + x^3 \to (x+1)x^2 + (x+1) \to x^3 + x^2 + x + 1 \to$$
$$(x+1) + x^2 + x + 1 \to x^2.$$

Hence, the remainder polynomial is found as $p(x) = x^2$.

3. In the final step, we construct the systematic code-word using the equation

$$c(x) = x^{n-k}d(x) + p(x)$$

as

$$c(x) = x^{7-4}(x^2 + 1) + x^2 \to c(x) = x^5 + x^3 + x^2.$$

We can express the code-word polynomial $c(x) = x^5 + x^3 + x^2$ in bit vector as

$$\mathbf{c} = \left[\underbrace{\mathbf{0101}}_{d}\,\underbrace{\mathbf{100}}_{p}\right]$$

where we see that the first 4-bits of the code-word are the data bits, and the next three bits are the parity bits.

**Exercise** The generator polynomial of $\boldsymbol{C}(n = 7, k = 4)$ cyclic code is given as

$$g(x) = x^3 + x + 1.$$

Systematically encode the data-word $\boldsymbol{d} = [1101]$ and obtain the code-word in systematic form.

## 4.6 Decoding of Cyclic Codes

Since cyclic codes are a class of linear block codes, they can be decoded using the syndrome tables which are used for the decoding of linear block codes. The syndromes for the cyclic codes can be expressed in polynomial forms, and similarly, the syndrome tables can be constructed using the polynomials.

For the code-word polynomial $c(x)$, we can write the received word polynomial $r(x)$ as

$$r(x) = c(x) + e(x) \tag{4.33}$$

where $e(x)$ is the error word polynomial. Substituting $c(x) = d(x)g(x)$ in Eq. (4.33), we obtain

$$r(x) = d(x)g(x) + e(x). \tag{4.34}$$

The syndrome polynomial for $r(x)$ is defined as

$$s(x) = R_{g(x)}[r(x)] \tag{4.35}$$

in which substituting Eq. (4.34), we obtain

$$s(x) = R_{g(x)}[d(x)g(x) + e(x)] \rightarrow s(x) = \underbrace{R_{g(x)}[d(x)g(x)]}_{=0} + R_{g(x)}[e(x)] \rightarrow$$

$$s(x) = R_{g(x)}[e(x)]. \tag{4.36}$$

The syndrome decoding operation for cyclic codes can be outlined as follows.

Let $c(x)$ be the code-word polynomial, and $r(x)$ is the received word polynomial such that $r(x) = c(x) + e(x)$ where $e(x)$ is the error polynomial.

1. Using $r(x)$, we determine the syndrome polynomial $s(x) = R_{g(x)}[r(x)]$.
2. Using the syndrome table, we determine the error polynomial $\widehat{e}(x)$ corresponding to the syndrome $s(x)$.
3. In the last step, the decoder's estimate about the transmitted code-word is calculated using

$$\widehat{c}(x) = r(x) + \widehat{e}(x). \tag{4.37}$$

**Example 4.12** The generator polynomial of single-error-correcting $\boldsymbol{C}(n = 7, k = 4)$ cyclic code is given as

$$g(x) = x^3 + x + 1.$$

Construct the syndrome table of this code.

**Solution 4.12** First, we list all the possible single-error patterns on the leftmost column as shown in Table 4.1.

In the next step, the polynomials for the error patterns are written in the second column as in Table 4.2.

In the third step, we calculate the syndrome using

$$s(x) = R_{g(x)}[e(x)].$$

**Table 4.1** Syndrome table construction for Example 4.12

| ***e*** | |
|---|---|
| 0000001 | |
| 0000010 | |
| 0000100 | |
| 0001000 | |
| 0010000 | |
| 0100000 | |
| 1000000 | |

**Table 4.2** Syndrome table construction for Example 4.12

| ***e*** | $e(x)$ |
|---|---|
| 0000001 | 1 |
| 0000010 | $x$ |
| 0000100 | $x^2$ |
| 0001000 | $x^3$ |
| 0010000 | $x^4$ |
| 0100000 | $x^5$ |
| 1000000 | $x^6$ |

**Table 4.3** Syndrome table construction for Example 4.12

| ***e*** | $e(x)$ | $s(x) = R_{g(x)}[e(x)]$ |
|---|---|---|
| 0000001 | 1 | 1 |
| 0000010 | $x$ | $x$ |
| 0000100 | $x^2$ | $x^2$ |
| 0001000 | $x^3$ | $x + 1$ |
| 0010000 | $x^4$ | $x^2 + x$ |
| 0100000 | $x^5$ | $x^2 + x + 1$ |
| 1000000 | $x^6$ | $x^2 + 1$ |

For the syndrome calculation, we use $g(x) = x^3 + x + 1 = 0 \rightarrow x^3 = x + 1$ equation to calculate the remainder polynomials; for instance, if $e(x) = x^4$, the remainder polynomial can be calculated as

$$x^4 \rightarrow x^3 x \rightarrow (x + 1)x \rightarrow x^2 + x.$$

The syndromes can be calculated as shown in Table 4.3.

## 4.7 Selection of Generator Polynomials of the Cyclic Codes

Assume that we want to design a cyclic code $\boldsymbol{C}(n, k)$. There are several questions to be answered for this design.

The first one is "How to determine the values of $n$ and $k$? Is there a relation between $n$ and $k$? Can we choose arbitrary values for $n$ and $k$?"

The second question is "How to determine the generator polynomial $g(x)$ whose degree is $n-k$?"

Let's answer all of these questions as follows. The generator polynomial $g(x)$ must be a divider of $x^n+1$. For this reason, there is a relation between $n$ and $k$, and we cannot select arbitrary value for $k$. Let's illustrate the concept by an example.

**Example 4.13** Find the generator polynomials of the cyclic codes $\boldsymbol{C}(n,k)$ where $n=7$.

**Solution 4.13** The generator polynomial $g(x)$ with degree $n-k$ should be a divider of

$$x^n+1 \to x^7+1.$$

We can factorize $x^7+1$ as in

$$x^7+1=(x+1)\left(x^3+x^2+1\right)\left(x^3+x+1\right).$$

We can choose the generator polynomial $g(x)$ considering the factors of $x^7+1$. We can select the generator polynomials, and accordingly, we can calculate the value of $k$ and determine the cyclic codes as

$$g(x)=x+1 \to n-k=1 \to 7-k=1 \to k=6 \to$$
$$\boldsymbol{C}(n=7,k=6) \text{ cyclic code exists}$$

$$g(x)=x^3+x^2+1 \to n-k=3 \to 7-k=3 \to k=4 \to$$
$$\boldsymbol{C}(n=7,k=4) \text{ cyclic code exists}$$

$$g(x)=x^3+x+1 \to n-k=3 \to 7-k=3 \to k=4 \to$$
$$\boldsymbol{C}(n=7,k=4) \text{ cyclic code exists}$$

$$g(x)=(x+1)\left(x^3+x^2+1\right) \to n-k=4 \to 7-k=4 \to k=3 \to$$
$$\boldsymbol{C}(n=7,k=3) \text{ cyclic code exists}$$

$$g(x)=\left(x^3+x^2+1\right)\left(x^3+x+1\right) \to n-k=6 \to 7-k=6 \to k=1 \to$$
$$\boldsymbol{C}(n=7,k=1) \text{ cyclic code exists}$$

**Table 4.4** Factorization of $x^n + 1$ for $n = 1, 2, \ldots, 31$

| |
|---|
| $x^1 + 1 \rightarrow (x + 1)$ |
| $x^2 + 1 \rightarrow (x + 1)^2$ |
| $x^3 + 1 \rightarrow (x + 1)(x^2 + x + 1)$ |
| $x^4 + 1 \rightarrow (x + 1)^4$ |
| $x^5 + 1 \rightarrow (x + 1)(x^4 + x^3 + x^2 + x + 1)$ |
| $x^6 + 1 \rightarrow (x + 1)^2(x^2 + x + 1)^2$ |
| $x^7 + 1 \rightarrow (x + 1)(x^3 + x + 1)(x^3 + x^2 + 1)$ |
| $x^8 + 1 \rightarrow (x + 1)^8$ |
| $x^9 + 1 \rightarrow (x + 1)(x^2 + x + 1)(x^6 + x^3 + 1)$ |
| $x^{10} + 1 \rightarrow (x + 1)^2(x^4 + x^3 + x^2 + x + 1)^2$ |
| $x^{11} + 1 \rightarrow (x + 1)(x^{10} + x^9 + x^8 + x^7 + x^6 + x^5 + x^4 + x^3 + x^2 + x + 1)$ |
| $x^{12} + 1 \rightarrow (x + 1)^4(x^2 + x + 1)^4$ |
| $x^{13} + 1 \rightarrow (x + 1)(x^{12} + x^{11} + x^{10} + x^9 + x^8 + x^7 + x^6 + x^5 + x^4 + x^3 + x^2 + x + 1)$ |
| $x^{14} + 1 \rightarrow (x + 1)^2(x^3 + x + 1)^2(x^3 + x^2 + 1)^2$ |
| $x^{15} + 1 \rightarrow (x + 1)(x^2 + x + 1)(x^4 + x + 1)(x^4 + x^3 + 1)(x^4 + x^3 + x^2 + x + 1)$ |
| $x^{16} + 1 \rightarrow (x + 1)^{16}$ |
| $x^{17} + 1 \rightarrow (x + 1)(x^8 + x^5 + x^4 + x^3 + 1)(x^8 + x^7 + x^6 + x^4 + x^2 + x + 1)$ |
| $x^{18} + 1 \rightarrow (x + 1)^2(x^2 + x + 1)^2(x^6 + x^3 + 1)^2$ |
| $x^{19} + 1 \rightarrow (x + 1)(x^{18} + x^{17} + x^{16} + x^{15} + x^{14} + \cdots + x + 1)$ |
| $x^{20} + 1 \rightarrow (x + 1)^4(x^4 + x^3 + x^2 + x + 1)^4$ |
| $x^{21} + 1 \rightarrow (x + 1)(x^2 + x + 1)(x^3 + x + 1)(x^3 + x^2 + 1)(x^6 + x^4 + x^2 + x + 1)(x^6 + x^5 + x^4 + x^2 + 1)$ |
| $x^{22} + 1 \rightarrow (x + 1)^2(x^{10} + x^9 + x^8 + x^7 + x^6 + x^5 + x^4 + x^3 + x^2 + x + 1)^2$ |
| $x^{23} + 1 \rightarrow (x + 1)(x^{11} + x^9 + x^7 + x^6 + x^5 + x + 1)(x^{11} + x^{10} + x^6 + x^5 + x^4 + x^2 + 1)$ |
| $x^{24} + 1 \rightarrow (x + 1)^8(x^2 + x + 1)^8$ |
| $x^{25} + 1 \rightarrow (x + 1)(x^4 + x^3 + x^2 + x + 1)(x^{20} + x^{15} + x^{10} + x^5 + 1)$ |
| $x^{26} + 1 \rightarrow (x + 1)^2(x^{12} + x^{11} + x^{10} + x^9 + x^8 + x^7 + x^6 + x^5 + x^4 + x^3 + x^2 + x + 1)^2$ |
| $x^{27} + 1 \rightarrow (x + 1)(x^2 + x + 1)(x^6 + x^3 + 1)(x^{18} + x^9 + 1)$ |
| $x^{28} + 1 \rightarrow (x + 1)^4(x^3 + x + 1)^4(x^3 + x^2 + 1)^4$ |
| $x^{29} + 1 \rightarrow (x + 1)(x^{28} + x^{27} + x^{26} + \cdots + x + 1)$ |
| $x^{30} + 1 \rightarrow (x + 1)^2(x^2 + x + 1)^2(x^4 + x + 1)^2(x^4 + x^3 + 1)^2(x^4 + x^3 + x^2 + x + 1)^2$ |
| $x^{31} + 1 \rightarrow (x + 1)(x^5 + x^2 + 1)(x^5 + x^3 + 1)(x^5 + x^3 + x^2 + x + 1)(x^5 + x^4 + x^2 + x + 1)$ $(x^5 + x^4 + x^3 + x + 1)(x^5 + x^4 + x^3 + x^2 + 1)$ |

$$g(x) = (x+1)\left(x^3+x+1\right) \rightarrow n-k=4 \rightarrow 7-k=4 \rightarrow k=3 \rightarrow$$

$$\boldsymbol{C}(n=7, k=3) \text{ cyclic code exists.}$$

Although we obtained more than one cyclic code with the same parameters $n$ and $k$, their code-words are different, since their generator polynomials are different.

Factorization of $x^n + 1$ for $n = 1, 2, \ldots, 31$ is tabulated in Table 4.4 which can be used for the design of cyclic codes.

**Example 4.14** The factorization of $x^9 + 1$ is given as

$$x^9 + 1 = (x + 1)\left(x^2 + x + 1\right)\left(x^6 + x^3 + 1\right).$$

Find the generator polynomials of the cyclic codes $\boldsymbol{C}(n, k)$ where $n = 9$, and determine the value of $k$ considering each generator polynomial found.

**Solution 4.14** We can choose the generator polynomial $g(x)$ considering the factors of $x^9 + 1$. We can select the generator polynomials, and accordingly, we can determine the value of $k$ and determine the cyclic code as

$$g(x) = x + 1 \rightarrow n - k = 1 \rightarrow 9 - k = 1 \rightarrow k = 8 \rightarrow$$
$$\boldsymbol{C}(n = 9, k = 8) \text{ cyclic code exists}$$

$$g(x) = x^2 + x + 1 \rightarrow n - k = 2 \rightarrow 9 - k = 2 \rightarrow k = 7 \rightarrow$$
$$\boldsymbol{C}(n = 9, k = 7) \text{ cyclic code exists}$$

$$g(x) = x^6 + x^3 + 1 \rightarrow n - k = 6 \rightarrow 9 - k = 6 \rightarrow k = 3 \rightarrow$$
$$\boldsymbol{C}(n = 9, k = 3) \text{ cyclic code exists}$$

$$g(x) = (x + 1)\left(x^2 + x + 1\right) \rightarrow n - k = 3 \rightarrow 9 - k = 3 \rightarrow k = 6 \rightarrow$$
$$\boldsymbol{C}(n = 9, k = 6) \text{ cyclic code exists}$$

$$g(x) = (x + 1)\left(x^6 + x^3 + 1\right) \rightarrow n - k = 7 \rightarrow 9 - k = 7 \rightarrow k = 2 \rightarrow$$
$$\boldsymbol{C}(n = 9, k = 2) \text{ cyclic code exists}$$

$$g(x) = \left(x^2 + x + 1\right)\left(x^6 + x^3 + 1\right) \rightarrow n - k = 8 \rightarrow 9 - k = 8 \rightarrow k = 1 \rightarrow$$
$$\boldsymbol{C}(n = 9, k = 1) \text{ cyclic code exists.}$$

## 4.8 Parity Check Polynomials of the Cyclic Codes

Let $g(x)$ be the generator polynomial of a cyclic code. The parity check polynomial $h(x)$ of a cyclic code is found from

$$g(x)h(x) = x^n + 1 \tag{4.38}$$

where the degree of the generator polynomial is $n - k$ and the degree of the parity check polynomial is $k$. The parity check polynomial can be written as

$$h(x) = x^k + b_{k-1}x^{k-1} + \ldots + b_2x^2 + b_1x^1 + x^0. \tag{4.39}$$

Let $d(x)$ be the data-word polynomial. The code-word polynomial for $d(x)$ can be obtained as

$$c(x) = d(x)g(x). \tag{4.40}$$

If we substitute Eq. (4.40) in

$$R_{x^n+1}[c(x)h(x)]. \tag{4.41}$$

we obtain

$$R_{x^n+1}[d(x)g(x)h(x)]$$

in which using Eq. (4.38), we get

$$R_{x^n+1}[d(x)(x^n + 1)] = \mathbf{0}.$$

Hence, we showed that

$$R_{x^n+1}[c(x)h(x)] = \mathbf{0}. \tag{4.42}$$

The Eq. (4.42) can be utilized for an alternative equation for the calculation of syndrome polynomial. The received word polynomial can be written as

$$r(x) = c(x) + e(x). \tag{4.43}$$

Substituting

$$g(x) = \frac{x^n + 1}{h(x)}$$

in

$$s(x) = R_{g(x)}[r(x)],$$

we obtain

$$s(x) = R_{\frac{x^n+1}{h(x)}}[r(x)]$$

which can be written as

$$s(x) = R_{x^n+1}[r(x)h(x)]. \tag{4.44}$$

If we use $r(x) = c(x) + e(x)$ in Eq. (4.44), we obtain

$$s(x) = R_{x^n+1}[e(x)h(x)]. \tag{4.45}$$

Hence, using the parity check polynomial, we can calculate the syndrome polynomial using either

$$s(x) = R_{x^n+1}[r(x)h(x)] \tag{4.46}$$

or

$$s(x) = R_{x^n+1}[e(x)h(x)]. \tag{4.47}$$

**Example 4.15** The generator polynomial of single-error-correcting $\boldsymbol{C}(n = 7, k = 4)$ cyclic code is given as

$$g(x) = x^3 + x^2 + 1.$$

Find the parity check polynomial.

**Solution 4.15** The parity check polynomial can be calculated as

$$h(x) = \frac{x^n + 1}{g(x)} \rightarrow h(x) = \frac{x^7 + 1}{x^3 + x^2 + 1} \rightarrow h(x) = x^4 + x^3 + x^2 + 1.$$

The encoding, syndrome calculation, and orthogonality formulas for the cyclic codes can be expressed using matrices and polynomials as shown in Table 4.5.

**Table 4.5** Encoding, syndrome calculation, and orthogonality formulas for cyclic codes

| Matrix form | Polynomial form |
|---|---|
| $\boldsymbol{c} = \boldsymbol{dG}$ | $c(x) = d(x)g(x)$ |
| $\boldsymbol{GH}^T = \boldsymbol{0}$ | $R_{x^n+1}(g(x)h(x)) = 0$ |
| $\boldsymbol{cH}^T = \boldsymbol{0}$ | $R_{x^n+1}(c(x)h(x)) = 0$ |
| $\boldsymbol{s} = \boldsymbol{rH}^T$ or $\boldsymbol{s} = \boldsymbol{eH}^T$ | $s(x) = R_{x^n+1}(r(x)h(x))$ or $s(x) = R_{g(x)}(r(x))$ |

## 4.9 Dual Cyclic Codes

The parity check polynomial of cyclic codes is NOT the generator polynomial of the dual cyclic code, but it is used to calculate the generator polynomial of the dual cyclic code. If $h(x)$ is the parity check polynomial of a cyclic code, then the generator polynomial of the dual cyclic code is calculated using

$$g_d(x) = x^m h(x^{-1})$$

where $m$ is the degree of $h(x)$.

**Example 4.16** The generator polynomial of single-error-correcting $\boldsymbol{C}(n = 7, k = 4)$ cyclic code is given as

$$g(x) = x^3 + x + 1.$$

Find the generator polynomial of the dual cyclic code.

**Solution 4.16** We can find the parity check polynomial of the given cyclic code as

$$h(x) = \frac{x^n + 1}{g(x)} \rightarrow h(x) = \frac{x^7 + 1}{x^3 + x + 1} \rightarrow h(x) = x^4 + x^2 + x + 1.$$

Using the parity check polynomial, we can calculate the generator polynomial of the dual code as

$$g_d(x) = x^m h(x^{-1}) \rightarrow g_d(x) = x^4 (x^{-4} + x^{-2} + x^{-1} + 1) \rightarrow$$
$$g_d(x) = x^4 + x^3 + x^2 + 1.$$

The bit vector representation of $h(x) = x^4 + x^2 + x + 1$ is

$$\boldsymbol{h} = [1\ 0\ 1\ 1\ 1] \tag{4.48}$$

and the bit vector representation of $g_d(x) = x^4 + x^3 + x^2 + 1$ is

$$\boldsymbol{g_d} = [1\ 1\ 1\ 0\ 1]. \tag{4.49}$$

When Eqs. (4.48) and (4.49) are compared to each other, we see that they are the reverse of each other.

**Exercise** The generator polynomial of $C(n = 9, k = 6)$ cyclic code is given as

$$g(x) = x^3 + 1.$$

Find the parity check polynomial of this code, and using the parity check polynomial, obtain the generator polynomial of the dual cyclic code.

## 4.10 Generator and Parity Check Matrices for Cyclic Codes

***Generator Matrices of the Cyclic Codes***

Cyclic codes are a type of linear block codes. For this reason, cyclic codes have generator and parity check matrices. The generator and parity check matrices can be constructed using the binary coefficients of generator and parity check polynomials. Let $g(x)$ be the generator polynomial of a cyclic code such that

$$g(x) = a_{n-k}x^{n-k} + a_{n-k-1}x^{n-k-1} + \ldots + a_1x^1 + a_0x^0$$

where $a_i \in F = \{0, 1\}$. Using the binary coefficients of the generator polynomial, we construct the generator matrix as

$$G = \begin{bmatrix} a_{n-k}\ a_{n-k-1} \ldots a_2\ a_1\ a_0\ 0\ 0 \ldots 0\ 0\ 0 \\ 0\ a_{n-k}\ a_{n-k-1} \ldots a_2\ a_1\ a_0\ 0 \ldots 0\ 0\ 0 \\ 0\ 0\ a_{n-k}\ a_{n-k-1} \ldots a_2\ a_1\ a_0\ 0 \ldots 0\ 0 \\ \vdots \\ 0\ 0\ 0 \ldots 0\ 0\ a_{n-k}\ a_{n-k-1} \ldots a_2\ a_1\ a_0 \end{bmatrix}_{k \times n} . \quad (4.50)$$

**Example 4.17** The generator polynomial of single-error-correcting $C(n = 7, k = 4)$ cyclic code is given as

$$g(x) = x^3 + x^2 + 1.$$

Find the generator matrix of this cyclic code.

**Solution 4.17** The generator polynomial $g(x) = x^3 + x^2 + 1$ can be expressed by a bit vector as

$$\boldsymbol{g} = [1\ 1\ 0\ 1].$$

The size of the generator matrix is $k \times n = 4 \times 7$. If we zero pad the vector $\boldsymbol{g} = [1\ 1\ 0\ 1]$ such that its length equals to 7, we obtain

$$\boldsymbol{g}_1 = [1\ 1\ 0\ 1\ \mathbf{0}\ \mathbf{0}\ \mathbf{0}].$$

If we rotate the bits of $\boldsymbol{g}_1$ to the right by one unit, we obtain

$$\boldsymbol{g}_2 = [\mathbf{0}\ 1\ 1\ 0\ 1\ \mathbf{0}\ \mathbf{0}].$$

If we rotate the bits of $\boldsymbol{g}_2$ to the right by one unit, we obtain

$$\boldsymbol{g}_3 = [\mathbf{0}\ \mathbf{0}\ 1\ 1\ 0\ 1\ \mathbf{0}].$$

If we rotate the bits of $\boldsymbol{g}_3$ to the right by one unit, we obtain

$$\boldsymbol{g}_4 = [\mathbf{0}\ \mathbf{0}\ \mathbf{0}\ 1\ 1\ 0\ 1].$$

The generator matrix of the cyclic code can be formed as

$$\boldsymbol{G} = \begin{bmatrix} \boldsymbol{g}_1 \\ \boldsymbol{g}_2 \\ \boldsymbol{g}_3 \\ \boldsymbol{g}_4 \end{bmatrix} \rightarrow \boldsymbol{G} = \begin{bmatrix} 1\ 1\ 0\ 1\ \mathbf{0}\ \mathbf{0}\ \mathbf{0} \\ \mathbf{0}\ 1\ 1\ 0\ 1\ \mathbf{0}\ \mathbf{0} \\ \mathbf{0}\ \mathbf{0}\ 1\ 1\ 0\ 1\ \mathbf{0} \\ \mathbf{0}\ \mathbf{0}\ \mathbf{0}\ 1\ 1\ 0\ 1 \end{bmatrix}. \tag{4.51}$$

To obtain the parity check matrix of the cyclic code, we can obtain the systematic form of Eq. (4.51), and using the systematic form of the generator matrix, we can construct the parity check matrix of the cyclic code.

***Parity Check Matrices of the Cyclic Codes***

Let $h(x)$ be the parity check polynomial of a cyclic code such that

$$h(x) = b_k x^k + b_{k-1} x^{k-1} + \ldots + b_0 x^1 + b_0 x^0 \tag{4.52}$$

where $a_i \in F = \{0, 1\}$. Using the binary coefficients of the parity check polynomial, we construct the parity check matrix as

$$H = \begin{bmatrix} b_0 & b_1 & b_2 & \dots & b_{k-1} & b_k & 0 & 0 & \dots & 0 & 0 & 0 \\ 0 & b_0 & b_1 & b_2 & \dots & b_{k-1} & b_k & 0 & 0 & \dots & 0 & 0 \\ 0 & 0 & b_0 & b_1 & b_2 & \dots & b_{k-1} & b_k & 0 & 0 & \dots & 0 \\ \vdots & \vdots & \vdots & \vdots & \vdots & \vdots & \vdots & \vdots & \vdots & \vdots & \vdots & \vdots \\ 0 & 0 & 0 & \dots & 0 & 0 & b_0 & b_1 & b_2 & \dots & b_{k-1} & b_k \end{bmatrix}_{(n-k)\times n}. \tag{4.53}$$

**Example 4.18** The generator polynomial of a single-error-correcting $\boldsymbol{C(n = 7, k = 4)}$ cyclic code is given as

$$g(x) = x^3 + x^2 + 1.$$

Find the parity check matrix of this cyclic code.

**Solution 4.18** We can find the parity check polynomial of the given cyclic code as

$$h(x) = \frac{x^n + 1}{g(x)} \to h(x) = \frac{x^7 + 1}{x^3 + x^2 + 1} \to h(x) = x^4 + x^3 + x^2 + 1.$$

The size of the parity check matrix is $(n - k) \times n = 3 \times 7$. The parity check polynomial $h(x) = x^4 + x^3 + x^2 + 1$ can be represented by a bit vector as

$$\boldsymbol{h} = [1\ 1\ 1\ 0\ 1]. \tag{4.54}$$

If we reverse the bit vector in Eq. (4.54), we get

$$\boldsymbol{h}_r = [1\ 0\ 1\ 1\ 1].$$

If we zero pad the vector $\boldsymbol{h}_r = [1\ 0\ 1\ 1\ 1]$ such that its length equals to 7, we obtain

$$\boldsymbol{h}_{r1} = [1\ 0\ 1\ 1\ 1\ \mathbf{0}\ \mathbf{0}].$$

If we rotate the bits of $\boldsymbol{h}_{r1}$ to the right by one unit, we obtain

$$\boldsymbol{h}_{r2} = [\mathbf{0}\ 1\ 0\ 1\ 1\ 1\ \mathbf{0}].$$

If we rotate the bits of $\boldsymbol{h}_{r2}$ to the right by one unit, we obtain

$$\boldsymbol{h}_{r3} = [\mathbf{0}\ \mathbf{0}\ 1\ 0\ 1\ 1\ 1].$$

The parity check matrix of the cyclic code can be formed as

$$\boldsymbol{H} = \begin{bmatrix} \boldsymbol{h}_{r1} \\ \boldsymbol{h}_{r2} \\ \boldsymbol{h}_{r3} \end{bmatrix} \rightarrow \boldsymbol{H} = \begin{bmatrix} 1\,0\,1\,1\,1\,\mathbf{0}\,\mathbf{0} \\ \mathbf{0}\,1\,0\,1\,1\,1\,\mathbf{0} \\ \mathbf{0}\,\mathbf{0}\,1\,0\,1\,1\,1 \end{bmatrix}. \tag{4.55}$$

The generator and parity check matrices given in Eqs. (4.51) and (4.55) satisfy

$$\boldsymbol{G}\boldsymbol{H}^T = \mathbf{0}.$$

That is,

$$\boldsymbol{G}\boldsymbol{H}^T = \begin{bmatrix} 1\,1\,0\,1\,\mathbf{0}\,\mathbf{0}\,\mathbf{0} \\ \mathbf{0}\,1\,1\,0\,1\,\mathbf{0}\,\mathbf{0} \\ \mathbf{0}\,\mathbf{0}\,1\,1\,0\,1\,\mathbf{0} \\ \mathbf{0}\,\mathbf{0}\,\mathbf{0}\,1\,1\,0\,1 \end{bmatrix} \times \begin{bmatrix} 1\,\mathbf{0}\,\mathbf{0} \\ 0\,1\,\mathbf{0} \\ 1\,0\,1 \\ 1\,1\,0 \\ 1\,1\,1 \\ \mathbf{0}\,1\,1 \\ \mathbf{0}\,\mathbf{0}\,1 \end{bmatrix} \rightarrow \boldsymbol{G}\boldsymbol{H}^T = \begin{bmatrix} 0\,0\,0 \\ 0\,0\,0 \\ 0\,0\,0 \\ 0\,0\,0 \end{bmatrix}.$$

## Problems

1. The polynomial $x^9 + 1$ can be factorized as

$$x^9 + 1 = (x+1)\left(x^2+x+1\right)\left(x^6+x^3+1\right).$$

The generator polynomial of a cyclic code with block length $n = 9$ is given as

$$g(x) = x^6 + x^3 + 1.$$

(a) Determine the value of $k$.
(b) Find the parity check polynomial and generator polynomial of the dual cyclic code.
(c) Find the generator and parity check matrices of this cyclic code.
(d) Find the minimum distance of this code.
(e) Construct the syndrome polynomial table of this cyclic code.

2. The generator polynomial of $\boldsymbol{C}(n=8, k=4)$ cyclic code is given as

$$g(x) = x^4 + 1.$$

(a) Find the parity check polynomial of this code.
(b) Express the data vector $\boldsymbol{d} = [1\ 0\ 1\ 1]$ in polynomial form, and encode the data polynomial using non-systematic and systematic encoding methods. Obtain the systematic and non-systematic code-word and determine the locations of both data and parity bits in each code-word.

3. A cyclic code is used to encode a data polynomial, and the code-word

$$c(x) = x^5 + x^4 + 1$$

is obtained. What can be the parameters of the cyclic code used, and determine a generator matrix for this code. After determining the generator polynomial, find the data-word polynomial which yields the given code-word after encoding operation.

4. Using the factorization

$$x^{15} + 1 = (x+1)\left(x^2+x+1\right)\left(x^4+x+1\right)\left(x^4+x^3+x^2+x+1\right)$$

(a) Determine the number of cyclic codes $\boldsymbol{C}(n=15, k)$.
(b) Find the generator polynomials of the cyclic codes $\boldsymbol{C}(n=15, k=11)$.
(c) Find the generator polynomial, parity check polynomial, generator matrix, and parity check matrix of the cyclic code $\boldsymbol{C}(n=15, k=7)$.

# Chapter 5
# Galois Fields

## 5.1 Equation Roots and Concept of Field Extension

A polynomial is defined with powers of a dummy parameter $x$ and coefficients selected from a field $(F, \oplus, \otimes)$. For instance, consider the field of real numbers for which ordinary addition and multiplication operations, i.e., mod-10 addition and multiplication operations, are defined, and a polynomial is given as

$$p(x) = x^2 - 2.5x + 1.5 \tag{5.1}$$

where the coefficients are selected from the real numbers, i.e., real number field $(R, +, \times)$.

Now consider the equation

$$p(x) = 0 \rightarrow x^2 - 2.5x + 1.5 = 0. \tag{5.2}$$

The roots of Eq. (5.1) can be found as $\alpha_1 = 1$ and $\alpha_2 = 1.5$. When 1 is substituted for $x$ in Eq. (5.1), we get

$$1^2 - 2.5 \times 1 + 1.5 \rightarrow 0.$$

In a similar manner, when 1.5 is substituted for $x$ in Eq. (5.1), we get

$$1.5^2 - 2.5 \times 1.5 + 1.5 \rightarrow 0.$$

That is, the roots satisfy the Eq. (5.1), and the roots are also real numbers, i.e., they are available in the real number field $(R, +, \times)$.

Now consider the polynomial



O. Gazi, *Forward Error Correction via Channel Coding*,
https://doi.org/10.1007/978-3-030-33380-5_5

$$p(x) = x^2 + 1$$

where we see that the coefficients are selected from the real numbers, i.e., real number field ($R$, + , ×).

Now, let's try to solve the equation

$$p(x) = 0 \rightarrow x^2 + 1 = 0. \tag{5.3}$$

The roots of Eq. (5.3) can NOT be found in the real number field, i.e., in ($R$, + , ×). That is, we cannot find an $\alpha \in R$ satisfying Eq. (5.3), i.e., there is no $\alpha \in R$ such that $\alpha^2 + 1 = 0$.

If the solution is not available in real number field, we should look for the solution in another field which includes the real number field as its subset. Such a field is the complex number field $C$, and assume that in this field we have $i \in C$ such that

$$i^2 + 1 = 0 \rightarrow i^2 = -1. \tag{5.4}$$

The elements of the complex number field can be obtained as

$$C = \{R\} \cup \{R + i \times R\}. \tag{5.5}$$

And defining two new operators (+, ×) for the addition and multiplication of two complex numbers, we can extend the real number field to the complex number field. This extension can be symbolically indicated by

$$R \rightarrow C.$$

Let's try to illustrate the field extension concept by some examples.

**Example 5.1** The finite field ($F_3$, ⊕ , ⊗) is defined as

$$F_3 = \{0, 1, 2\}$$

$$\oplus \rightarrow \text{Mod} - 3 \text{ additon operation}$$

$$\otimes \rightarrow \text{Mod} - 3 \text{ multiplication operation.}$$

Consider the polynomial

$$p(x) = x^2 + 2x + 2. \tag{5.6}$$

We see that the coefficients of Eq. (5.6) are selected from $F = \{0, 1, 2\}$.

Now consider the solution of the equation

$$p(x) = 0 \rightarrow x^2 + 2x + 2 = 0. \tag{5.7}$$

We should look for the solution from the field elements $F = \{0, 1, 2\}$. Let's try each field element in the Eq. (5.7), and decide the ones that satisfy Eq. (5.7) as follows:

$$x = 0 \rightarrow x^2 + 2x + 2 = 0 \rightarrow 0^2 + 2 \times 0 + 2 = 0 \rightarrow 2 = 0 \rightarrow \text{incorrect} \tag{5.8}$$

$$x = 1 \rightarrow x^2 + 2x + 2 = 0 \rightarrow 1^2 + 2 \times 1 + 2 = 0 \rightarrow 2 = 0 \rightarrow \text{incorrect} \tag{5.9}$$

$$x = 2 \rightarrow x^2 + 2x + 2 = 0 \rightarrow 2^2 + 2 \times 2 + 2 = 0 \rightarrow 1 = 0 \rightarrow \text{incorrect.} \tag{5.10}$$

Thus, considering Eqs. (5.8), (5.9) and (5.10), we can say that there is no root of Eq. (5.7) in the field $F_3 = \{0, 1, 2\}$. We should look for the roots in another field which contains the field $F_3 = \{0, 1, 2\}$ as its subset. Assume that we have such a field, and let's show this field as

$$E = \{0, 1, 2, \alpha, \ldots\} \tag{5.11}$$

where the element $\alpha$ is the root of Eq. (5.7), i.e.,

$$x = \alpha \rightarrow x^2 + 2x + 2 = 0 \rightarrow \alpha^2 + 2\alpha + 2 = 0. \surd$$

From Eq. (5.11), it is clear that $F_3 \subset E$. Since $E$ is a field, then it should satisfy all the properties of a field. To check whether such a field exists or not, we need to find all the other elements of $E$. Then, we ask the question "How can we determine all the other elements of $E$ ?"

This question can also be stated as "How can we extend the finite field $F_3$ to another finite field $E$ such that the extended field contains an element $\alpha$ which is a root of Eq. (5.7)?"

In fact, this is the main topic of this chapter, and we will first explain the extension of finite fields in details.

### 5.1.1 Extension of Finite Fields

In this section, we will only study the extension of binary field, i.e., Galois field, indicated by $F$, $F_2$, or $GF(2)$. We have stated in Chap. 1 that a set of polynomials may satisfy the properties of a field, and the coefficients of the polynomials are selected from a number field.

For instance, consider the binary field $F = \{0, 1\}$ and the set of all the polynomials with degrees less than three. The coefficients of the polynomials are selected from the field $F = \{0, 1\}$. The polynomial set can be written as

$$G = \{0, 1, \alpha, \alpha + 1, \alpha^2, \alpha^2 + 1, \alpha^2 + \alpha, \alpha^2 + \alpha + 1\}. \tag{5.12}$$

If we make a multiplication and addition table for the elements of Eq. (5.12), we can see from the tables that all the properties of a field are satisfied. Thus, we can say that $G$ is a field under polynomial multiplication and addition operations.

Now consider the polynomial $p(x) = x^3 + x + 1$ whose coefficients are selected from the elements of the binary field $F = \{0, 1\}$. Let's try to find the solution of the equation

$$p(x) = 0 \rightarrow x^3 + x + 1 = 0 \tag{5.13}$$

in the binary field. If we try the elements 0 and 1 in Eq. (5.13), we see that neither of them are the solutions of Eq. (5.13), i.e.,

$$x = 0 \rightarrow x^3 + x + 1 = 0 \rightarrow 0^3 + 0 + 1 = 0 \rightarrow 1 = 0 \rightarrow \text{incorrect}$$

$$x = 1 \rightarrow x^3 + x + 1 = 0 \rightarrow 1^3 + 1 + 1 = 0 \rightarrow 1 = 0 \rightarrow \text{incorrect.}$$

Then we should look for the solution in another field extended from $F = \{0, 1\}$. Assume that $\alpha$ is a root of the equation, i.e.,

$$x = \alpha \rightarrow \alpha^3 + \alpha + 1 = 0$$

and $\alpha$ is available in another field which includes the binary field as its subset, then the extended field can written as

$$GF = \{0, 1, \alpha, \ldots\}.$$

We wonder the other elements of $GF$. However, we previously showed that the set of polynomials in Eq. (5.12) is a field. Then, we can use it as our extended field. The extended field in Eq. (5.12) includes eight elements, and for this reason, we can show the extended field as $GF(8)$. That is, our extended field which includes the solution of the equation

$$x^3 + x + 1 = 0 \tag{5.14}$$

can be written as

$$GF(8) = \{0, 1, \alpha, \alpha + 1, \alpha^2, \alpha^2 + 1, \alpha^2 + \alpha, \alpha^2 + \alpha + 1\}. \tag{5.15}$$

There are three roots of the equation

$$x^3 + x + 1 = 0.$$

One of the roots is $\alpha$; this means that $\alpha^3 + \alpha + 1 = 0$ from which we can write that

$$\alpha^3 = \alpha + 1. \quad (5.16)$$

What about the other roots? Are they available in Eq. (5.15)? If we check the elements of Eq. (5.15) one by one, we see that $\alpha^2$ and $\alpha^2 + \alpha$ are also roots of Eq. (5.13). That is, if we put $x = \alpha^2$ in Eq. (5.13), we get

$$\left(\alpha^2\right)^3 + \alpha^2 + 1 = 0$$

where, employing Eq. (5.16), we obtain

$$(\alpha + 1)^2 + \alpha^2 + 1 = 0 \rightarrow \alpha^2 + 1 + \alpha^2 + 1 = 0 \rightarrow 0 = 0\surd$$

which is a correct equality. In a similar manner, if we use $x = \alpha^2 + \alpha$ in Eq. (5.13), we see that it is also a root of the given polynomial.

### 5.1.2 *Irreducible Polynomial*

The polynomial $p(x)$ with degree $r$ is an irreducible polynomial if $p(x)$ divides the polynomial

$$x^n + 1 \quad (5.17)$$

where

$$n = 2^r - 1 \quad (5.18)$$

and $p(x)$ divides at least one polynomial

$$x^m + 1 \quad (5.19)$$

where

$$m < 2^r - 1. \quad (5.20)$$

An irreducible polynomial cannot be factorized.

**Example 5.2** The polynomial $p(x) = x^4 + x^3 + x^2 + x + 1$ divides $x^{15} + 1$ where $15 = 2^4 - 1$, and $p(x)$ also divides $x^5 + 1$. Thus, we can conclude that the polynomial

$$p(x) = x^4 + x^3 + x^2 + x + 1$$

is an irreducible polynomial, and it cannot be factorized.

### 5.1.3 *Primitive Polynomial*

The polynomial $p(x)$ with degree $r$ is a primitive polynomial if $p(x)$ divides the polynomial

$$x^n + 1$$

where

$$n = 2^r - 1$$

and $p(x)$ does NOT divide a polynomial of the form

$$x^m + 1$$

where

$$m < 2^r - 1.$$

**Example 5.3** Let's show that the polynomial

$$p(x) = x^3 + x + 1$$

is a primitive polynomial.

The degree of $p(x)$ is $r = 3$. If $p(x)$ is a primitive polynomial, then it should divide $x^7 + 1, 7 = 2^3 - 1$, and it should not divide all the polynomials $x^m + 1$ where $m < 7$. We can show that

$$x^3 + x + 1 \text{ divides } x^7 + 1$$

$$x^3 + x + 1 \text{ does NOT divide } x^6 + 1$$

$$x^3 + x + 1 \text{ does NOT divide } x^5 + 1$$

$$x^3 + x + 1 \text{ does NOT divide } x^4 + 1$$

$$x^3 + x + 1 \text{ does NOT divide } x^3 + 1.$$

Thus, we can conclude that $p(x) = x^3 + x + 1$ is a primitive polynomial.

**Exercise** Show that the polynomials

$$p_1(x) = x^4 + x + 1$$

and

$$p_2(x) = x^4 + x^3 + 1$$

are primitive polynomials.

***Note*** Let $p(x)$ be a polynomial with degree $m$. If $2^m - 1$ is a prime number, then we can say that $p(x)$ is a primitive polynomial. On the other hand, if $2^m - 1$ is NOT a prime number, then $p(x)$ may or may not be a primitive polynomial.

**Example 5.4** The polynomial $p(x) = x^3 + x + 1$ is a primitive polynomial, since $2^3 - 1 = 7$ is a prime number.

The polynomial $p(x) = x^4 + x + 1$ is a primitive polynomial, although $2^4 - 1 = 15$ is NOT a prime number.

The polynomial $p(x) = x^4 + x^3 + x^2 + x + 1$ is NOT a primitive polynomial, and $2^4 - 1 = 15$ is NOT a prime number.

## 5.2 Construction of Extended Finite Fields

Let $p(x)$ be a primitive polynomial with degree $m$. Using the primitive polynomial, we can generate all the elements of the extended field $GF(2^m)$ via a recursive computation. The generation of the elements of the extended field is outlined as follows:

1. We assume that $\alpha$ is a root of $p(x)$ such that

$$p(\alpha) = 0. \tag{5.21}$$

2. From Eq. (5.21), we write a recursive statement in the form

$$\alpha^m = c_1\alpha^{m-1} + c_2\alpha^{m-2} + \ldots + c_m\alpha^0 \text{ such that } c_i \in F = \{0, 1\}, i = 1, \ldots, m. \tag{5.22}$$

3. The elements of the extended Galois field $GF(2^m)$ can be written as

$$GF(2^m) = \{0, 1, \alpha, \alpha^2, \ldots, \alpha^r\} \text{ where } r = 2^m - 2. \tag{5.23}$$

If we want to write the polynomial expressions for the elements of extended field $GF(2^m)$, then we use Eq. (5.22) to simplify the $\alpha^i$ expressions in Eq. (5.23).

**Example 5.5** The primitive polynomial $p(x)$ is given as

$$p(x) = x^3 + x + 1.$$

Using the given primitive polynomial, construct the extended field $GF(2^3)$.

**Solution 5.5** Let's follow the steps mentioned in the previous paragraphs for the determination of elements of the extended field.

1. Let $\alpha$ be a root of $p(x) = x^3 + x + 1$; then, we can write that

$$\alpha^3 + \alpha + 1 = 0. \tag{5.24}$$

2. From Eq. (5.24), we obtain the recursive statement

$$\alpha^3 = \alpha + 1. \tag{5.25}$$

3. The elements of the extended field $GF(2^3)$ can be written as

$$GF(2^3) = \{0, 1, \alpha, \alpha^2, \ldots, \alpha^6\} \text{ where } 6 = 2^3 - 2. \tag{5.26}$$

The $\alpha^i$ terms in Eq. (5.26) can be expressed in polynomial form using Eq. (5.25) as in

$$\begin{aligned}
&\alpha^3 \to \alpha + 1 \\
&\alpha^4 \to \alpha \times \alpha^3 \to \alpha(\alpha + 1) \to \alpha^2 + \alpha \\
&\alpha^5 \to \alpha \times \alpha^4 \to \alpha \times (\alpha^2 + \alpha) \to \alpha^3 + \alpha^2 \to \alpha^2 + \alpha + 1 \\
&\alpha^6 \to \alpha \times \alpha^5 \to \alpha \times (\alpha^2 + \alpha + 1) \to \alpha^3 + \alpha^2 + \alpha \to \alpha^2 + 1.
\end{aligned} \tag{5.27}$$

Using the polynomial equivalents of $\alpha^i$ terms, we can write the extended field in Eq. (5.26) as in

$$GF(2^3) = \{0, 1, \alpha, \alpha^2, \alpha^2 + \alpha, \alpha^2 + 1, \alpha^2 + \alpha + 1\}.$$

***Construction of the Extended Field Using an Irreducible Polynomial***

Let $p(x)$ be an irreducible polynomial with order $m$. We can construct the extended field $GF(2^m)$ with the help of $p(x)$ as follows.

1. We assume that $\alpha$ is a root of $p(x)$ such that

$$p(\alpha) = 0. \tag{5.28}$$

2. From Eq. (5.28), we write a recursive statement in the form

$$\alpha^m = c_1\alpha^{m-1} + c_2\alpha^{m-2} + \ldots + c_m\alpha^0 \ \text{ such that } c_i \in F = \{0, 1\}, i = 1, \ldots, m. \tag{5.29}$$

3. There exists a polynomial of $\alpha$, i.e., $\beta = f(\alpha)$, with degree less than $m$, and using this polynomial, we can construct the extended field as

$$GF(2^m) = \left\{0, 1, \beta, \beta^2, \ldots, \beta^r\right\} \text{ where } r = 2^m - 2. \tag{5.30}$$

Using the recursive statement in Eq. (5.29), we can write the field elements of Eq. (5.30) as polynomials.

**Example 5.6** The irreducible polynomial $p(x)$ is given as

$$p(x) = x^4 + x^3 + x^2 + x + 1. \tag{5.31}$$

Using the given irreducible polynomial, construct the extended field $GF(2^4)$ (Table 5.1).

**Solution 5.6** 1. We assume that $\alpha$ is a root of $p(x)$ such that

$$p(\alpha) = 0. \tag{5.32}$$

2. From Eq. (5.32), we write a recursive statement in the form

$$\alpha^4 = \alpha^3 + \alpha^2 + \alpha + 1. \tag{5.33}$$

3. Let's choose a polynomial of $\alpha$ as

**Table 5.1** List of irreducible and primitive polynomials with their hexadecimal representations

| $n$ | Primitive polynomials | Irreducible polynomials |
|---|---|---|
| 1 | $x \rightarrow 2$<br>$x+1 \rightarrow 3$ | – |
| 2 | $x^2+x+1 \rightarrow 7$ | – |
| 3 | $x^3+x+1 \rightarrow B$<br>$x^3+x^2+1 \rightarrow D$ | – |
| 4 | $x^4+x+1 \rightarrow 13$ $x^4+x^3+1 \rightarrow 19$ | $x^4+x^3+x^2+x+1 \rightarrow 1F$ |
| 5 | $x^5+x^4+x^3+x^2+1 \rightarrow 3D$<br>$x^5+x^4+x^2+x+1 \rightarrow 37$<br>$x^5+x^3+x^2+x+1 \rightarrow 2F$<br>$x^5+x^4+x^3+x+1 \rightarrow 3B$<br>$x^5+x^3+1 \rightarrow 29$<br>$x^5+x^2+1 \rightarrow 25$ | – |
| 6 | $x^6+x+1 \rightarrow 43$<br>$x^6+x^5+1 \rightarrow 61$<br>$x^6+x^5+x^3+x^2+1 \rightarrow 49$<br>$x^6+x^5+x^4+x+1 \rightarrow 73$<br>$x^6+x^4+x^3+x+1 \rightarrow 5B$<br>$x^6+x^5+x^2+x+1 \rightarrow 57$ | $x^6+x^2+1 \rightarrow 45$ |

$$f(\alpha) = \alpha + 1 \rightarrow \beta = \alpha + 1.$$

Using Eq. (5.33), we can calculate the powers of $\beta$ as in

$$\beta^1 \rightarrow \alpha + 1$$

$$\beta^2 \rightarrow \beta\beta \rightarrow \alpha^2 + 1$$

$$\beta^3 \rightarrow \beta^2\beta \rightarrow (\alpha^2+1)(\alpha+1) \rightarrow \alpha^3+\alpha^2+\alpha+1$$

$$\beta^4 \rightarrow \beta^3\beta \rightarrow (\alpha^3+\alpha^2+\alpha+1)(\alpha+1) \rightarrow \alpha^4+\alpha^3+\alpha^2+\alpha+\alpha^3+\alpha^2+\alpha+1$$
$$\rightarrow \alpha^3+\alpha^2+\alpha$$

$$\beta^5 \rightarrow \beta^4\beta \rightarrow (\alpha^3+\alpha^2+\alpha)(\alpha+1) \rightarrow \alpha^4+\alpha^3+\alpha^2+\alpha^3+\alpha^2+\alpha$$
$$\rightarrow \alpha^3+\alpha^2+1$$

$$\beta^6 \rightarrow \beta^5\beta \rightarrow (\alpha^3+\alpha^2+1)(\alpha+1) \rightarrow \alpha^4+\alpha^3+\alpha+\alpha^3+\alpha^2+1 \rightarrow \alpha^3$$

$$\beta^7 \rightarrow \beta^6\beta \rightarrow (\alpha^3)(\alpha+1) \rightarrow \alpha^4+\alpha^3 \rightarrow \alpha^2+\alpha+1$$

$$\beta^8 \rightarrow \beta^7\beta \rightarrow (\alpha^2+\alpha+1)(\alpha+1) \rightarrow \alpha^3+\alpha^2+\alpha+\alpha^2+\alpha+1 \rightarrow \alpha^3+1$$

$$\beta^9 \rightarrow \beta^8\beta \rightarrow (\alpha^3+1)(\alpha+1) \rightarrow \alpha^4+\alpha+\alpha^3+1 \rightarrow \alpha^2$$

$$\beta^{10} \rightarrow \beta^9\beta \rightarrow (\alpha^2)(\alpha+1) \rightarrow \alpha^3+\alpha^2$$

$$\beta^{11} \to \beta^{10}\beta \to (\alpha^3+\alpha^2)(\alpha+1) \to \alpha^4+\alpha^3+\alpha^3+\alpha^2 \to \alpha^3+\alpha+1$$
$$\beta^{12} \to \beta^{11}\beta \to (\alpha^3+\alpha+1)(\alpha+1) \to \alpha^4+\alpha^2+\alpha+\alpha^3+\alpha+1 \to \alpha$$
$$\beta^{13} \to \beta^{12}\beta \to (\alpha)(\alpha+1) \to \alpha^2+\alpha$$
$$\beta^{14} \to \beta^{13}\beta \to (\alpha^2+\alpha)(\alpha+1) \to \alpha^3+\alpha.$$

The extended field $GF(2^4)$ can be written as

$$GF(2^4) = \{0, 1, \beta, \beta^2, \beta^3, \beta^4, \beta^5, \beta^6, \beta^7, \beta^8, \beta^9, \beta^{10}, \beta^{11}, \beta^{12}, \beta^{13}, \beta^{14}\}$$

or using the polynomial form of $\beta^i$, we can express the extended field $GF(2^4)$ as

$$\begin{aligned} GF(2^4) = \{&0, 1, \alpha+1, \alpha^2+1, \alpha^3+\alpha^2+\alpha+1,\\ &\alpha^3+\alpha^2+\alpha, \alpha^3+\alpha^2+1,\\ &\alpha^3, \alpha^2+\alpha+1, \alpha^3+1,\\ &\alpha^2, \alpha^3+\alpha^2, \alpha^3+\alpha+1,\\ &\alpha, \alpha^2+\alpha, \alpha^3+\alpha.\} \end{aligned}$$

**Exercise** The polynomial $p(x) = x^5 + x^2 + 1$ is a primitive polynomial. Using $p(x)$, generate all the elements of $GF(2^5)$, i.e., construct the extended field $GF(2^5)$ from the base field $F = \{0, 1\}$, i.e., binary field.

## 5.3 Conjugate Classes

Consider the solution of the equation

$$x^2 - 2x + 2 = 0 \tag{5.34}$$

in complex number field. There are two roots of this equation, and the roots are

$$1 + j \text{ and } 1 - j. \tag{5.35}$$

The roots in Eq. (5.35) are said to be conjugates of each other, and we can make a set consisting of the roots which are conjugate of each other as

$$\{1 + j, 1 - j\}$$

which can be called as a conjugate class.

A similar concept can be pursued for the solution of equations in finite field. The roots of the equation $p(x) = 0$ in a finite field appear as conjugates of each other. For instance, the equation

$$x^3 + x + 1 = 0$$

has three roots in $GF(2^8)$, and the roots are conjugates of each other. In fact, in $GF(2^8)$, every element is a root of a polynomial equation, and conjugates of this root are also available in $GF(2^8)$. In fact, three elements of $GF(2^8)$ are the roots of

$$x^3 + x + 1 = 0$$

and another three are the roots of

$$x^3 + x^2 + 1 = 0$$

and the remaining two are the roots of

$$x = 0 \text{ and } x = 1.$$

Let $\beta$ be an element of the extended field $GF(2^m)$. We know that $\beta$ is a root of an equation $p(x) = 0$. Assume that there are $r$ roots of $p(x) = 0$. The other roots of $p(x) = 0$, i.e., conjugates of $\beta$, can be calculated as

$$\beta^2, \beta^4, \beta^8, \ldots, \beta^{2^r} \tag{5.36}$$

or in short we can write Eq. (5.36) as

$$\beta^{2^i}, i = 1, \ldots, r. \tag{5.37}$$

and we have

$$\beta^{2^r+1} = \beta. \tag{5.38}$$

The conjugate class including $\beta$ can be written as

$$\{\beta, \beta^2, \beta^4, \beta^8, \ldots, \beta^{2^r}\} \tag{5.39}$$

such that $\beta^{2^r+1} = \beta$.

**Example 5.7** Using the primitive polynomial $p(x) = x^3 + x + 1$, construct the extended field $GF(2^3)$. Find the conjugates of $\alpha^3$, and determine the polynomial for which $\alpha^3$ is a root.

**Solution 5.7** If $\alpha$ is a root of

$$p(x) = 0 \rightarrow x^3 + x + 1 = 0$$

then we have

$$\alpha^3 + \alpha + 1 = 0 \rightarrow \alpha^3 = \alpha + 1. \tag{5.40}$$

The extended field $GF(2^8)$ can be written as

$$GF\left(2^8\right) = \left\{0, 1, \alpha, \alpha^2, \alpha^3, \alpha^4, \alpha^5, \alpha^6\right\}$$

where $\alpha^7 = 1$ and $\alpha^i$, $i = 1, \ldots, 6$ can be converted to polynomial expressions using the recursive statement in Eq. (5.40). The conjugates of $\beta = \alpha^3$ can be calculated as follows.

The first conjugate of $\beta$ is

$$\beta^2 \rightarrow \left(\alpha^3\right)^2 \rightarrow \alpha^6. \tag{5.41}$$

The second conjugate of $\beta$ is

$$\beta^4 \rightarrow \left(\beta^2\right)^2 \rightarrow \left(\alpha^6\right)^2 \rightarrow \alpha^{12} \rightarrow \underbrace{\alpha^7}_{=1}\alpha^5 \rightarrow \alpha^5. \tag{5.42}$$

The third conjugate of $\beta$ is

$$\beta^8 \rightarrow \left(\beta^4\right)^2 \rightarrow \left(\alpha^5\right)^2 \rightarrow \alpha^{10} \rightarrow \underbrace{\alpha^7}_{=1}\alpha^3 \rightarrow \alpha^3 \tag{5.43}$$

which is the same as $\beta$. Then, there is no third conjugate. There are only two conjugates of $\beta = \alpha^3$, and the conjugates are

$$\left\{\beta^2, \beta^4\right\} \rightarrow \left\{\alpha^6, \alpha^5\right\}.$$

The conjugate class including $\beta$ can be written as

$$\left\{\beta, \beta^2, \beta^4\right\} \rightarrow \left\{\alpha^3, \alpha^6, \alpha^5\right\}.$$

Since the elements $\beta$, $\beta^2$, $\beta^4$ are the roots of an equation

$$q(x) = 0$$

the polynomial $q(x)$ can be written as

$$q(x) = (x+\beta)(x+\beta^2)(x+\beta^4)$$

in which, inserting $\alpha^3$, $\alpha^6$, $\alpha^5$ for $\beta$, $\beta^2$, $\beta^4$, we obtain

$$q(x) = (x+\alpha^3)(x+\alpha^6)(x+\alpha^5). \tag{5.44}$$

***Note*** There is no "–" operator defined in $GF(2)$; only mod-2 addition and multiplication operations are defined. For this reason, it is not correct to write the polynomial $q(x)$ as

$$q(x) = (x-\alpha^3)(x-\alpha^6)(x-\alpha^5).$$

Expanding Eq. (5.44) and using the recursive statement in Eq. (5.40) we can simplify Eq. (5.44) as

$$(x+\alpha^3)(x+\alpha^6)(x+\alpha^5) \rightarrow (x^2+\alpha^6x+\alpha^3x+\alpha^9)(x+\alpha^5)$$

$$\rightarrow x^3+\alpha^6x^2+\alpha^3x^2+\underbrace{\alpha^9}_{=\alpha^2}x+\alpha^5x^2+\underbrace{\alpha^{11}}_{=\alpha^4}x+\underbrace{\alpha^8}_{=\alpha}x+\underbrace{\alpha^{14}}_{=1}$$

$$\rightarrow x^3+\alpha^6x^2+\alpha^3x^2+\alpha^2x+\alpha^5x^2+\alpha^4x+\alpha x+1$$

which can be simplified as

$$x^3+(\alpha^6+\alpha^3+\alpha^5)x^2+(\alpha^2+\alpha^4+\alpha)x+1. \tag{5.45}$$

It is a convention to express the coefficients of the polynomials as powers of $\alpha$. The coefficients in Eq. (5.45) can be expressed as powers of $\alpha$. For this purpose, we first write the polynomial equivalents of each $\alpha^i$ term; next, we sum the polynomials; and, lastly, we express the summation results using powers of $\alpha$. Accordingly, the coefficients of Eq. (5.45) can be converted to the powers of $\alpha$ as in

$$\underbrace{\alpha^6}_{\alpha^2+1}+\underbrace{\alpha^3}_{\alpha+1}+\underbrace{\alpha^5}_{\alpha^2+\alpha+1} \rightarrow \alpha^2+1+\alpha+1+\alpha^2+\alpha+1 \rightarrow 1$$

$$\alpha^2+\underbrace{\alpha^4}_{\alpha^2+\alpha}+\alpha \rightarrow \alpha^2+\alpha^2+\alpha+\alpha \rightarrow 0.$$

Thus, the polynomials in Eq. (5.45) happen to be as

$$x^3+x^2+1$$

which has no coefficients involving $\alpha$ terms. All the coefficients are available in the base field, i.e., binary field $GF(2)=\{0,1\}$.

*Note* The equation whose roots are the complex numbers $1 + j$ and $1 - j$ can be determined as

$$x^2 - 2x + 2 = 0$$

where we see that the coefficients of the polynomials $x^2 - 2x + 2$ are all real numbers belonging to the real number field, i.e., belonging to the base field, and the roots are available in the complex number field, i.e., available in the extended field.

**Example 5.8** Find all the conjugates of $\alpha^3$ in the extended field $GF(2^4)$.

**Solution 5.8** For $GF(2^4)$, we have $\alpha^{15} = 1$. If $\beta \in GF(2^4)$, then the conjugates of $\beta$ are generated according to

$$\beta^2, \beta^4, \beta^8, \ldots$$

until a repetition occurs in the generated elements. Accordingly, we can generate the conjugates of $\beta = \alpha^3$ as follows:

$$\beta^2 \rightarrow \left(\alpha^3\right)^2 \rightarrow \alpha^6$$

$$\beta^4 \rightarrow \left(\alpha^6\right)^2 \rightarrow \alpha^{12}$$

$$\beta^8 \rightarrow \left(\alpha^{12}\right)^2 \rightarrow \alpha^{24} \rightarrow \alpha^9$$

$$\beta^{16} \rightarrow \left(\alpha^9\right)^2 \rightarrow \alpha^{18} \rightarrow \alpha^3 \text{ is the same as } \beta\text{; stop here.}$$

Thus, the conjugate class including $\beta = \alpha^3$ can be written as

$$\{\beta, \beta^2, \beta^4, \beta^8\} \rightarrow \{\alpha^3, \alpha^6, \alpha^{12}, \alpha^9\}.$$

**Example 5.9** For the previous example, find the conjugates of $\alpha^6$.

**Solution 5.9** Let $\beta = \alpha^6$; the conjugates of $\beta$ can be found as

$$\beta^2 \rightarrow \left(\alpha^6\right)^2 \rightarrow \alpha^{12}$$

$$\beta^4 \rightarrow \left(\alpha^{12}\right)^2 \rightarrow \alpha^{24} \rightarrow \alpha^9$$

$$\beta^8 \rightarrow \left(\alpha^9\right)^2 \rightarrow \alpha^{18} \rightarrow \alpha^3$$

$$\beta^{16} \rightarrow \left(\alpha^3\right)^2 \rightarrow \alpha^6 \rightarrow \alpha^6 \text{ is the same as } \beta\text{; stop here.}$$

Hence we found the conjugates of $\alpha^6$ as

$$\{\alpha^3, \alpha^{12}, \alpha^9\}.$$

## 5.4 Order of a Finite Field Element

Let $\beta$ be an element of the extended field $GF(2^m)$. The order of $\beta$ is an integer $n$ such that

$$\beta^n = 1.$$

**Example 5.10** Let $p(x)$ be a primitive polynomial with degree 4. If $\alpha$ is a root of the equation $p(x) = 0$, then the powers of $\alpha$ can be used to construct the extended field $GF(2^4)$ as

$$GF\left(2^4\right) = \left\{0, 1, \alpha, \alpha^2, \alpha^3, \ldots, \alpha^{14}\right\}$$

and we have

$$\alpha^{15} = 1. \tag{5.46}$$

The order of the elements $\beta_1 = \alpha^5$ and $\beta_2 = \alpha^{10}$ is 3, since we have

$$(\beta_1)^3 = \left(\alpha^5\right)^3 \rightarrow (\beta_1)^3 = \alpha^{15} \rightarrow 1$$

$$(\beta_2)^3 = \left(\alpha^{10}\right)^3 \rightarrow (\beta_2)^3 = \alpha^{30} \rightarrow 1.$$

**Remark** In a conjugate class, all the elements have the same order.

**Remark** Let $\beta$ be an element of the extended field $GF(2^m)$. If the order of $\beta$ is an integer $r$, i.e., $\beta^r = 1$, then we can say that $r$ is a divider of $2^m - 1$.

**Remark** The element $\beta$ of the extended field $GF(2^m)$ is a primitive element, if the order of $\beta$ is $2^m - 1$. Primitive element is the element such that by taking its successive powers we can generate all the other field elements.

**Example 5.11** What can be the order of the elements in the extended field $GF(2^6)$?

**Solution 5.11** The orders of elements are the dividers of $2^6 - 1 = 63$. The integer 63 can be written as

$$63 = 3 \times 3 \times 7$$

from which we can conclude that the order of an element in $GF(2^6)$ can be one of the numbers

$$3, 9, 7, 21, \text{and } 63.$$

**Example 5.12** In $GF(2^4)$, the order of the elements $\alpha$, $\alpha^2$, $\alpha^4$, $\alpha^7$, $\alpha^8$, $\alpha^{11}$, $\alpha^{13}$, $\alpha^{14}$ is the same, and it is 1. This means that they are all primitive elements.

**Remark** Using the powers of primitive elements, we can generate all the field elements.

## 5.5 Minimal Polynomials

Minimal polynomials are defined for conjugate classes. Each conjugate class has its own minimal polynomial. If a conjugate class is given as

$$J = \{\beta_1, \beta_2, \ldots \beta_k\} \tag{5.47}$$

then the minimal polynomial of the conjugate class $J$ is calculated as

$$m(x) = (x + \beta_1)(x + \beta_2) \ldots (x + \beta_k). \tag{5.48}$$

When the equation Eq. (5.48) is expanded, we get a polynomial with binary coefficients, i.e., extended field elements other than the binary ones do not appear as the coefficients of the polynomial. In other words, although the roots are in the extended field, the coefficients of the polynomial are all in the base field.

***Note*** Minimal polynomials are irreducible polynomials, and they can also be primitive polynomials.

**Example 5.13** Assume that the polynomial $p(x) = x^3 + x + 1$ is used to construct the extended field $GF(2^3)$. Find the conjugates of $\beta = \alpha^3$, and find the minimal polynomial of the conjugate class.

**Solution 5.13** The conjugates of $\alpha^3$ can be evaluated as

$$\beta^2 \rightarrow \left(\alpha^3\right)^2 \rightarrow \alpha^6$$

$$\beta^4 \rightarrow \left(\alpha^6\right)^2 \rightarrow \alpha^{12} \rightarrow \alpha^5$$

$$\beta^8 \rightarrow \left(\alpha^5\right)^2 \rightarrow \alpha^{10} \rightarrow \alpha^3 \text{ is the same as } \beta.$$

Thus the conjugate class including $\beta$ can be written as

$$J = \{\beta, \beta^2, \beta^4\} \rightarrow J = \{\alpha^3, \alpha^6, \alpha^5\}.$$

The minimal polynomial of $J$ can be formed as

$$m(x) = (x+\beta)\left(x+\beta^2\right)\left(x+\beta^4\right) \rightarrow m(x) = \left(x+\alpha^3\right)\left(x+\alpha^6\right)\left(x+\alpha^5\right)$$

which can be simplified using as the recursive equation $\alpha^3 = \alpha + 1$, $\alpha^7 = 1$, and extended field elements

$$GF\left(2^3\right) = \left\{0, 1, \alpha, \alpha^2, \underbrace{\alpha+1}_{\alpha^3}, \underbrace{\alpha^2+\alpha}_{\alpha^4}, \underbrace{\alpha^2+\alpha+1}_{\alpha^5}, \underbrace{\alpha^2+1}_{\alpha^6}\right\} \quad (5.49)$$

as

$$\begin{aligned} m(x) &= \left(x+\alpha^3\right)\left(x+\alpha^6\right)\left(x+\alpha^5\right) \\ &= \left(x^2 + (\alpha^6+\alpha^3)x + \underbrace{\alpha^9}_{=\alpha^2}\right)\left(x+\alpha^5\right) \\ &= x^3 + \left(\underbrace{\alpha^6}_{\alpha^2+1} + \underbrace{\alpha^5}_{\alpha^2+\alpha+1} + \underbrace{\alpha^3}_{\alpha+1}\right)x^2 + ((\alpha^6+\alpha^3)\alpha^5 + \alpha^2)x + \underbrace{\alpha^{14}}_{=1} \\ &= x^3 + x^2 + \left(\underbrace{\alpha^4}_{\alpha^2+\alpha} + \alpha + \alpha^2\right)x + 1 \\ &= x^3 + x^2 + 1. \end{aligned}$$

Hence, we found the minimal polynomial of the conjugate class $J = \{\alpha^3, \alpha^6, \alpha^5\}$ as

$$m(x) = x^3 + x^2 + 1.$$

The conjugate classes and the corresponding minimal polynomials for $GF(2^3)$ are depicted in Table 5.2.

If we multiply the minimal polynomials $m_1(x)$, $m_2(x)$, and $m_3(x)$, we get

$$(x+1)\left(x^3+x+1\right)\left(x^3+x^2+1\right) \rightarrow x^7 + 1.$$

We used the polynomial $p(x) = x^3 + x + 1$ to generate the $GF(2^3)$. When we inspect the Table 5.2, we see that the polynomial used to generate the extended field is one of the minimal polynomials of conjugate classes.

**Table 5.2** The conjugate classes and the corresponding minimal polynomials for $GF(2^3)$

| Conjugate classes in $GF(2^3)$ | Minimal polynomials |
|---|---|
| $J_0 = \{0\}$ | $m_0(x) = x$ |
| $J_1 = \{1\}$ | $m_1(x) = x + 1$ |
| $J_2 = \{\alpha, \alpha^2, \alpha^4\}$ | $m_2(x) = x^3 + x + 1$ |
| $J_3 = \{\alpha^3, \alpha^5, \alpha^6\}$ | $m_3(x) = x^3 + x^2 + 1$ |

**Table 5.3** Generation of $GF(2^3)$ by different primitive polynomials

| $p(x) = x^3 + x + 1$ | $p(x) = x^3 + x^2 + 1$ |
|---|---|
| $0 \rightarrow 0$ | $0 \rightarrow 0$ |
| $1 \rightarrow 1$ | $1 \rightarrow 1$ |
| $\alpha \rightarrow \alpha$ | $\alpha \rightarrow \alpha$ |
| $\alpha^2 \rightarrow \alpha^2$ | $\alpha^2 \rightarrow \alpha^2$ |
| $\alpha^3 \rightarrow \alpha + 1$ | $\alpha^3 \rightarrow \alpha^2 + 1$ |
| $\alpha^4 \rightarrow \alpha^2 + \alpha$ | $\alpha^4 \rightarrow \alpha^2 + \alpha + 1$ |
| $\alpha^5 \rightarrow \alpha^2 + \alpha + 1$ | $\alpha^5 \rightarrow \alpha + 1$ |
| $\alpha^6 \rightarrow \alpha^2 + 1$ | $\alpha^6 \rightarrow \alpha^2 + \alpha$ |

If we use the polynomial $p(x) = x^3 + x^2 + 1$ to generate the field elements, we obtain the extended field $GF(2^3)$ as

$$GF\left(2^3\right) = \left\{0, 1, \alpha, \alpha^2, \underbrace{\alpha^2 + 1}_{\alpha^3}, \underbrace{\alpha^2 + \alpha + 1}_{\alpha^4}, \underbrace{\alpha + 1}_{\alpha^5}, \underbrace{\alpha^2 + \alpha}_{\alpha^6}\right\}. \tag{5.50}$$

If we compare Eqs. (5.50) to (5.49), we see that, although the same set of polynomials are generated, the corresponding $\alpha^i$ terms of the polynomials are different from each other. This comparison is depicted in Table 5.3.

The minimal polynomial of the conjugate class $J = \{\alpha^3, \alpha^6, \alpha^5\}$ using $\alpha^3 = \alpha^2 + 1$, $\alpha^7 = 1$ and extended field elements

$$GF\left(2^3\right) = \left\{0, 1, \alpha, \alpha^2, \underbrace{\alpha^2 + 1}_{\alpha^3}, \underbrace{\alpha^2 + \alpha + 1}_{\alpha^4}, \underbrace{\alpha + 1}_{\alpha^5}, \underbrace{\alpha^2 + \alpha}_{\alpha^6}\right\}$$

can be calculated as

$$m(x) = \left(x + \alpha^3\right)\left(x + \alpha^6\right)\left(x + \alpha^5\right)$$

$$= \left(x^2 + \left(\alpha^6 + \alpha^3\right)x + \underbrace{\alpha^9}_{=\alpha^2}\right)\left(x + \alpha^5\right)$$

$$= x^3 + \left( \underbrace{\alpha^6}_{\alpha^2+\alpha} + \underbrace{\alpha^5}_{\alpha+1} + \underbrace{\alpha^3}_{\alpha^2+1} \right) x^2 + ((\alpha^6 + \alpha^3)\alpha^5 + \alpha^2)x + \underbrace{\alpha^{14}}_{=1}$$

$$= x^3 + 0x^2 + \left( \underbrace{\alpha^4}_{\alpha^2+\alpha+1} + \alpha + \alpha^2 \right) x + 1$$

$$= x^3 + x + 1.$$

Thus, we see that if $p(x) = x^3 + x + 1$ is used for the generation of extended field $GF(2^3)$, then the minimal polynomial of the conjugate class $J = \{\alpha^3, \alpha^6, \alpha^5\}$ happens to be

$$m(x) = x^3 + x^2 + 1.$$

On the other hand, if $p(x) = x^3 + x^2 + 1$ is used for the generation of extended field $GF(2^3)$, then the minimal polynomial of the conjugate class $J = \{\alpha^3, \alpha^6, \alpha^5\}$ happens to be

$$m(x) = x^3 + x + 1.$$

**Example 5.14** Assume that $p(x)$ is a primitive polynomial with degree 4, and it is used for the generation of the extended field $GF(2^4)$. Write all the conjugate classes, and show that the extended field $GF(2^4)$ can be written as the union of all conjugate classes.

**Solution 5.14** In $GF(2^4)$, we have $\alpha^{15} = 1$. Using the equality $\alpha^{15} = 1$, we can form the conjugates of $\beta = \alpha$ as in

$$\beta^2 \rightarrow \alpha^2$$

$$\beta^4 \rightarrow \left(\alpha^2\right)^2 \rightarrow \alpha^4$$

$$\beta^8 \rightarrow \left(\alpha^4\right)^2 \rightarrow \alpha^8$$

$$\beta^{16} \rightarrow \left(\alpha^8\right)^2 \rightarrow \alpha^{16} \rightarrow \alpha \text{ which is the same as } \beta.$$

Thus, the conjugate class including $\alpha$ can be written as

$$J_2 = \left\{\alpha, \alpha^2, \alpha^4, \alpha^8\right\}.$$

To find another conjugate class, first, we look for an element of $GF(2^4)$ which is not available in the previously constructed conjugate classes. Such an element is $\alpha^3$, since $\alpha^3 \in GF(2^4)$ and $\alpha^3 \notin J_2$. The conjugates of $\beta = \alpha^3$ can be determined as

$$\beta^2 \rightarrow \left(\alpha^3\right)^2 \rightarrow \alpha^6$$

$$\beta^4 \rightarrow \left(\alpha^6\right)^2 \rightarrow \alpha^{12}$$

$$\beta^8 \rightarrow \left(\alpha^{12}\right)^2 \rightarrow \alpha^{24} \rightarrow \alpha^9$$

$$\beta^{16} \rightarrow \left(\alpha^9\right)^2 \rightarrow \alpha^{18} \rightarrow \alpha^3 \text{ which is the same as } \beta.$$

Thus, the conjugate class including $\alpha^3$ can be written as

$$J_3 = \left\{\alpha^3, \alpha^6, \alpha^{12}, \alpha^9\right\}.$$

To find another conjugate class, we look for an element of $GF(2^4)$ which is not available in $J_2$ and $J_3$. Such an element is $\alpha^5$, since $\alpha^5 \in GF(2^4)$ and $\alpha^5 \notin J_2$, $\alpha^5 \notin J_3$. The conjugates of $\beta = \alpha^5$ can be determined as

$$\beta^2 \rightarrow \left(\alpha^5\right)^2 \rightarrow \alpha^{10}$$

$$\beta^4 \rightarrow \left(\alpha^{10}\right)^2 \rightarrow \alpha^5 \text{ which is the same as } \beta.$$

Thus, the conjugate class including $\alpha^5$ can be written as

$$J_4 = \left\{\alpha^5, \alpha^{10}\right\}.$$

To find another conjugate class, we look for an element of $GF(2^4)$ which is not available in $J_2$, $J_3$, and $J_4$. Such an element is $\alpha^7$, since $\alpha^7 \in GF(2^4)$ and $\alpha^7 \notin J_2$, $\alpha^7 \notin J_3$, $\alpha^7 \notin J_4$.

The conjugates of $\beta = \alpha^7$ can be determined as

$$\beta^2 \rightarrow \left(\alpha^7\right)^2 \rightarrow \alpha^{14}$$

$$\beta^4 \rightarrow \left(\alpha^{14}\right)^2 \rightarrow \alpha^{28} \rightarrow \alpha^{13}$$

$$\beta^8 \rightarrow \left(\alpha^{13}\right)^2 \rightarrow \alpha^{26} \rightarrow \alpha^{11}$$

**Table 5.4** Conjugate classes for $GF(2^4)$

| |
|---|
| $J_0 = \{0\}$ |
| $J_1 = \{1\}$ |
| $J_2 = \{\alpha, \alpha^2, \alpha^4, \alpha^8\}$ |
| $J_3 = \{\alpha^3, \alpha^6, \alpha^{12}, \alpha^9\}$ |
| $J_4 = \{\alpha^5, \alpha^{10}\}$ |
| $J_5 = \{\alpha^7, \alpha^{14}, \alpha^{13}, \alpha^{11}\}$ |

$$\beta^{16} \rightarrow \left(\alpha^{11}\right)^2 \rightarrow \alpha^7 \text{ which is the same as } \beta.$$

Thus, the conjugate class including $\alpha^7$ can be written as

$$J_5 = \left\{\alpha^7, \alpha^{14}, \alpha^{13}, \alpha^{11}\right\}.$$

Considering the base field elements, we can write all the conjugate classes as in Table 5.4.

The extended field $GF(2^4)$ can be written as the union of conjugate classes, i.e., we have

$$GF\left(2^4\right) = J_0 \cup J_1 \cup J_2 \cup J_3 \cup J_4 \cup J_5.$$

**Example 5.15** Using the primitive polynomial $p(x) = x^4 + x + 1$, generate the extended field $GF(2^4)$, and obtain the minimal polynomials of all the conjugate classes.

**Solution 5.15** If $\alpha$ is a root of the equation $p(x) = 0$, we can write the recursive expression

$$\alpha^4 = \alpha + 1. \tag{5.51}$$

The extended field elements using the powers of $\alpha$ can be written as

$$GF\left(2^4\right) = \left\{0, 1, \alpha, \alpha^2, \alpha^3, \alpha^4, \alpha^5, \alpha^6, \alpha^7, \alpha^8, \alpha^9, \alpha^{10}, \alpha^{11}, \alpha^{12}, \alpha^{13}, \alpha^{14}\right\}. \tag{5.52}$$

We can obtain the polynomial equivalents of power of $\alpha$ in Eq. (5.52) using Eq. (5.51) as in

$$
\begin{aligned}
&0 \to 0\\
&1 \to 1\\
&\alpha \to \alpha\\
&\alpha^2 \to \alpha^2\\
&\alpha^3 \to \alpha^3\\
&\alpha^4 \to \alpha + 1\\
&\alpha^5 \to \alpha^2 + \alpha\\
&\alpha^6 \to \alpha^3 + \alpha^2\\
&\alpha^7 \to \alpha^3 + \alpha + 1\\
&\alpha^8 \to \alpha^2 + 1\\
&\alpha^9 \to \alpha^3 + \alpha\\
&\alpha^{10} \to \alpha^2 + \alpha + 1\\
&\alpha^{11} \to \alpha^3 + \alpha^2 + \alpha\\
&\alpha^{12} \to \alpha^3 + \alpha^2 + \alpha + 1\\
&\alpha^{13} \to \alpha^3 + \alpha^2 + 1\\
&\alpha^{14} \to \alpha^3 + 1.
\end{aligned} \tag{5.53}
$$

The conjugate classes of the extended field $GF(2^4)$ are evaluated in the previous example, and they are given in Table 5.4 as

$$
\begin{aligned}
J_0 &= \{0\}\\
J_1 &= \{1\}\\
J_2 &= \{\alpha, \alpha^2, \alpha^4, \alpha^8\}\\
J_3 &= \{\alpha^3, \alpha^6, \alpha^{12}, \alpha^9\}\\
J_4 &= \{\alpha^5, \alpha^{10}\}\\
J_5 &= \{\alpha^7, \alpha^{14}, \alpha^{13}, \alpha^{11}\}.
\end{aligned}
$$

The minimal polynomials of the conjugate classes can be formed as

$$
\begin{aligned}
J_0 = \{0\} &\to m_0(x) = x + 0\\
J_1 = \{1\} &\to m_1(x) = x + 1\\
J_2 = \{\alpha, \alpha^2, \alpha^4, \alpha^8\} &\to m_2(x) = (x+\alpha)(x+\alpha^2)(x+\alpha^4)(x+\alpha^8)\\
J_3 = \{\alpha^3, \alpha^6, \alpha^{12}, \alpha^9\} &\to m_3(x) = (x+\alpha^3)(x+\alpha^6)(x+\alpha^{12})(x+\alpha^9)\\
J_4 = \{\alpha^5, \alpha^{10}\} &\to m_4(x) = (x+\alpha^5)(x+\alpha^{10})
\end{aligned}
$$

$$J_5 = \{\alpha^7, \alpha^{14}, \alpha^{13}, \alpha^{11}\} \rightarrow m_5(x) = (x + \alpha^7)(x + \alpha^{14})(x + \alpha^{134})(x + \alpha^{11})$$

where the minimal polynomials can be simplified using the polynomial equivalents of powers of $\alpha$ in Eq. (5.53) and the equality $\alpha^{15} = 1$ as

$$J_0 = \{0\} \rightarrow m_0(x) = x + 0 \rightarrow m_0(x) = x$$
$$J_1 = \{1\} \rightarrow m_1(x) = x + 1 \rightarrow m_1(x) = x + 1$$
$$J_2 = \{\alpha, \alpha^2, \alpha^4, \alpha^8\} \rightarrow m_2(x) = x^4 + x + 1$$
$$J_3 = \{\alpha^3, \alpha^6, \alpha^{12}, \alpha^9\} \rightarrow m_3(x) = x^4 + x^3 + x^2 + x + 1$$
$$J_4 = \{\alpha^5, \alpha^{10}\} \rightarrow m_4(x) = x^2 + x + 1$$
$$J_5 = \{\alpha^7, \alpha^{14}, \alpha^{13}, \alpha^{11}\} \rightarrow m_5(x) = x^4 + x^3 + 1$$

in which we see that a minimal polynomial includes coefficients only from the base field.

**Example 5.16** Assume that, using a primitive polynomial $p(x)$ of degree 5, we generate the extended field $GF(2^5)$. The minimal polynomial of a conjugate class is given as

$$m(x) = x^3 + \alpha^2 x + x + \alpha^3.$$

Comment on the calculated minimal polynomial.

**Comment** The minimal polynomial given in the question contains coefficients from the extended field elements. Such a minimal polynomial cannot exist. It is a fake minimal polynomial.

## 5.6 Polynomials in Extended Fields

Polynomials can be constructed using the extended field $GF(2^m)$ elements for the coefficients of $x^i$ terms. For instance, using the elements of

$$GF(2^3) = \{0, 1, \alpha, \alpha^2, \alpha^3, \alpha^4, \alpha^5, \alpha^6\}$$

we can define the polynomials

$$p_1(x) = \alpha^3 x^5 + \alpha x^3 + \alpha^3 x^2 + x + \alpha^4$$

$$p_2(x) = \alpha^6 x^6 + \alpha^3 x^4 + \alpha^5 x^2 + 1.$$

While adding or multiplying two polynomials in $GF(2^m)$, we use the recursive statement obtained from $p(\alpha) = 0$, where $p(x)$ is a primitive polynomial and $\alpha$ is a root of $p(x)$, and $\alpha^{m-1} = 1$ for the simplification of the coefficients in the resulting polynomial.

**Example 5.17** Using the primitive polynomial $p(x) = x^3 + x + 1$, we generate the extended field $GF(2^3)$. Two polynomials over $GF(2^3)$ are defined as

$$p_1(x) = \alpha x^6 + \alpha^2 x^4 + \alpha^3 x^2 + \alpha$$

$$p_2(x) = \alpha^6 x^6 + \alpha^3 x^4 + \alpha^5 x^2 + 1.$$

Find $p_1(x) + p_2(x)$ and $p_1(x) \times p_2(x)$.

**Solution 5.17** The extended field $GF(2^3)$ can be written as

$$GF(2^3) = \{0, 1, \alpha, \alpha^2, \alpha^3, \alpha^4, \alpha^5, \alpha^6\}$$

where the $\alpha^i$ terms can be expressed by polynomials as

$$0 \to 0$$

$$1 \to 1$$

$$\alpha \to \alpha$$

$$\alpha^2 \to \alpha^2$$

$$\alpha^3 \to \alpha + 1$$

$$\alpha^4 \to \alpha^2 + \alpha$$

$$\alpha^5 \to \alpha^2 + \alpha + 1$$

$$\alpha^6 \to \alpha^2 + 1.$$

In addition, we have

$$\alpha^7 = 1 \text{ and } \alpha^3 = \alpha + 1.$$

The addition of polynomials $p_1(x)$ and $p_1(x)$ results in

$$p(x) = p_1(x) + p_2(x) \to$$

$$p(x) = (\alpha + \alpha^6)x^6 + (\alpha^2 + \alpha^3)x^4 + (\alpha^3 + \alpha^5)x^2 + \alpha + 1$$

where the coefficients can be simplified as

$$\alpha + \alpha^6 \rightarrow \alpha + \alpha^2 + 1 \rightarrow \alpha^5$$

$$\alpha^2 + \alpha^3 \rightarrow \alpha^2 + \alpha + 1 \rightarrow \alpha^5$$

$$\alpha^3 + \alpha^5 \rightarrow \alpha + 1 + \alpha^2 + \alpha + 1 \rightarrow \alpha^2$$

$$\alpha + 1 \rightarrow \alpha^3$$

leading to the simplified expression

$$p(x) = \alpha^5 x^6 + \alpha^5 x^4 + \alpha^2 x^2 + \alpha^3.$$

The multiplication of $p_1(x)$ and $p_2(x)$ can be performed as

$$\begin{aligned} p_1(x) \times p_2(x) &= \left(\alpha x^6 + \alpha^2 x^4 + \alpha^3 x^2 + \alpha\right)\left(\alpha^6 x^6 + \alpha^3 x^4 + \alpha^5 x^2 + 1\right) \\ &= \alpha^7 x^{12} + \alpha^4 x^{10} + \alpha^6 x^8 + \alpha x^6 + \alpha^8 x^{10} + \alpha^5 x^8 + \alpha^7 x^6 + \alpha^2 x^4 + \alpha^9 x^8 + \alpha^6 x^6 \\ &\quad + \alpha^8 x^4 + \alpha^3 x^2 + \alpha^7 x^6 + \alpha^4 x^4 + \alpha^6 x^2 + \alpha \end{aligned}$$

in which, adding the confidents of the same powers of $x$, we obtain

$$\begin{aligned} p_1(x) \times p_2(x) = \alpha^7 x^{12} &+ (\alpha^4 + \alpha^8)x^{10} + (\alpha^6 + \alpha^5 + \alpha^9)x^8 \\ &+ (\alpha + \alpha^7 + \alpha^6 + \alpha^7)x^6 + (\alpha^2 + \alpha^8 + \alpha^4)x^4 + (\alpha^3 + \alpha^6)x^2 + \alpha \end{aligned}$$

in which, using $\alpha^7 = 1$ and polynomial forms in the powers of $\alpha$, we obtain the expression

$$\begin{aligned} p_1(x) \times p_2(x) &= x^{12} + (\alpha^2 + \alpha + \alpha)x^{10} + (\alpha^2 + 1 + \alpha^2 + \alpha + 1 + \alpha^2)x^8 \\ &+ (\alpha + \alpha^2 + 1)x^6 + (\alpha^2 + \alpha + \alpha^2 + \alpha)x^4 + (\alpha + 1 + \alpha^2 + 1)x^2 + \alpha \end{aligned}$$

leading to

$$\begin{aligned} p_1(x) \times p_2(x) = x^{12} &+ (\alpha^2)x^{10} + (\alpha + \alpha^2)x^8 + (\alpha + \alpha^2 + 1)x^6 + (0)x^4 \\ &+ (\alpha + \alpha^2)x^2 + \alpha \end{aligned}$$

in which, expressing the coefficients as powers of $\alpha$, we get the simplified expression

$$p_1(x) \times p_2(x) = x^{12} + \alpha^2 x^{10} + \alpha^4 x^8 + \alpha^5 x^6 + \alpha^4 x^2 + \alpha.$$

## 5.7 Binary Representation of Extended Field Elements

In Chap. 1, we stated that the polynomials can be represented by number vectors, and the vector elements are the coefficients of the polynomials. This rule is valid for the polynomials in finite extended fields. However, in this case, since the coefficients of the polynomials are also polynomials, we can also represent the coefficients by binary vectors.

**Example 5.18** We can represent the polynomial

$$p_1(x) = \alpha^5 x^6 + \alpha^4 x^4 + \alpha^3 x^2 + \alpha$$

whose coefficients are selected from the extended finite field $GF(2^3)$, which is constructed using the primitive polynomial $p(x) = x^3 + x + 1$, in vector form as

$$\boldsymbol{p_1} = \left[\alpha^5 \; 0 \; \alpha^4 \; 0 \; \alpha^3 \; 0 \; \alpha\right]. \tag{5.54}$$

The powers of $\alpha$ in Eq. (5.54) can be represented in polynomial forms as

$$\begin{aligned} \alpha^5 &\rightarrow \alpha^2 + \alpha + 1 \\ \alpha^4 &\rightarrow \alpha^2 + \alpha \\ \alpha^3 &\rightarrow \alpha + 1 \\ \alpha^1 &\rightarrow \alpha \end{aligned}$$

where the polynomials can be expressed using bit groups consisting of three bits as

$$\begin{aligned} \alpha^5 &\rightarrow \alpha^2 + \alpha + 1 \rightarrow 111 \\ \alpha^4 &\rightarrow \alpha^2 + \alpha \rightarrow 110 \\ \alpha^3 &\rightarrow \alpha + 1 \rightarrow 011 \\ \alpha^1 &\rightarrow \alpha \rightarrow 010. \end{aligned} \tag{5.55}$$

Using the binary representation of the polynomials in Eq. (5.55), we obtain its vector form as

$$\boldsymbol{p_1} = [111 \;\; 000 \;\; 110 \;\; 000 \;\; 011 \;\; 000 \;\; 010]. \tag{5.56}$$

Note that 0 polynomial is represented as 000.

## 5.8 Equations in Extended Fields

We can define equations having coefficients from an extended finite field. For instance, using some elements of the extended field $GF(2^3)$ for coefficients, we can define the equation pair

$$\alpha^2 x + \alpha y = \alpha^4$$
$$\alpha x + \alpha^3 y = \alpha^5.$$

To find the solutions of the equations with coefficients from the extended field, we follow the same method while finding the solutions of equations having coefficients in the real number field. We try to eliminate one of the variables, in case that there is more than one variable.

**Example 5.19** Assume that, using the primitive polynomial $p(x) = x^3 + x + 1$, we construct the extended field $GF(2^3)$, and two equations in $GF(2^3)$ are given as

$$\begin{aligned} \alpha^2 x + \alpha^3 y &= \alpha^4 \\ \alpha^4 x + \alpha^2 y &= \alpha^5. \end{aligned} \tag{5.57}$$

Find the solution of Eq. (5.57).

**Solution 5.19** The elements of the extended field $GF(2^3)$ can be written as

$$0 \rightarrow 0$$
$$1 \rightarrow 1$$
$$\alpha \rightarrow \alpha$$
$$\alpha^2 \rightarrow \alpha^2$$
$$\alpha^3 \rightarrow \alpha + 1$$
$$\alpha^4 \rightarrow \alpha^2 + \alpha$$
$$\alpha^5 \rightarrow \alpha^2 + \alpha + 1$$
$$\alpha^6 \rightarrow \alpha^2 + 1.$$

Using the polynomial form of the extended field elements and the equality $\alpha^7 = 1$, we can find the solution of Eq. (5.57) as follows:

We multiply the first equation by $\alpha^2$ and sum the two equations, i.e.,

$$\alpha^2 \left( \alpha^2 x + \alpha^3 y = \alpha^4 \right)$$

$$\alpha^4 x + \alpha^2 y = \alpha^5$$

leading to

$$(\alpha^5 + \alpha^2)y = \alpha^6 + \alpha^5 \rightarrow (\alpha^2 + \alpha + 1 + \alpha^2)y = \alpha^2 + 1 + \alpha^2 + \alpha + 1 \rightarrow$$
$$(\alpha + 1)y = \alpha \rightarrow \alpha^3 y = \alpha \rightarrow \alpha^4\alpha^3 y = \alpha^4\alpha \rightarrow y = \alpha^5.$$

Putting the root $y = \alpha^5$ into $\alpha^2 x + \alpha^3 y = \alpha^4$, we get

$$\alpha^2 x + \alpha^3\alpha^5 = \alpha^4 \rightarrow \alpha^2 x + \alpha = \alpha^4 \rightarrow$$
$$\alpha^2 x = \alpha^4 + \alpha \rightarrow \alpha^2 x = \alpha^2 + \alpha + \alpha \rightarrow x = 1.$$

Thus, the roots of the equation pair are

$$(x, y) = (1, \alpha^5).$$

**Exercise** Assume that, using the primitive polynomial $p(x) = x^3 + x + 1$, we construct the extended field $GF(2^3)$ and two equations in $GF(2^3)$ are given as

$$\begin{aligned} \alpha^4 x + \alpha^2 y &= \alpha^3 \\ \alpha^2 x + \alpha^3 y &= \alpha^2 \end{aligned} \tag{5.58}$$

Find the solution of Eq. (5.58).

## 5.9 Matrices in Extended Fields

As in real number field, we can define matrices with elements in finite fields. Once we define a matrix, we can calculate its determinant or inverse.

**Example 5.20** In $GF(2^3)$ constructed using $p(x) = x^3 + x + 1$, a square matrix is given as

$$A = \begin{bmatrix} \alpha^2 & \alpha^4 \\ \alpha & \alpha^2 \end{bmatrix}. \tag{5.59}$$

Find the inverse of Eq. (5.59)

**Solution 5.20** The determinant of Eq. (5.59) is calculated as

$$\begin{vmatrix} \alpha^2 & \alpha^4 \\ \alpha^2 & \alpha^2 \end{vmatrix} = \alpha^2\alpha^2 + \alpha^2\alpha^4 \rightarrow \alpha^4 + \alpha^6 = \alpha^2 + \alpha + \alpha^2 + 1 \rightarrow \alpha + 1 \rightarrow \alpha^3.$$

Using the determinant, we can calculate the inverse of Eq. (5.59) as in

$$\begin{bmatrix} \alpha^2 & \alpha^4 \\ \alpha & \alpha^2 \end{bmatrix}^{-1} = \frac{1}{\alpha^3}\begin{bmatrix} \alpha^2 & \alpha \\ \alpha^4 & \alpha^2 \end{bmatrix} \rightarrow \frac{1}{\alpha^3}\begin{bmatrix} \alpha^9 & \alpha^8 \\ \alpha^4 & \alpha^9 \end{bmatrix} \rightarrow \begin{bmatrix} \alpha^6 & \alpha^5 \\ \alpha & \alpha^6 \end{bmatrix}.$$

**Properties**
Consider the polynomial

$$p(x) = a_n x^n + a_{n-1}x^{n-1} + a_1 x + a_0$$

defined in the extended field $GF(2^m)$. The square of the polynomial is calculated as

$$[p(x)]^2 = (a_n)^2 x^{2n} + (a_{n-1})^2 x^{2(n-1)} + (a_1)^2 x^2 + (a_0)^2.$$

In general, for $k = 2^r$, we can write that

$$[p(x)]^k = (a_n)^k x^{kn} + (a_{n-1})^k x^{k(n-1)} + (a_1)^k x^k + (a_0)^k. \tag{5.60}$$

**Example 5.21** We can write that

$$(x^3 + x^2)^2 = x^6 + x^4$$

$$(x^3 + x^2 + x)^2 = ([x^3 + x^2] + x)^2 \rightarrow (x^3 + x^2)^2 + x^2 \rightarrow x^6 + x^4 + x^2.$$

Another example is

$$(x^5 + x^3 + x + 1)^4 = x^{20} + x^{12} + x^4 + 1.$$

In $GF(2^4)$, we can write that

$$(\alpha^2 x^3 + x + \alpha^5)^8 = \alpha^{16} x^{24} + x^8 + \alpha^{40} \rightarrow (\alpha^2 x^3 + x + \alpha^5)^8 = x^{24} + x^8 + \alpha^8$$

where we used the equality $\alpha^{16} = 1$.

**Example 5.22** Simplify the following mathematical term in $GF(2^4)$

$$\left(\alpha^3x^2+\alpha^2x+\alpha\right)^8.$$

**Solution 5.22** Using the property in Eq. (5.60) and equality $\alpha^{15}=1$, we can evaluate $(\alpha^3x^2+\alpha^2x+\alpha)^8$ as in

$$\left(\alpha^3x^2+\alpha^2x+\alpha\right)^8\rightarrow\left(\alpha^3x^2\right)^8+\left(\alpha^2x\right)^8+\alpha^8\rightarrow\alpha^9x^{16}+\alpha x+\alpha^8.$$

**Theorem** The roots of $x^n+1$ where $m=2^m-1$ are all the nonzero elements of the extended field $GF(2^m)$.

**Theorem** The minimal polynomials of the conjugate classes are either irreducible or primitive polynomials.

Note that primitive polynomials are already irreducible polynomials.

**Property** A polynomial in $GF(2^m)$ can be represented by an $m$-bit vector.

**Notation** The conjugates of $\beta=\alpha^i$ are evaluated according to

$$\{\beta,\beta^2,\beta^4,\beta^8,\ldots\}. \tag{5.61}$$

The minimal polynomial of the conjugate class can be denoted by $m(x)$, or considering the powers of the elements in Eq. (5.61), we can denote the same minimal polynomial by

$$m_i(x),m_{2i}(x),m_{4i}(x),\ \ldots$$

that is,

$$m(x)=m_i(x)=m_{2i}(x)=m_{4i}(x)\ldots$$

Although the minimal polynomial is defined for a conjugate class, we can label the minimal polynomial considering its roots which are the elements of the conjugate class.

## Problems

1. Decide whether the polynomials

$$x^4 + x + 1 \qquad x^4 + x^2 + x + 1$$

are irreducible polynomials or not.

2. Construct the extended field $GF(2^4)$ using the primitive polynomial $p(x) = x^4 + x^3 + 1$. Repeat the construction process using the primitive polynomial $p(x) = x^4 + x + 1$. Comment on the field elements for both constructions.

3. Decide whether the polynomial

$$p(x) = x^5 + x^4 + x^2 + x + 1$$

is a primitive polynomial or not.

4. Find all the conjugate classes of $GF(2^5)$. How many minimal polynomials do we have in $GF(2^5)$?

5. Obtain the binary representation of the polynomial

$$p(\alpha) = \alpha^2 + \alpha$$

in $GF(2^3)$ and $GF(2^5)$.

6. Assume that the extended field $GF(2^5)$ is constructed using the primitive polynomial $p(x) = x^5 + x^3 + 1$. Obtain the conjugates of $\alpha^3$ in $GF(2^5)$, and find the minimal polynomial for the obtained conjugate class.
7. Factorize $x^{15} + 1$.
8. Expand $(x^4 + \alpha^2x^3 + \alpha^5x + \alpha^3)^4$ in $GF(2^3)$.
9. Obtain the binary representation of the polynomial $p(x) = x^4 + \alpha^4x^3 + \alpha^{12}x + \alpha^9$ in $GF(2^3)$ and $GF(2^5)$. Assume that the primitive polynomials $p(x) = x^3 + x + 1$ and $p(x) = x^5 + x^3 + x^2 + x + 1$ are used for the construction of the extended fields $GF(2^3)$ and $GF(2^5)$.
10. Evaluate the inverses of $\alpha^2$, $\alpha^2 + 1$ in $GF(2^3)$. Use the primitive polynomial $p(x) = x^3 + x^2 + 1$.
11. Calculate $\sqrt[3]{\alpha^4}, \sqrt[2]{\alpha^3}, \sqrt[5]{\alpha^2}$ in $GF(2^3)$.
12. Find the roots of $x^3 + \alpha^3x^2 + \alpha^5x + \alpha = 0$ in $GF(2^4)$. Determine the $p(x)$ used to construct the extended field $GF(2^4)$ by yourself.
13. Using $p(x) = x^2 + x + 1$, construct $GF(2^2)$. Find all the conjugate classes; calculate the minimal polynomials.
14. Solve the equation set

$$\alpha^3 x + \alpha y = \alpha^2$$
$$\alpha^5 x + \alpha^2 y = \alpha^4$$

in $GF(2^3)$. Use $p(x) = x^3 + x + 1$ as your primitive polynomial.

15. Find the determinant of the matrix

$$A = \begin{bmatrix} \alpha & \alpha^2 \\ \alpha^3 & \alpha^8 \end{bmatrix}$$

in $GF(2^4)$. Use $p(x) = x^4 + x + 1$ as your primitive polynomial.

16. Calculate the inverse of the matrix

$$A = \begin{bmatrix} \alpha^4 & \alpha^2 \\ \alpha^5 & \alpha^8 \end{bmatrix}$$

in $GF(2^4)$. Use $p(x) = x^4 + x + 1$ as your primitive polynomial.

17. Calculate the inverse of the matrix

$$A = \begin{bmatrix} \alpha^2 & \alpha^6 & \alpha^5 \\ \alpha^3 & \alpha^4 & \alpha^4 \\ \alpha^7 & \alpha^9 & \alpha^{12} \end{bmatrix}$$

in $GF(2^4)$. Use $p(x) = x^4 + x + 1$ as your primitive polynomial.

18. Find the solution of the equation set

$$\alpha^3 x + \alpha^9 y + \alpha z = \alpha^5$$
$$\alpha^7 x + \alpha^2 y + \alpha^2 z = \alpha^{13}$$
$$\alpha^3 x + \alpha^4 y + \alpha^3 z = \alpha^6$$

in $GF(2^4)$. Use $p(x) = x^4 + x + 1$ as your primitive polynomial.

# Chapter 6
# BCH Codes

## 6.1 BCH Codes and Generator Polynomials of BCH Codes

Consider that using a primitive polynomial, we generate the extended field $GF(2^m)$ which is written as

$$GF(2^m) = \{0, 1, \alpha, \alpha^2, \ldots, \alpha^l\} \quad \text{where } l = 2^m - 2. \tag{6.1}$$

Let's denote the minimal polynomial of the conjugate class in which $\alpha^i$ appears by $m_i(x)$; in other words, $m_i(x)$ is the minimal polynomial of $\alpha^i$.

The generator polynomial of a $t$-error-correcting $BCH$ code is obtained as

$$g(x) = LCM\{m_1(x), m_3(x), m_5(x), \ldots, m_{2t-1}\} \tag{6.2}$$

where $LCM$ denotes the least common multiple operation. The parameters of a $BCH$ code can be specified as

$$BCH(n, k) = BCH(2^m - 1, 2^m - 1 - r) \tag{6.3}$$

where $r$ is the degree of the generator polynomial $g(x)$ given in Eq. (6.2). $BCH$ codes are cyclic codes.

**Example 6.1** Obtain the generator polynomial of the single-error-correcting $BCH$ code. Use $GF(2^3)$ which is constructed using the primitive polynomial $p(x) = x^3 + x + 1$.

**Solution 6.1** The conjugate classes and the corresponding minimal polynomials can be obtained as in Table 6.1.

The generator polynomial of the single-error-correcting $BCH$ code can be calculated as


O. Gazi, *Forward Error Correction via Channel Coding*,
https://doi.org/10.1007/978-3-030-33380-5_6

**Table 6.1** Conjugate classes and minimal polynomials for $GF(2^3)$

| Conjugate classes | Minimal polynomials |
|---|---|
| 0 | $x$ |
| 1 | $x + 1$ |
| $\alpha, \alpha^2, \alpha^4$ | $m_1(x) = x^3 + x + 1$ |
| $\alpha^3, \alpha^5, \alpha^6$ | $m_3(x) = x^3 + x^2 + 1$ |

$$g(x) = LCM\{m_1(x), m_3(x), \ldots, m_{2t-1}\} \to g(x) = LCM\{m_1(x)\} \to$$
$$g(x) = m_1(x) \to g(x) = x^3 + x + 1.$$

The degree of the generator polynomial is $r = 3$. Then, the parameters of the *BCH* code can be calculated using $m = 3$ and $r = 6$ as

$$BCH(n, k) = BCH(2^m - 1, 2^m - 1 - r) \to BCH(7, 4).$$

**Example 6.2** Obtain the generator polynomial of the double-error-correcting *BCH* code. Use $GF(2^3)$ which is constructed using the primitive polynomial $p(x) = x^3 + x + 1$.

**Solution 6.2** Using Table 6.1, the generator polynomial of the double-error-correcting *BCH* code can be calculated as

$$g(x) = LCM\{m_1(x), m_3(x), \ldots, m_{2t-1}\} \to g(x) = LCM\{m_1(x), m_3(x)\} \to$$
$$g(x) = m_1(x)m_3(x) \to g(x) = x^6 + x^5 + x^4 + x^3 + x^2 + x + 1.$$

The degree of the generator polynomial is $r = 6$. Then, the parameters of the *BCH* code can be calculated using $m = 3$ and $r = 6$ as

$$BCH(n, k) = BCH(2^m - 1, 2^m - 1 - r) \to BCH(7, 1)$$

which is a repetition code.

If we count the nonzero coefficients of the generator polynomial, we see that the minimum distance of the code is $d_{\min} = 7$, and this means that the designed code can correct up to three-bit errors. That is, although we started to design a *BCH* code that can correct two-bit errors, we came up with a code that can correct up to three-bit errors.

**Example 6.3** Obtain the generator polynomial of the single-error-correcting *BCH* code. Use $GF(2^4)$ which is constructed using the primitive polynomial $p(x) = x^4 + x + 1$.

**Solution 6.3** Using the minimal polynomials of Table 6.2, the generator polynomial of the single-error-correcting *BCH* code can be calculated as

**Table 6.2** Conjugate classes and minimal polynomials for $GF(2^4)$

| Conjugate classes | Minimal polynomials |
|---|---|
| 0 | $x$ |
| 1 | $x + 1$ |
| $\alpha, \alpha^2, \alpha^4, \alpha^6$ | $m_1(x) = x^4 + x + 1$ |
| $\alpha^3, \alpha^6, \alpha^9, \alpha^{12}$ | $m_3(x) = x^4 + x^3 + x^2 + x + 1$ |
| $\alpha^5, \alpha^{10}$ | $m_5(x) = x^2 + x + 1$ |
| $\alpha^7, \alpha^{11}, \alpha^{13}, \alpha^{14}$ | $m_7(x) = x^4 + x^3 + 1$ |

$$g(x) = LCM\{m_1(x), m_3(x), \ldots, m_{2t-1}\} \rightarrow g(x) = LCM\{m_1(x)\} \rightarrow$$
$$g(x) = m_1(x) \rightarrow g(x) = x^4 + x + 1.$$

The degree of the generator polynomial is $r = 4$. Then, the parameters of the *BCH* code can be calculated as

$$BCH(n, k) = BCH(2^m - 1, 2^m - 1 - r) \rightarrow BCH\left(2^4 - 1, 2^4 - 1 - 4\right)$$
$$\rightarrow BCH(15, 11).$$

**Exercise** For the previous example, calculate the generator matrix using the generator polynomial.

**Example 6.4** Obtain the generator polynomial of the double-error-correcting *BCH* code. Use $GF(2^4)$ which is constructed using the primitive polynomial $p(x) = x^4 + x + 1$.

**Solution 6.4** The conjugate classes and the corresponding minimal polynomials are given in Table 6.2. The generator polynomial of the double-error-correcting *BCH* code can be calculated as

$$g(x) = LCM\{m_1(x), m_3(x), \ldots, m_{2t-1}\} \rightarrow g(x) = LCM\{m_1(x), m_3(x)\} \rightarrow$$
$$g(x) = m_1(x)m_3(x) \rightarrow g(x) = \left(x^4 + x + 1\right)\left(x^4 + x^3 + x^2 + x + 1\right) \rightarrow$$
$$g(x) = \left(x^4 + x + 1\right)\left(x^4 + x^3 + x^2 + x + 1\right) \rightarrow g(x) = x^8 + x^7 + x^6 + x^4 + 1.$$

The degree of the generator polynomial is $r = 8$. Then, the parameters of the *BCH* code can be calculated as

$$BCH(n, k) = BCH(2^m - 1, 2^m - 1 - r) \rightarrow BCH\left(2^4 - 1, 2^4 - 1 - 8\right)$$
$$\rightarrow BCH(15, 7).$$

**Exercise** For the previous example, using the generator polynomial, construct the generator matrix of the $BCH(15, 7)$ cyclic code.

**Example 6.5** Obtain the generator polynomial of the triple-error-correcting *BCH* code. Use $GF(2^4)$ which is constructed using the primitive polynomial $p(x) = x^4 + x + 1$.

**Solution 6.5** The conjugate classes and the corresponding minimal polynomials can be obtained as in Table 6.2. The generator polynomial of the double-error-correcting *BCH* code can be calculated as

$$g(x) = LCM\{m_1(x), m_3(x), \ldots, m_{2t-1}\} \rightarrow g(x) = LCM\{m_1(x), m_3(x), m_5(x)\} \rightarrow$$
$$g(x) = m_1(x)m_3(x)m_5(x) \rightarrow$$
$$g(x) = \left(x^4 + x + 1\right)\left(x^4 + x^3 + x^2 + x + 1\right)\left(x^2 + x + 1\right) \rightarrow$$
$$g(x) = x^{10} + x^8 + x^5 + x^4 + x^2 + x + 1.$$

The degree of the generator polynomial is $r = 10$. Then, the parameters of the *BCH* code can be calculated as

$$BCH(n,k) = BCH(2^m - 1, 2^m - 1 - r) \rightarrow BCH\left(2^4 - 1, 2^4 - 1 - 10\right)$$
$$\rightarrow BCH(15,5).$$

**Example 6.6** Obtain the generator polynomial of the four-bit error-correcting *BCH* code. Use $GF(2^4)$ which is constructed using the primitive polynomial $p(x) = x^4 + x + 1$.

**Solution 6.6** The conjugate classes and the corresponding minimal polynomials are given in Table 6.2. The generator polynomial of the double-error-correcting *BCH* code can be calculated as

$$g(x) = LCM\{m_1(x), m_3(x), \ldots, m_{2t-1}\} \rightarrow$$
$$g(x) = LCM\{m_1(x), m_3(x), m_5(x), m_7(x)\} \rightarrow$$
$$g(x) = m_1(x)m_3(x)m_5(x)m_7(x) \rightarrow$$
$$g(x) = \left(x^4 + x + 1\right)\left(x^4 + x^3 + x^2 + x + 1\right)\left(x^2 + x + 1\right)\left(x^4 + x^3 + 1\right) \rightarrow$$
$$g(x) = x^{14} + x^{13} + x^{12} + x^{11} + x^{10} + x^9 + x^8 + x^7 + x^6 + x^5 + x^4 + x^3 + x^2 + x$$
$$+ 1.$$

The degree of the generator polynomial is $r = 14$. Then, the parameters of the *BCH* code can be calculated as

$$BCH(n,k) = BCH(2^m - 1, 2^m - 1 - r) \rightarrow BCH\left(2^4 - 1, 2^4 - 1 - 14\right)$$
$$\rightarrow BCH(15,1)$$

**Table 6.3** BCH codes generated over $GF(2^4)$

| Code designed | Generator polynomial | Minimum distance and error correction capability |
|---|---|---|
| $BCH(15,11)$ | $g(x) = x^4 + x + 1$ | $d_{\min} = 3 \quad t_c = 1$ |
| $BCH(15,7)$ | $g(x) = x^8 + x^7 + x^6 + x^4 + 1$ | $d_{\min} = 5 \quad t_c = 2$ |
| $BCH(15,5)$ | $g(x) = x^{10} + x^8 + x^5 + x^4 + x^2 + x + 1$ | $d_{\min} = 7 \quad t_c = 3$ |
| $BCH(15,1)$ | $g(x) = x^{14} + x^{13} + x^{12} + x^{11} + x^{10} + x^9 + x^8 + x^7 + x^6 + x^5 + x^4 + x^3 + x^2 + x + 1$ | $d_{\min} = 15 \quad t_c = 7$ |

**Table 6.4** Conjugate classes and minimal polynomials for $GF(2^5)$

| Conjugate classes | Minimal polynomials |
|---|---|
| 0 | $x$ |
| 1 | $x + 1$ |
| $\alpha, \alpha^2, \alpha^4, \alpha^8, \alpha^{16}$ | $m_1(x) = x^5 + x^2 + 1$ |
| $\alpha^3, \alpha^6, \alpha^{12}, \alpha^{24}, \alpha^{17}$ | $m_3(x) = x^5 + x^4 + x^3 + x^2 + 1$ |
| $\alpha^5, \alpha^{10}, \alpha^{20}, \alpha^9, \alpha^{18}$ | $m_5(x) = x^5 + x^4 + x^2 + x + 1$ |
| $\alpha^7, \alpha^{14}, \alpha^{28}, \alpha^{25}, \alpha^{19}$ | $m_7(x) = x^5 + x^3 + x^2 + x + 1$ |
| $\alpha^{11}, \alpha^{22}, \alpha^{13}, \alpha^{26}, \alpha^{21}$ | $m_{11}(x) = x^5 + x^4 + x^3 + x + 1$ |
| $\alpha^{15}, \alpha^{30}, \alpha^{29}, \alpha^{27}, \alpha^{23}$ | $m_{15}(x) = x^5 + x^3 + 1$ |

which is a repetition code. If we count the nonzero coefficients of the generator polynomial, we see that the minimum distance of the code is $d_{\min} = 15$, and this means that the designed code can correct up to seven-bit errors. That is, although we started to design a *BCH* code that can correct five-bit errors, we came up with a code that can correct up to seven-bit errors.

The $BCH(n, k)$ codes designed in $GF(2^4)$, constructed using the private polynomial, can be summarized in Table 6.3.

**Example 6.7** Obtain the generator polynomial of the single-error-correcting *BCH* code. Use $GF(2^5)$ which is constructed using the primitive polynomial $p(x) = x^5 + x^2 + 1$.

**Solution 6.7** Using the minimal polynomials of Table 6.4, the generator polynomial of the single-error-correcting *BCH* code can be calculated as

$$g(x) = LCM\{m_1(x), m_3(x), \ldots, m_{2t-1}\} \to g(x) = LCM\{m_1(x)\} \to$$
$$g(x) = m_1(x) \to g(x) = x^5 + x^2 + 1.$$

The degree of the generator polynomial is $r = 5$. Then, the parameters of the *BCH* code can be calculated as

$$BCH(n, k) = BCH(2^m - 1, 2^m - 1 - r) \to BCH(2^5 - 1, 2^5 - 1 - 5)$$
$$\to BCH(31, 26).$$

**Example 6.8** Obtain the generator polynomial of the double-error-correcting *BCH* code. Use $GF(2^5)$ which is constructed using the primitive polynomial $p(x) = x^5 + x^2 + 1$.

**Solution 6.8** Using the minimal polynomials of Table 6.4, the generator polynomial of the double-error-correcting *BCH* code can be calculated as

$$g(x) = LCM\{m_1(x), m_3(x), \ldots, m_{2t-1}\} \rightarrow g(x) = LCM\{m_1(x), m_3(x)\} \rightarrow$$
$$g(x) = m_1(x)m_3(x) \rightarrow$$
$$g(x) = \left(x^5 + x^2 + 1\right)\left(x^5 + x^4 + x^3 + x^2 + 1\right) \rightarrow$$
$$g(x) = x^{10} + x^9 + x^8 + x^6 + x^5 + x^3 + 1.$$

The degree of the generator polynomial is $r = 10$. Then, the parameters of the *BCH* code can be calculated as

$$BCH(n,k) = BCH(2^m - 1, 2^m - 1 - r) \rightarrow BCH\left(2^5 - 1, 2^5 - 1 - 10\right)$$
$$\rightarrow BCH(31, 21).$$

Considering the nonzero coefficients of the generator polynomial, we can calculate the minimum distance of the code as $d_{\min} = 7$. This means that the designed code can correct up to three bits.

**Example 6.9** Obtain the generator polynomial of the triple-error-correcting *BCH* code. Use $GF(2^5)$ which is constructed using the primitive polynomial $p(x) = x^5 + x^2 + 1$.

**Solution 6.9** Using the minimal polynomials of Table 6.4, the generator polynomial of the triple-error-correcting *BCH* code can be calculated as

$$g(x) = LCM\{m_1(x), m_3(x), \ldots, m_{2t-1}\} \rightarrow g(x) = LCM\{m_1(x), m_3(x), m_5(x)\} \rightarrow$$
$$g(x) = m_1(x)m_3(x)m_5(x) \rightarrow$$
$$g(x) = \left(x^5 + x^2 + 1\right)\left(x^5 + x^4 + x^3 + x^2 + 1\right)\left(x^5 + x^4 + x^2 + x + 1\right) \rightarrow$$
$$g(x) = x^{15} + x^{11} + x^{10} + x^9 + x^8 + x^7 + x^5 + x^3 + x^2 + x + 1.$$

The degree of the generator polynomial is $r = 15$. Then, the parameters of the *BCH* code can be calculated as

$$BCH(n,k) = BCH(2^m - 1, 2^m - 1 - r) \rightarrow BCH\left(2^5 - 1, 2^5 - 1 - 15\right)$$
$$\rightarrow BCH(31, 16).$$

Considering the nonzero coefficients of the generator polynomial, we can calculate the minimum distance of the code as $d_{\min} = 11$. This means that the designed code can correct up to five bits.

**Example 6.10** Obtain the generator polynomial of the four-bit error-correcting *BCH* code. Use $GF(2^5)$ which is constructed using the primitive polynomial $p(x) = x^5 + x^2 + 1$.

**Solution 6.10** Using the minimal polynomials of Table 6.4, the generator polynomial of the four-bit error-correcting *BCH* code can be calculated as

$$g(x) = LCM\{m_1(x), m_3(x), \ldots, m_{2t-1}\} \rightarrow$$
$$g(x) = LCM\{m_1(x), m_3(x), m_5(x), m_7(x)\} \rightarrow$$
$$g(x) = m_1(x)m_3(x)m_5(x)m_7(x) \rightarrow$$
$$g(x) = (x^5 + x^2 + 1)(x^5 + x^4 + x^3 + x^2 + 1)(x^5 + x^4 + x^2 + x + 1) \times (x^5 + x^3 + x^2 + x + 1) \rightarrow$$
$$g(x) = x^{20} + x^{18} + x^{17} + x^{13} + x^{10} + x^9 + x^8 + x^7 + x^4 + x^2 + 1.$$

The degree of the generator polynomial is $r = 20$. Then, the parameters of the *BCH* code can be calculated as

$$BCH(n, k) = BCH(2^m - 1, 2^m - 1 - r) \rightarrow BCH(2^5 - 1, 2^5 - 1 - 20) \rightarrow BCH(31, 11).$$

Considering the nonzero coefficients of the generator polynomial, we can calculate the minimum distance of the code as $d_{\min} = 11$. This means that the designed code can correct up to five-bit errors.

**Exercise** Repeat the previous example for $GF(2^5)$ which is constructed using the primitive polynomial $p(x) = x^5 + x^3 + 1$.

***Note*** The elements of $GF(2^5)$ generated by $p(x) = x^5 + x^2 + 1$ are given in Table 6.5.

## 6.2 Parity Check and Generator Matrices of BCH Codes

The parity and generator matrices of BCH codes can be obtained via two methods. Let's now explain these two methods.

**Method 1**

Since the BCH codes are binary cyclic codes, the methods presented in Chap. 4 for the construction of the generator and parity check matrices of the cyclic codes are also valid for BCH codes. For the reminder, if the generator polynomial of BCH code is given as

$$g(x) = a^{n-k}x^{n-k} + a_{n-k-1}x^{n-k-1} + \ldots + a_2x^2 + a_1x^1 + a_0x^0 \tag{6.4}$$

then the generator matrix of the BCH code is formed as

**Table 6.5** Generation of $GF(2^5)$ using the primitive polynomial $p(x) = x^5 + x^2 + 1$

| $\alpha^i$ form | Vector form | Polynomial form |
|---|---|---|
| 0 | 00000 | 0 |
| 1 | 00001 | 1 |
| $\alpha$ | 00010 | $\alpha$ |
| $\alpha^2$ | 001000 | $\alpha^2$ |
| $\alpha^3$ | 01000 | $\alpha^3$ |
| $\alpha^4$ | 10000 | $\alpha^4$ |
| $\alpha^5$ | 00101 | $\alpha^2 + 1$ |
| $\alpha^6$ | 01010 | $\alpha^3 + \alpha$ |
| $\alpha^7$ | 10100 | $\alpha^4 + \alpha^2$ |
| $\alpha^8$ | 01101 | $\alpha^3 + \alpha^2 + 1$ |
| $\alpha^9$ | 11010 | $\alpha^4 + \alpha^3 + \alpha$ |
| $\alpha^{10}$ | 10001 | $\alpha^4 + 1$ |
| $\alpha^{11}$ | 00111 | $\alpha^2 + \alpha + 1$ |
| $\alpha^{12}$ | 01110 | $\alpha^3 + \alpha^2 + \alpha$ |
| $\alpha^{13}$ | 11100 | $\alpha^4 + \alpha^3 + \alpha^2$ |
| $\alpha^{14}$ | 11101 | $\alpha^4 + \alpha^3 + \alpha^2 + 1$ |
| $\alpha^{15}$ | 11111 | $\alpha^4 + \alpha^3 + \alpha^2 + \alpha + 1$ |
| $\alpha^{16}$ | 11011 | $\alpha^4 + \alpha^3 + \alpha + 1$ |
| $\alpha^{17}$ | 10011 | $\alpha^4 + \alpha + 1$ |
| $\alpha^{18}$ | 00011 | $\alpha + 1$ |
| $\alpha^{19}$ | 00110 | $\alpha^2 + \alpha$ |
| $\alpha^{20}$ | 01100 | $\alpha^3 + \alpha^2$ |
| $\alpha^{21}$ | 11000 | $\alpha^4 + \alpha^3$ |
| $\alpha^{22}$ | 10101 | $\alpha^4 + \alpha^2 + 1$ |
| $\alpha^{23}$ | 01111 | $\alpha^3 + \alpha^2 + \alpha + 1$ |
| $\alpha^{24}$ | 11110 | $\alpha^4 + \alpha^3 + \alpha^2 + 1$ |
| $\alpha^{25}$ | 11001 | $\alpha^4 + \alpha^3 + 1$ |
| $\alpha^{26}$ | 10111 | $\alpha^4 + \alpha^2 + \alpha + 1$ |
| $\alpha^{27}$ | 01011 | $\alpha^3 + \alpha + 1$ |
| $\alpha^{28}$ | 10110 | $\alpha^4 + \alpha^2 + \alpha$ |
| $\alpha^{29}$ | 01001 | $\alpha^3 + 1$ |
| $\alpha^{30}$ | 10010 | $\alpha^4 + \alpha$ |

$$\boldsymbol{G} = \begin{bmatrix} a_{n-k}\, a_{n-k-1} \ldots a_2\, a_1\, a_0\, 0\, 0 \ldots 0\, 0\, 0 \\ 0\, a_{n-k}\, a_{n-k-1} \ldots a_2\, a_1\, a_0\, 0 \ldots 0\, 0\, 0 \\ 0\, 0\, a_{n-k}\, a_{n-k-1} \ldots a_2\, a_1\, a_0\, 0 \ldots 0\, 0 \\ \vdots \\ 0\, 0\, 0 \ldots 0\, 0\, a_{n-k}\, a_{n-k-1} \ldots a_2\, a_1\, a_0 \end{bmatrix}_{k \times n} \quad (6.5)$$

On the other hand, if $h(x)$ is the parity check polynomial of a cyclic code such that

$$h(x) = b_k x^k + b_{k-1} x^{k-1} + \ldots + b_0 x^1 + b_0 x^0 \tag{6.6}$$

where $b_i \in F = \{0, 1\}$, then, using the binary coefficients of the parity check polynomial, we construct the parity check matrix as

$$\boldsymbol{H} = \begin{bmatrix} b_0\, b_1\, b_2 \ldots b_{k-1}\, b_k\, 0\, 0 \ldots 0\, 0\, 0 \\ 0\, b_0\, b_1\, b_2 \ldots b_{k-1}\, b_k\, 0 \ldots 0\, 0\, 0 \\ 0\, 0\, b_0\, b_1\, b_2 \ldots b_{k-1}\, b_k\, 0 \ldots 0\, 0 \\ \vdots \\ 0\, 0\, 0 \ldots 0\, 0\, b_0\, b_1\, b_2 \ldots b_{k-1}\, b_k \end{bmatrix}_{(n-k)\times n} . \tag{6.7}$$

**Example 6.11** The generator polynomial of $BCH(15, 7)$ code is given as

$$g(x) = x^8 + x^7 + x^6 + x^4 + 1.$$

Find the generator and parity check matrices of $BCH(15, 7)$ code.

**Solution 6.11** The coefficients of the generator matrix can be written in bit vector form as

$$[\mathbf{111010001}].$$

Using the coefficients of the generator polynomial, the first row of the generator matrix including 15 bits is formed as

$$[\mathbf{111010001}000000]$$

and rotating the first row to the right by one bit, we get the second row of the generator matrix as

$$\begin{bmatrix} \mathbf{111010001}000000 \\ 0\mathbf{111010001}00000 \end{bmatrix}$$

and rotating the second row to the right by one bit, we get the third row of the generator matrix as

$$\begin{bmatrix} \mathbf{111010001}000000 \\ 0\mathbf{111010001}00000 \\ 00\mathbf{111010001}0000 \end{bmatrix}$$

and rotating the third row to the right by one bit, we get the fourth row of the generator matrix as

$$\begin{bmatrix} \mathbf{111010001}000000 \\ 0\mathbf{111010001}00000 \\ 00\mathbf{111010001}0000 \\ 000\mathbf{111010001}000 \end{bmatrix}$$

and rotating the fourth row to the right by one bit, we get the fifth row of the generator matrix as

$$\begin{bmatrix} \mathbf{111010001}000000 \\ 0\mathbf{111010001}00000 \\ 00\mathbf{111010001}0000 \\ 000\mathbf{111010001}000 \\ 0000\mathbf{111010001}00 \end{bmatrix}$$

and rotating the fifth row to the right by one bit, we get the sixth row of the generator matrix as

$$\begin{bmatrix} \mathbf{111010001}000000 \\ 0\mathbf{111010001}00000 \\ 00\mathbf{111010001}0000 \\ 000\mathbf{111010001}000 \\ 0000\mathbf{111010001}00 \\ 00000\mathbf{111010001}0 \end{bmatrix}$$

and rotating the sixth row to the right by one bit, we get the generator matrix as

$$\boldsymbol{G} = \begin{bmatrix} \mathbf{111010001}000000 \\ 0\mathbf{111010001}00000 \\ 00\mathbf{111010001}0000 \\ 000\mathbf{111010001}000 \\ 0000\mathbf{111010001}00 \\ 00000\mathbf{111010001}0 \\ 000000\mathbf{111010001} \end{bmatrix}.$$

For the generation of the parity check matrix, we need the parity check polynomial of the given code. The parity check polynomial of the given code can be calculated as

$$h(x) = \frac{x^n + 1}{g(x)} \rightarrow h(x) = \frac{x^{15} + 1}{x^8 + x^7 + x^6 + x^4 + 1} \rightarrow h(x) = x^7 + x^6 + x^4 + 1.$$

The parity check polynomial can be expressed in bit vector form as

$$[11010001]. \tag{6.8}$$

When the bit vector in Eq. (6.8) is reversed, we obtain

$$[\mathbf{10001011}]. \tag{6.9}$$

Using the reversed bit vector in Eq. (6.9), the first row of the parity check matrix including 15 bits is formed as

$$[\mathbf{10001011}0000000]$$

and rotating the first row to the right by one bit, we get the second row of the parity check matrix as

$$\begin{bmatrix} \mathbf{10001011}0000000 \\ 0\mathbf{10001011}000000 \end{bmatrix}$$

and rotating the second row to the right by one bit, we get the third row of the parity check matrix as

$$\begin{bmatrix} \mathbf{10001011}0000000 \\ 0\mathbf{10001011}000000 \\ 00\mathbf{10001011}00000 \end{bmatrix}$$

and rotating the third row to the right by one bit, we get the fourth row of the parity check matrix as

$$\begin{bmatrix} \mathbf{10001011}0000000 \\ 0\mathbf{10001011}000000 \\ 00\mathbf{10001011}00000 \\ 000\mathbf{10001011}0000 \end{bmatrix}$$

and rotating the fourth row to the right by one bit, we get the fifth row of the parity check matrix as

$$\begin{bmatrix} \mathbf{10001011}0000000 \\ 0\mathbf{10001011}000000 \\ 00\mathbf{10001011}00000 \\ 000\mathbf{10001011}0000 \\ 0000\mathbf{10001011}000 \end{bmatrix}$$

and rotating the fifth row to the right by one bit, we get the sixth row of the parity check matrix as

$$\begin{bmatrix} \mathbf{10001011}0000000 \\ 0\mathbf{10001011}000000 \\ 00\mathbf{10001011}00000 \\ 000\mathbf{10001011}0000 \\ 0000\mathbf{10001011}000 \\ 00000\mathbf{10001011}00 \end{bmatrix}$$

and rotating the sixth row to the right by one bit, we get the seventh row of the parity check matrix as

$$\begin{bmatrix} \mathbf{10001011}0000000 \\ 0\mathbf{10001011}000000 \\ 00\mathbf{10001011}00000 \\ 000\mathbf{10001011}0000 \\ 0000\mathbf{10001011}000 \\ 00000\mathbf{10001011}00 \\ 000000\mathbf{10001011}0 \end{bmatrix}$$

and rotating the seventh row to the right by one bit, we get the parity check matrix as

$$
\boldsymbol{H} = \begin{bmatrix} 100010110000000 \\ 010001011000000 \\ 001000101100000 \\ 000100010110000 \\ 000010001011000 \\ 000001000101100 \\ 000000100010110 \\ 000000010001011 \end{bmatrix}.
$$

It can be verified that the obtained matrices $\boldsymbol{G}$ and $\boldsymbol{H}$ satisfy

$$
\boldsymbol{G}\boldsymbol{H}^T = \boldsymbol{0}. \tag{6.10}
$$

### 6.2.1 Second Method to Obtain the Generator and Parity Check Matrices of BCH Codes

The second method to obtain the generator and parity check matrices of the *BCH* codes rely on the use of polynomials which are the elements of the extended field *GF* $(2^m)$. For a $t$-error-correcting $BCH(n, k)$ code, the parity check matrix is formed as

$$
\boldsymbol{H} = \begin{bmatrix} 1 & \alpha & \alpha^2 & \cdots & \alpha^{n-1} \\ 1 & (\alpha)^3 & (\alpha^2)^3 & \cdots & (\alpha^{n-1})^3 \\ 1 & (\alpha)^5 & (\alpha^2)^5 & \cdots & (\alpha^{n-1})^5 \\ \vdots & \vdots & \vdots & \vdots & \vdots \\ 1 & (\alpha)^{2t-1} & (\alpha^2)^{2t-1} & \cdots & (\alpha^{n-1})^{2t-1} \end{bmatrix}. \tag{6.11}
$$

The generator matrix for a $t$-error-correcting $BCH(n, k)$ code is obtained as

$$G = \begin{bmatrix} g_1 \\ g_2 \\ g_3 \\ g_4 \\ \vdots \\ g_l \end{bmatrix} = \begin{bmatrix} 1 & 1 & 1 & 1 & 1 \\ 1 & \alpha & \alpha^2 & \cdots & \alpha^{n-1} \\ 1 & (\alpha)^3 & (\alpha^2)^3 & \cdots & (\alpha^{n-1})^3 \\ 1 & (\alpha)^5 & (\alpha^2)^5 & \cdots & (\alpha^{n-1})^5 \\ \vdots & \vdots & \vdots & \vdots & \vdots \\ 1 & (\alpha)^m & (\alpha^2)^m & \cdots & (\alpha^{n-1})^m \end{bmatrix} \tag{6.12}$$

where $m = 2l - 3$, $l \geq 2$, and $l < k$.

**Example 6.12** Find the generator and parity check matrices of the double-error-correcting $BCH(15, 7)$ code. Use $GF(2^4)$ constructed using the primitive polynomial $p(x) = x^4 + x + 1$.

**Solution 6.12** The parity check matrix can be formed for $t = 2$ using Eq. (6.11) as

$$H = \begin{bmatrix} 1 & \alpha & \alpha^2 & \alpha^3 & \alpha^4 & \alpha^5 & \alpha^6 & \alpha^7 & \alpha^8 & \alpha^9 \\ 1 & (\alpha)^3 & (\alpha^2)^3 & (\alpha^3)^3 & (\alpha^4)^3 & (\alpha^5)^3 & (\alpha^6)^3 & (\alpha^7)^3 & (\alpha^8)^3 & (\alpha^9)^3 \end{bmatrix.$$
$$\begin{matrix} \alpha^{10} & \alpha^{11} & \alpha^{12} & \alpha^{13} & \alpha^{14} \\ (\alpha^{10})^3 & (\alpha^{11})^3 & (\alpha^{12})^3 & (\alpha^{13})^3 & (\alpha^{14})^3 \end{matrix}\right]$$

which can be simplified using $\alpha^{15} = 1$ as

$$H = \begin{bmatrix} 1 & \alpha & \alpha^2 & \alpha^3 & \alpha^4 & \alpha^5 & \alpha^6 & \alpha^7 & \alpha^8 & \alpha^9 & \alpha^{10} & \alpha^{11} & \alpha^{12} & \alpha^{13} & \alpha^{14} \\ 1 & \alpha^3 & \alpha^6 & \alpha^9 & \alpha^{12} & 1 & \alpha^3 & \alpha^6 & \alpha^9 & \alpha^{12} & 1 & \alpha^3 & \alpha^6 & \alpha^9 & \alpha^{12} \end{bmatrix}. \tag{6.13}$$

The $\alpha^i$, $i = 1, \ldots, 14$ terms in Eq. (6.13) can be represented by polynomials in $GF(2^4)$, and these polynomials can be expressed by three-bit column vectors as in

$$0 \rightarrow 0 \rightarrow \begin{bmatrix} 0 \\ 0 \\ 0 \\ 0 \end{bmatrix} \quad 1 \rightarrow 1 \rightarrow \begin{bmatrix} 1 \\ 0 \\ 0 \\ 0 \end{bmatrix} \quad \alpha \rightarrow \alpha \rightarrow \begin{bmatrix} 0 \\ 1 \\ 0 \\ 0 \end{bmatrix} \quad \alpha^2 \rightarrow \alpha^2 \rightarrow \begin{bmatrix} 0 \\ 0 \\ 1 \\ 0 \end{bmatrix}$$

$$\alpha^3 \rightarrow \alpha^3 \rightarrow \begin{bmatrix} 0 \\ 0 \\ 0 \\ 1 \end{bmatrix} \quad \alpha^4 \rightarrow \alpha + 1 \rightarrow \begin{bmatrix} 1 \\ 1 \\ 0 \\ 0 \end{bmatrix} \quad \alpha^5 \rightarrow \alpha^2 + \alpha \rightarrow \begin{bmatrix} 0 \\ 1 \\ 1 \\ 0 \end{bmatrix} \quad \alpha^6 \rightarrow \alpha^3 + \alpha^2 \rightarrow \begin{bmatrix} 0 \\ 0 \\ 1 \\ 1 \end{bmatrix}$$

$$\alpha^7 \to \alpha^3 + \alpha + 1 \to \begin{bmatrix} 1 \\ 1 \\ 0 \\ 1 \end{bmatrix} \quad \alpha^8 \to \alpha^2 + 1 \to \begin{bmatrix} 1 \\ 0 \\ 1 \\ 0 \end{bmatrix} \quad \alpha^9 \to \alpha^3 + \alpha \to \begin{bmatrix} 0 \\ 1 \\ 0 \\ 1 \end{bmatrix}$$

$$\alpha^{10} \to \alpha^2 + \alpha + 1 \to \begin{bmatrix} 1 \\ 1 \\ 1 \\ 0 \end{bmatrix} \quad \alpha^{11} \to \alpha^3 + \alpha^2 + \alpha \to \begin{bmatrix} 0 \\ 1 \\ 1 \\ 1 \end{bmatrix}$$

$$\alpha^{12} \to \alpha^3 + \alpha^2 + \alpha + 1 \to \begin{bmatrix} 1 \\ 1 \\ 1 \\ 1 \end{bmatrix} \quad \alpha^{13} \to \alpha^3 + \alpha^2 + 1 \to \begin{bmatrix} 1 \\ 0 \\ 1 \\ 1 \end{bmatrix}$$

$$\alpha^{14} \to \alpha^3 + 1 \to \begin{bmatrix} 1 \\ 0 \\ 0 \\ 1 \end{bmatrix}.$$

Replacing the $\alpha^i$ terms by their column vector equivalents, we get the parity check matrix in bit form as

$$\boldsymbol{H} = \begin{bmatrix} 100010011010111 \\ 010011010111100 \\ 001001101011110 \\ 000100110101111 \\ ----------- \\ \mathbf{10001}1000110001 \\ \mathbf{00011}0001100011 \\ \mathbf{00101}0010100101 \\ \mathbf{01111}0111101111 \end{bmatrix}.$$

**Example 6.13** The extended field $GF(2^4)$ is constructed using the primitive polynomial $p(x) = x^4 + x + 1$. The generator polynomial of triple-error-correcting *BCH* (15, 5) code over $GF(2^4)$ is evaluated as

$$g(x) = x^{10} + x^8 + x^5 + x^4 + x^2 + x + 1.$$

Find the parity check and generator matrices of $BCH(15, 5)$ code.

**Solution 6.13** Using Eq. (6.11), we can form the parity check matric of $BCH(15, 5)$ as in

$$\boldsymbol{H} = \begin{bmatrix} 1 & \alpha & \alpha^2 & \alpha^3 & \alpha^4 & \alpha^5 & \alpha^6 & \alpha^7 & \alpha^8 & \alpha^9 & \alpha^{10} & \alpha^{11} & \alpha^{12} & \alpha^{13} & \alpha^{14} \\ 1 & \alpha^3 & \alpha^6 & \alpha^9 & \alpha^{12} & \alpha^{15} & \alpha^{18} & \alpha^{21} & \alpha^{24} & \alpha^{27} & \alpha^{30} & \alpha^{33} & \alpha^{36} & \alpha^{39} & \alpha^{42} \\ 1 & \alpha^5 & \alpha^{10} & \alpha^{15} & \alpha^{20} & \alpha^{25} & \alpha^{30} & \alpha^{35} & \alpha^{40} & \alpha^{45} & \alpha^{50} & \alpha^{55} & \alpha^{60} & \alpha^{65} & \alpha^{70} \end{bmatrix}$$

which can be simplified using $\alpha^{15} = 1$ as

$$\boldsymbol{H} = \begin{bmatrix} 1 & \alpha & \alpha^2 & \alpha^3 & \alpha^4 & \alpha^5 & \alpha^6 & \alpha^7 & \alpha^8 & \alpha^9 & \alpha^{10} & \alpha^{11} & \alpha^{12} & \alpha^{13} & \alpha^{14} \\ 1 & \alpha^3 & \alpha^6 & \alpha^9 & \alpha^{12} & 1 & \alpha^3 & \alpha^6 & \alpha^9 & \alpha^{12} & 1 & \alpha^3 & \alpha^6 & \alpha^9 & \alpha^{12} \\ 1 & \alpha^5 & \alpha^{10} & 1 & \alpha^5 & \alpha^{10} & 1 & \alpha^5 & \alpha^{10} & 1 & \alpha^5 & \alpha^{10} & 1 & \alpha^5 & \alpha^{10} \end{bmatrix}$$

in which, expressing the $\alpha^i$ terms by their polynomial representations and representing each polynomial by its column bit vector equivalent, we get the parity check matrix

$$\boldsymbol{H} = \begin{bmatrix} 100010011010111 \\ 010011010111100 \\ 001001101011110 \\ 000100110101111 \\ ------------ \\ 100011000110001 \\ 000110001100011 \\ 001010010100101 \\ 011110111101111 \\ ------------ \\ 101101101101101 \\ 011011011011011 \\ 011011011011011 \\ 000000000000000 \end{bmatrix} \tag{6.14}$$

where it is seen that the last row of the matrix is zero, and the rows 10 and 11 are the same. For this reason, omitting the last two rows of Eq. (6.14), we obtain the final form of the parity check matrix as in

$$\boldsymbol{H} = \begin{bmatrix} 100010011010111 \\ 010011010111100 \\ 001001101011110 \\ 000100110101111 \\ - - - - - - - - - \\ 100011000110001 \\ 000110001100011 \\ 001010010100101 \\ 011110111101111 \\ - - - - - - - - - \\ 101101101101101 \\ 011011011011011 \end{bmatrix}. \tag{6.15}$$

The size of the matrix in Eq. (6.14) is $12 \times 15$ which is NOT correct, and on the other hand, the size of the matrix in Eq. (6.15) is $10 \times 15$ which is correct, since the size of the parity check matrix should be $(n - k) \times k$, and for the $BCH(15, 5)$ code, we have $n = 15$ and $k = 5$.

The generator matrix can be formed using Eq. (6.12) as

$$\boldsymbol{G} = \begin{bmatrix} 1 & 1 & 1 & 1 & 1 & 1 & 1 & 1 & 1 & 1 & 1 & 1 & 1 & 1 & 1 \\ 1 & \alpha & \alpha^2 & \alpha^3 & \alpha^4 & \alpha^5 & \alpha^6 & \alpha^7 & \alpha^8 & \alpha^9 & \alpha^{10} & \alpha^{11} & \alpha^{12} & \alpha^{13} & \alpha^{14} \end{bmatrix}$$

where, expressing each term by a column bit vector including five bits and eliminating the redundant rows, we obtain

$$\boldsymbol{G} = \begin{bmatrix} 111111111111111 \\ 100010011010111 \\ 010011010111100 \\ 001001101011110 \\ 000100110101111 \end{bmatrix}.$$

We can verify that $\boldsymbol{GH}^T = 0$, i.e., we have

$$
GH^T = 0 \rightarrow \begin{bmatrix} 111111111111111 \\ 100010011010111 \\ 010011010111100 \\ 001001101011110 \\ 000100110101111 \end{bmatrix} \left[\begin{array}{c|c|c} 1000 & 1000 & 10 \\ 0100 & 0001 & 01 \\ 0010 & 0011 & 11 \\ 0001 & 0101 & 10 \\ 1100 & 1111 & 01 \\ 0110 & 1000 & 11 \\ 0011 & 0001 & 10 \\ 1101 & 0011 & 01 \\ 1010 & 0101 & 11 \\ 0101 & 1111 & 10 \\ 1110 & 1000 & 01 \\ 0111 & 0001 & 11 \\ 1111 & 0011 & 10 \\ 1011 & 0101 & 01 \\ 1001 & 1111 & 11 \end{array}\right] = \begin{bmatrix} 0000000000 \\ 0000000000 \\ 0000000000 \\ 0000000000 \\ 0000000000 \end{bmatrix}.
$$

**Example 6.14** Find the parity check and generator matrices of double-error-correcting $BCH(31, 16)$ code designed using the elements of $GF(2^5)$ which is constructed using the primitive polynomial $p(x) = x^5 + x^2 + 1$.

**Solution 6.14** Using Eq. (6.11), we obtain the parity check matrix of the $BCH$ $(31, 16)$ code as

$$
\boldsymbol{H} = \begin{bmatrix} 1 & \alpha & \alpha^2 & \cdots & \alpha^{30} \\ 1 & (\alpha)^3 & (\alpha^2)^3 & \cdots & (\alpha^{30})^3 \\ 1 & (\alpha)^5 & (\alpha^2)^5 & \cdots & (\alpha^{30})^5 \end{bmatrix}
$$

in which, simplifying the $\alpha^i$, $i = 31, \ldots, 150$ using $\alpha^{31} = 1$, expressing each $\alpha^i$, $i = 1, \ldots, 30$ in terms of polynomials, and finally converting each polynomial to a column vector involving five bits, removing redundant rows if necessary, we obtain the parity check matrix as

$$H=\left[\begin{array}{c}
10000100101100111110001101111010\\
01000010010110011111000110111101\\
00100101100111110001101110101000\\
00010010110011111000110111010100\\
00001001011001111100011011101010\\
\hline
10000110010011111011100010101100\\
00111110111000101011010000110010\\
00001100100111110111000101011010\\
01111101110001010110100001100100\\
00010101101000011001001111101110\\
\hline
11110100010010101100000111001101\\
00010010101100000111001101111101\\
01011000011100110111110100010010\\
00011100110111110100010010101100\\
00110111110100010010101100000111
\end{array}\right].$$

## 6.3 Syndrome Calculation for BCH Codes

The generator polynomial of a *t*-error-correcting *BCH* code can be obtained using

$$g(x)=LCM\{m_1(x),m_3(x),\ \ldots,m_{2t-1}\}. \tag{6.16}$$

It is possible to show that $\alpha$, $\alpha^2$, $\alpha^3$, …, $\alpha^{2t}$ are the roots of $g(x)$, and this means that

$$g\left(\alpha^i\right)=0\ \ i=1,2,\ \ldots,2t. \tag{6.17}$$

Since the *BCH* codes are cyclic codes, for a given data-word polynomial $d(x)$, the code-word polynomial can be written as

$$c(x)=d(x)g(x) \tag{6.18}$$

in which, employing Eq. (6.17), we get

$$c\left(\alpha^i\right)=0\ \ i=1,2,\ \ldots,2t. \tag{6.19}$$

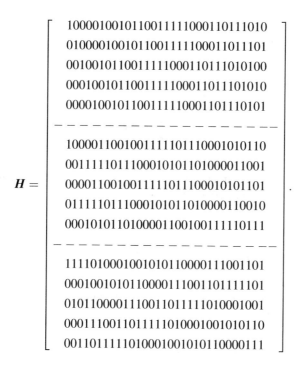

**Table 6.6** Encoding, syndrome calculation, and orthogonality formulas for linear block, linear cyclic, and BCH cyclic codes

| Linear block | Cyclic | BCH cyclic |
|---|---|---|
| $\mathbf{c} = \mathbf{d}\mathbf{G}$ | $c(x) = d(x)g(x)$ | $c(x) = d(x)g(x)$ |
| $\mathbf{G}\mathbf{H}^T = \mathbf{0}$ | $R_{x^n+1}(g(x)h(x)) = 0$ | $R_{x^n+1}(g(x)h(x)) = 0$ |
| $\mathbf{c}\mathbf{H}^T = \mathbf{0}$ | $R_{x^n+1}(c(x)h(x)) = 0$ | $R_{x^n+1}(c(x)h(x)) = 0$ |
| $\mathbf{s} = \mathbf{r}\mathbf{H}^T$ or $\mathbf{s} = \mathbf{e}\mathbf{H}^T$ | $s(x) = R_{x^n+1}(r(x)h(x))$ or $s(x) = R_{g(x)}(r(x))$ | $S_i = r(\alpha^i)$ $i = 1, 2, \ldots, 2t$ |

Now, let's consider the transmission of a code-word $c(x)$, and at the receiver side, we have the received word polynomial

$$r(x) = c(x) + e(x) \tag{6.20}$$

where $e(x)$ corresponds to the error word polynomial.

***Syndromes for BCH Codes***

For a received word polynomial $r(x)$, there are $2t$ syndromes, and these syndromes are calculated using

$$S_i = r(\alpha^i) \quad \text{where } i = 1, 2, \ldots, 2t. \tag{6.21}$$

Note that error syndromes are the elements of an extended field $GF(2^m)$.

In Table 6.6, encoding, syndrome calculation, and orthogonality formulas for linear block, linear cyclic, and BCH cyclic codes are summarized.

**Example 6.15** The generator polynomial of the double-error-correcting $BCH(15, 7)$ cyclic code is given as

$$g(x) = x^8 + x^7 + x^6 + x^4 + 1.$$

The data-word polynomial is $d(x) = x^5 + x^2 + 1$. Obtain the systematic and non-systematic code-words for $d(x)$.

**Solution 6.15** The non-systematic code-word resulting in the encoding of $d(x)$ can be calculated as

$$c(x) = d(x)g(x) \rightarrow c(x) = \left(x^5 + x^2 + 1\right)\left(x^8 + x^7 + x^6 + x^4 + 1\right)$$

leading to

$$c(x) = x^{13} + x^{12} + x^{11} + x^{10} + x^7 + x^5 + x^4 + x^2 + 1.$$

The systematic form of the code-word can be obtained as follows. First, we multiply $d(x)$ by $x^{n-k} \rightarrow x^8$ resulting in

$$x^8 d(x) \to x^{13} + x^{10} + x^8.$$

Next, we calculate the remainder polynomial using

$$r(x) = R_{g(x)}\left(x^{x-k} d(x)\right)$$

as

$$r(x) = x^6 + x + 1.$$

Finally, the systematic code-word is obtained as

$$c(x) = x^{n-k} d(x) + r(x) \to c(x) = x^{13} + x^{10} + x^8 + x^6 + x + 1.$$

**Example 6.16** The double-error-correcting $BCH(15,7)$ cyclic code is designed in $GF(2^4)$ which is constructed using the primitive polynomial $p(x) = x^4 + x + 1$. A code-word polynomial of $BCH(15,7)$ is given as

$$c(x) = x^{13} + x^{10} + x^8 + x^6 + x + 1.$$

Show that $c(\alpha^i) = 0$ for $i = 1, 2, 3, 4$.

**Solution 6.16** For the reminder, first, let's sort the extended field $GF(2^4)$ elements generated by the primitive polynomial $p(x) = x^4 + x + 1$ as in

$$\begin{aligned}
&0 \to 0 \\
&1 \to 1 \\
&\alpha \to \alpha \\
&\alpha^2 \to \alpha^2 \\
&\alpha^3 \to \alpha^3 \\
&\alpha^4 \to \alpha + 1 \\
&\alpha^5 \to \alpha^2 + \alpha \\
&\alpha^6 \to \alpha^3 + \alpha^2 \\
&\alpha^7 \to \alpha^3 + \alpha + 1 \\
&\alpha^8 \to \alpha^2 + 1 \\
&\alpha^9 \to \alpha^3 + \alpha \\
&\alpha^{10} \to \alpha^2 + \alpha + 1 \\
&\alpha^{11} \to \alpha^3 + \alpha^2 + \alpha \\
&\alpha^{12} \to \alpha^3 + \alpha^2 + \alpha + 1 \\
&\alpha^{13} \to \alpha^3 + \alpha^2 + 1 \\
&\alpha^{14} \to \alpha^3 + 1.
\end{aligned} \tag{6.22}$$

Using the field elements and the identity $\alpha^{15} = 1$, we can evaluate

$$c(x) = x^{13} + x^{10} + x^8 + x^6 + x + 1$$

for $x = \alpha^i$, $i = 1, 2, 3, 4$ as

$$x = \alpha \rightarrow c(\alpha) = \alpha^{13} + \alpha^{10} + \alpha^8 + \alpha^6 + \alpha + 1 \rightarrow$$
$$c(\alpha) = (\alpha^3 + \alpha^2 + 1) + (\alpha^2 + \alpha + 1) + (\alpha^2 + 1) + (\alpha^3 + \alpha^2) + \alpha + 1 \rightarrow$$
$$c(\alpha) = 0$$
$$x = \alpha^2 \rightarrow c(\alpha^2) = \alpha^{26} + \alpha^{20} + \alpha^{16} + \alpha^{12} + \alpha^2 + 1 \rightarrow$$
$$c(\alpha^2) = \alpha^{11} + \alpha^5 + \alpha^1 + \alpha^{12} + \alpha^2 + 1$$
$$c(\alpha^2) = (\alpha^3 + \alpha^2 + \alpha) + (\alpha^2 + \alpha) + \alpha^1 + (\alpha^3 + \alpha^2 + \alpha + 1) + \alpha^2 + 1 \rightarrow$$
$$c(\alpha^2) = 0$$
$$x = \alpha^3 \rightarrow c(\alpha^3) = \alpha^{39} + \alpha^{30} + \alpha^{24} + \alpha^{18} + \alpha^3 + 1 \rightarrow$$
$$c(\alpha^3) = \alpha^9 + \alpha^0 + \alpha^9 + \alpha^3 + \alpha^3 + 1 \rightarrow c(\alpha^3) = 0$$
$$x = \alpha^4 \rightarrow c(\alpha^4) = \alpha^{52} + \alpha^{40} + \alpha^{32} + \alpha^{24} + \alpha^4 + 1 \rightarrow$$
$$c(\alpha^4) = \alpha^7 + \alpha^{10} + \alpha^2 + \alpha^9 + \alpha^4 + 1$$
$$c(\alpha^4) = (\alpha^3 + \alpha + 1) + (\alpha^2 + \alpha + 1) + \alpha^2 + (\alpha^3 + \alpha) + (\alpha + 1) + 1 \rightarrow$$
$$c(\alpha^4) = 0.$$

Thus, we showed that

$$c(\alpha) = c(\alpha^2) = c(\alpha^3) = c(\alpha^4) = 0.$$

In fact, if $c(\alpha) = 0$, then using the property

$$(x_1 + x_2 + \ldots + x_i)^2 = x_1^2 + x_2^2 + \ldots + x_i^2$$

we can write that

$$c(\alpha^2) = c(\alpha^4) = 0.$$

## 6.4 Syndrome Equations and Syndrome Decoding

The error pattern that corrupts the transmitted code-word can be expressed in polynomial form. If there are $v$ bit errors during the communication, then we can express the error word in polynomial form as

$$e(x) = x^{p_1} + x^{p_2} + \ldots + x^{p_v} \tag{6.23}$$

where the exponential terms $p_1, p_2, \ldots, p_v$ indicate the positions of the errors.

**Example 6.17** For the code-word length $n = 8$, an error word is given as

$$\boldsymbol{e} = [01000101] \tag{6.24}$$

where the bits "1's" indicate the errors at the received word. The positions of "1's" are explicitly shown in

$$\boldsymbol{e} = \left[\underbrace{0}_{7}\;\underbrace{\mathbf{1}}_{\mathbf{6}}\;\underbrace{0}_{5}\;\underbrace{0}_{4}\;\underbrace{0}_{3}\;\underbrace{\mathbf{1}}_{\mathbf{2}}\;\underbrace{0}_{1}\;\underbrace{\mathbf{1}}_{\mathbf{0}}\right]$$

from which we see that the errors occur at the bit positions 0, 2, and 6. The polynomial form of $\boldsymbol{e}$ can be written as

$$e(x) = x^6 + x^2 + x^0$$

where it is clear that the exponents of $x$ correspond to the positions of the bit errors.

**Syndrome Equations**

Considering the transmission of the code-word polynomial $c(x)$, the received word polynomial can be written as

$$r(x) = c(x) + e(x).$$

where the error polynomial $e(x)$ can be expressed as in Eq. (6.23). Using $r(x)$, we can calculate the syndromes as

$$S_i = r(\alpha^i) \rightarrow S_i = \underbrace{c(\alpha^i)}_{=0} + e(\alpha^i) \rightarrow S_i = e(\alpha^i) \quad i = 1, 2, \ldots, 2t$$

in which, employing Eq. (6.23), we obtain the syndrome equations

$$
\begin{aligned}
&S_1 = e(\alpha) \to S_1 = \alpha^{p_1} + \alpha^{p_2} + \cdots + \alpha^{p_v} \\
&S_2 = e(\alpha^2) \to S_2 = \alpha^{2p_1} + \alpha^{2p_2} + \cdots + \alpha^{2p_v} \\
&S_3 = e(\alpha^3) \to S_3 = \alpha^{3p_1} + \alpha^{3p_2} + \cdots + \alpha^{3p_v} \\
&S_4 = e(\alpha^4) \to S_4 = \alpha^{4p_1} + \alpha^{4p_2} + \cdots + \alpha^{4p_v} \\
&\quad\vdots \\
&S_{2t} = e(\alpha^{2t}) \to S_{2t} = \alpha^{2tp_1} + \alpha^{2tp_2} + \cdots + \alpha^{2tp_v}
\end{aligned}
\tag{6.25}
$$

where, using the property $S_{2i} = S_i^2$, we get the reduced set of syndrome equations

$$
\begin{aligned}
&S_1 = e(\alpha) \to S_1 = \alpha^{p_1} + \alpha^{p_2} + \cdots + \alpha^{p_v} \\
&S_3 = e(\alpha^3) \to S_3 = \alpha^{3p_1} + \alpha^{3p_2} + \cdots + \alpha^{3p_v} \\
&S_5 = e(\alpha^5) \to S_5 = \alpha^{5p_1} + \alpha^{5p_2} + \cdots + \alpha^{5p_v} \\
&\quad\vdots \\
&S_{2t-1} = e(\alpha^{2t-1}) \to S_{2t-1} = \alpha^{(2t-1)p_1} + \cdots + \alpha^{(2t-1)p_v}
\end{aligned}
\tag{6.26}
$$

In Eq. (6.25), making use of the parameter change as $X_i = \alpha^{p_i}$, we express Eq. (6.25) as

$$
\begin{aligned}
&S_1 = X_1 + X_2 + \cdots + X_v \\
&S_2 = X_1^2 + X_2^2 + \cdots + X_v^2 \\
&S_3 = X_1^3 + X_2^3 + \cdots + X_v^3 \\
&\quad\vdots \\
&S_{2t} = X_1^{2t} + X_2^{2t} + \cdots + X_v^{2t}
\end{aligned}
\tag{6.27}
$$

and Eq. (6.26) as

$$
\begin{aligned}
&S_1 = X_1 + X_2 + \cdots + X_v \\
&S_3 = X_1^3 + X_2^3 + \cdots + X_v^3 \\
&S_5 = X_1^5 + X_2^5 + \cdots + X_v^5 \\
&\quad\vdots \\
&S_{2t-1} = X_1^{2t-1} + X_2^{2t-1} + \cdots + X_v^{2t-1}
\end{aligned}
\tag{6.28}
$$

**Example 6.18** For $t = v = 3$, Eq. (6.28) reduces to

$$
\begin{aligned}
&S_1 = X_1 + X_2 + X_3 \\
&S_3 = X_1^3 + X_2^3 + X_3^3 . \\
&S_5 = X_1^5 + X_2^5 + X_3^5
\end{aligned}
$$

In Eq. (6.28), there are $t$ equations and $v$ unknowns. If $v \leq t$, then there exists a solution of Eq. (6.28). This means that the number of bit errors occurred is less than the error correction capability of the code. That is, all the bit error positions can be determined, and error correction can be performed successfully. If $v \geq t$, then the number of unknowns in Eq. (6.26) becomes greater than the number of equations, i.e., $t$. This means that the solution of Eq. (6.26) does not exist.

**Example 6.19** The syndrome equation set for a *BCH* code is given as

$$\begin{aligned} S_1 &= X_1 + X_2 + X_3 + X_4 \\ S_2 &= X_1^2 + X_2^2 + X_3^2 + X_4^2 . \\ S_3 &= X_1^3 + X_2^3 + X_3^3 + X_4^3 \end{aligned} \tag{6.29}$$

(a) What is the error correction capability of this code?
(b) What is the assumed number of errors occurred?
(c) Can this equation set be solved?

**Solution 6.19** If we compare the given equation set to the Eq. (6.28), we see that

$$t = 2 \text{ and } v = 4.$$

Since the assumed number of errors occurred is greater than the error correction capability of the code, i.e., $v > t$, in other words, the number of unknowns is greater than the number of equations in (6.29), the equation set does not have a solution. This result can also be interpreted as the code is not capable of correcting these many bit errors.

**Example 6.20** The extended field $GF(2^4)$ is constructed using the primitive polynomial $p(x) = x^4 + x + 1$. The double-error-correcting $BCH(15,7)$ code is designed over $GF(2^4)$. A code-word polynomial of $BCH(15,7)$ is given as

$$c(x) = x^{13} + x^{10} + x^8 + x^6 + x + 1.$$

After transmission of $c(x)$, the received word polynomial incurs the error word polynomial

$$e(x) = x^2 + x.$$

Find the received word polynomial, and obtain the syndrome equations. Solve the syndrome equations and determine the erroneous bit positions.

**Solution 6.20** The received word polynomial can be calculated as

$$r(x) = c(x) + e(x) \rightarrow r(x) = x^{13} + x^{10} + x^8 + x^6 + x + 1 + x^2 + x$$

leading to

$$r(x) = x^{13} + x^{10} + x^8 + x^6 + x^2 + 1.$$

Since our code is a double-error-correcting code, i.e., $t = 2$, we need to calculate the syndromes

$$S_1, S_3, \ldots, S_{2t-1} \rightarrow S_1, S_3.$$

The syndromes $S_1$ and $S_3$ can be calculated using the field elements in Eq. (6.22) and the equality $\alpha^{15} = 1$ as

$$S_1 = r(\alpha) \rightarrow S_1 = \alpha^{13} + \alpha^{10} + \alpha^8 + \alpha^6 + \alpha^2 + 1 \rightarrow$$

$$S_1 = \left(\alpha^3 + \alpha^2 + 1\right) + \left(\alpha^2 + \alpha + 1\right) + \left(\alpha^2 + 1\right) + \left(\alpha^3 + \alpha^2\right) + \alpha^2 + 1 \rightarrow$$

$$S_1 = \alpha^2 + \alpha \rightarrow S_1 = \alpha^5.$$

$$S_3 = r\left(\alpha^3\right) \rightarrow S_3 = \alpha^{39} + \alpha^{30} + \alpha^{24} + \alpha^{18} + \alpha^6 + 1 \rightarrow$$

$$S_3 = \alpha^9 + \alpha^0 + \alpha^9 + \alpha^3 + \alpha^6 + 1 \rightarrow$$

$$S_3 = \alpha^3 + \alpha^6 \rightarrow S_3 = \alpha^3 + \alpha^3 + \alpha^2 \rightarrow S_3 = \alpha^2.$$

We assume that $v = t$, i.e., the number of bit errors equals to the number of correctable bit errors. Referring to Eq. (6.28), we can write the syndrome equations as

$$S_1 = X_1 + X_2$$

$$S_3 = X_1^3 + X_2^3$$

in which, using the calculated syndromes, we get

$$\begin{aligned} X_1 + X_2 &= \alpha^5 \\ X_1^3 + X_2^3 &= \alpha^2 \end{aligned} \quad (6.30)$$

Now, let's solve the equation pair in Eq. (6.30). If we take the cube of the first equation, we get

$$X_1^3 + X_2^3 + X_1^2 X_2 + X_1 X_2^2 = \alpha^{15}$$

which can be written as

$$X_1^3 + X_2^3 + X_1 X_2 (X_1 + X_2) = \alpha^{15}$$

in which, using the identities in Eq. (6.30), we get

$$\alpha^2 + X_1X_2(\alpha^5) = \alpha^{15}$$

from which we obtain

$$X_1X_2 = \frac{1+\alpha^2}{\alpha^5} \to X_1X_2 = \frac{\alpha^8}{\alpha^5} \to X_1X_2 = \alpha^3. \tag{6.31}$$

Using $X_1 + X_2 = \alpha^5$, we can write that

$$X_1 = X_2 + \alpha^5. \tag{6.32}$$

Substituting Eq. (6.32) into Eq. (6.31), we obtain

$$(X_2 + \alpha^5)X_2 = \alpha^3$$

from which we obtain

$$X_2^2 + \alpha^5 X_2 + \alpha^3 = 0. \tag{6.33}$$

Making use of the parameter change, the equation in Eq. (6.33) can be written as

$$X^2 + \alpha^5 X + \alpha^3 = 0. \tag{6.34}$$

The roots of Eq. (6.34) can be found in a trivial manner trying the field elements $\alpha, \alpha^2, \alpha^3, \ldots, \alpha^{14}$ in Eq. (6.34) and deciding on those satisfying the equality. The two roots of Eq. (6.12) are $X_1$ and $X_2$. If we try $X = \alpha$ in Eq. (6.34), we get

$$\alpha^2 + \alpha^5\alpha + \alpha^3 = 0$$

which can be simplified as

$$\alpha^2 + (\alpha^3 + \alpha^2) + \alpha^3 = 0 \to 0 = 0.$$

Thus, we understand that $X = \alpha$ is a root of Eq. (6.34). Now, if we try $X = \alpha^2$ in Eq. (6.34), we get

$$\alpha^4 + \alpha^5\alpha^2 + \alpha^3 = 0$$

which can be simplified as

$$(\alpha + 1) + (\alpha^3 + \alpha + 1) + \alpha^3 = 0 \to 0 = 0$$

from which we understand that $X = \alpha^2$ is a root of Eq. (6.34). Then, we can conclude that

$$X_1 = \alpha \quad X_2 = \alpha^2$$

where the powers of $\alpha$ indicate the positions of errors. Hence, we can write the error polynomial as

$$e(x) = x^2 + x.$$

Using the error polynomial, we can get the decoded code-word as

$$\widehat{c}(x) = r(x) + e(x)$$

leading to

$$\widehat{c}(x) = x^{13} + x^{10} + x^8 + x^6 + x + 1.$$

**Exercise** For the previous example, if the received word polynomial is

$$r(x) = x^{13} + x^{10} + x^7 + x^6 + x + 1$$

find the error polynomial using syndrome equations.

**Exercise** The extended field $GF(2^5)$ is constructed using the primitive polynomial $p(x) = x^5 + x^2 + 1$. The generator polynomial of $BCH(31, 16)$ code over this field can be calculated as

$$g(x) = x^{15} + x^{11} + x^{10} + x^9 + x^8 + x^7 + x^5 + x^3 + x^2 + x + 1.$$

Encode the data-word polynomial $d(x) = x^{11} + x^9 + x^8 + x^4 + x^2 + x + 1$ with this code. Let $c(x)$ be the code-word polynomial obtained after encoding operation. The error polynomial is given as

$$e(x) = x^7 + x^5 + 1.$$

Write the received word polynomial, and using syndrome decoding, calculate the error polynomial.

### 6.4.1 The Error Location Polynomial

In the previous example, we found that for the double-error-correcting *BCH* code, the location of the errors can be determined considering the roots of

$$X^2 + \alpha^5 X + \alpha^3 = 0.$$

We can generalize this issue for the *t*-error-correcting *BCH* codes. We can state that, for a *t*-error-correcting *BCH* code, the error locations can be determined considering the roots of

$$x^v + \sigma_1 x^{v-1} + \ldots + \sigma_{v-1} x + \sigma_v = 0 \tag{6.35}$$

where $v$ is the bit error number such that $v \leq t$. The polynomial whose roots are the reciprocal of the roots of Eq. (6.35), i.e., reciprocal of the error locations, is called the error location polynomial, and it is obtained by substituting $1/x$ in Eq. (6.35) as

$$\left(\frac{1}{x}\right)^v + \sigma_1 \left(\frac{1}{x}\right)^{v-1} + \ldots + \sigma_{v-1}\left(\frac{1}{x}\right) + \sigma_v = 0 \rightarrow$$

$$\sigma_0 + \sigma_1 x + \ldots + \sigma_{v-1} x^{v-1} + \sigma_v x^v = 0. \tag{6.36}$$

The error location polynomial is denoted by $\sigma(x)$, i.e.,

$$\sigma(x) = \sigma_0 + \sigma_1 x + \ldots + \sigma_{v-1} x^{v-1} + \sigma_v x^v. \tag{6.37}$$

If we denote the roots of Eq. (6.37) by $X_1, X_2, \ldots, X_v$, then the error location polynomial is $\sigma(x)$ which can be written as

$$\sigma(x) = \left(x + \frac{1}{X_1}\right)\left(x + \frac{1}{X_2}\right)\cdots\left(x + \frac{1}{X_v}\right). \tag{6.38}$$

Equating Eq. (6.38) to zero, we get

$$\left(x + \frac{1}{X_1}\right)\left(x + \frac{1}{X_2}\right)\cdots\left(x + \frac{1}{X_v}\right) = 0 \rightarrow (xX_1 + 1)(xX_2 + 1)\cdots(xX_v + 1) = 0. \tag{6.39}$$

Now, equating Eqs. (6.36) and (6.39), we get

$$\sigma_0 + \sigma_1 x + \ldots + \sigma_{v-1} x^{v-1} + \sigma_v x^v = (xX_1 + 1)(xX_2 + 1)\cdots(xX_v + 1) \tag{6.40}$$

where, expanding the polynomial on the right and equating the coefficients of terms $x^i$ with the same exponential value, we obtain the set of equations

$$\begin{aligned}
\sigma_0 &= 1 \\
\sigma_1 &= X_1 + X_2 + \cdots + X_{v-1} + X_v \\
\sigma_2 &= X_1X_2 + X_2X_3 + \cdots + X_{v-1}X_v \\
&\vdots \\
\sigma_v &= X_1X_2X_3 \ldots X_{v-1}X_v
\end{aligned} \tag{6.41}$$

**Example 6.21** For $v = 4$, Eq. (6.41) reduces to

$$\begin{aligned}
\sigma_0 &= 1 \\
\sigma_1 &= X_1 + X_2 + X_3 + X_4 \\
\sigma_2 &= X_1X_2 + X_2X_3 + X_3X_4 \, . \\
\sigma_3 &= X_1X_2X_3 + X_2X_3X_4 \\
\sigma_4 &= X_1X_2X_3X_4
\end{aligned}$$

Now, let's express the syndromes in Eq. (6.27) in terms of the coefficients appearing in Eq. (6.41). The syndrome equations in Eq. (6.27) and the coefficient equations in Eq. (6.41) are given below for reminder

$$\begin{aligned}
S_1 &= X_1 + X_2 + \cdots + X_v \\
S_2 &= X_1^2 + X_2^2 + \cdots + X_v^2 \\
S_3 &= X_1^3 + X_2^3 + \cdots + X_v^3 \\
&\vdots \\
S_{2t} &= X_1^{2t} + X_2^{2t} + \cdots + X_v^{2t} \\
\\
\sigma_0 &= 1 \\
\sigma_1 &= X_1 + X_2 + \cdots + X_{v-1} + X_v \\
\sigma_2 &= X_1X_2 + X_2X_3 + \cdots + X_{v-1}X_v \\
&\vdots \\
\sigma_v &= X_1X_2X_3 \ldots X_{v-1}X_v .
\end{aligned} \tag{6.42}$$

Assuming that $v \leq t$, the syndromes $S_i$, $i = 1, \ldots, 2v$ in Eq. (6.42) can be expressed in terms of the coefficients $\sigma_j$, $j = 1, \ldots, v$ as

$$\begin{aligned}
S_1 &= \sigma_1 \\
S_2 &= \sigma_1 S_1 \\
S_3 &= \sigma_1 S_2 + \sigma_2 S_1 + \sigma_3 \\
S_4 &= \sigma_1 S_3 + \sigma_2 S_2 + \sigma_3 S_1 \\
&\vdots \\
S_v &= \sigma_1 S_{v-1} + \sigma_2 S_{v-2} + \cdots + \sigma_{v-2} S_2 + \sigma_{v-1} S_1 + v\sigma_v \\
S_{v+1} &= \sigma_1 S_v + \sigma_2 S_{v-1} + \cdots + \sigma_{v-1} S_2 + \sigma_v S_1 \\
S_{v+2} &= \sigma_1 S_{v+2} + \sigma_2 S_v + \cdots + \sigma_{v-1} S_3 + \sigma_v S_2 \\
S_{2v} &= \sigma_1 S_{2v-1} + \sigma_2 S_{2v-1} + \cdots + \sigma_{v-1} S_{v+1} + \sigma_v S_v
\end{aligned} \tag{6.43}$$

which can be written in matrix form as

$$\begin{bmatrix} S_1 \\ S_2 \\ S_3 \\ S_4 \\ \vdots \\ S_v \\ S_{v+1} \\ S_{v+2} \\ \vdots \\ S_{2v} \end{bmatrix} = \begin{bmatrix} 1 & 0 & 0 & \cdots & 0 & 0 \\ S_1 & 0 & 0 & \cdots & 0 & 0 \\ S_2 & S_1 & 1 & \cdots & 0 & 0 \\ S_3 & S_2 & S_1 & \cdots & 0 & 0 \\ \vdots & \vdots & \vdots & \cdots & \vdots & \vdots \\ S_{v-1} & S_{v-2} & S_{v-3} & \cdots & S_1 & v \\ S_v & S_{v-1} & S_{v-2} & \cdots & S_2 & S_1 \\ S_{v+1} & S_v & S_{v-1} & \cdots & S_3 & S_2 \\ \vdots & \vdots & \vdots & \cdots & \vdots & \vdots \\ S_{2v-1} & S_{2v-2} & S_{2v-3} & \cdots & S_{v+1} & S_v \end{bmatrix} \begin{bmatrix} \sigma_1 \\ \sigma_2 \\ \sigma_3 \\ \sigma_4 \\ \vdots \\ \vdots \\ \sigma_{v-3} \\ \sigma_{v-2} \\ \sigma_{v-1} \\ \sigma_v \end{bmatrix} \tag{6.44}$$

where, employing the property $S_{2i} = S_i^2$, we get the simplified form of Eq. (6.44) as

$$\begin{bmatrix} S_1 \\ S_3 \\ S_5 \\ \vdots \\ S_{2v-1} \end{bmatrix} = \begin{bmatrix} 1 & 0 & 0 & \cdots & 0 & 0 \\ S_2 & S_1 & 1 & \cdots & 0 & 0 \\ S_4 & S_3 & S_2 & \cdots & 0 & 0 \\ \vdots & \vdots & \vdots & \cdots & \vdots & \vdots \\ S_{2v-1} & S_{2v-2} & S_{2v-3} & \cdots & S_{v+1} & S_v \end{bmatrix} \begin{bmatrix} \sigma_1 \\ \sigma_2 \\ \vdots \\ \sigma_{v-1} \\ \sigma_v \end{bmatrix} \tag{6.45}$$

**Example 6.22** For $v = 4$, construct the matrix form of syndrome equations given in Eqs. (6.44) and (6.45).

**Solution 6.22** If we inspect Eq. (6.44), we see that the syndrome $S_j$, $j \leq v$ is calculated according to

$$S_v = [S_{v-1}\ S_{v-2} \ldots S_2\ S_1\ v] \begin{bmatrix} \sigma_1 \\ \sigma_2 \\ \vdots \\ \sigma_{v-1} \\ \sigma_v \end{bmatrix} \tag{6.46}$$

and the syndrome $S_j$, $j > v$ is calculated using

$$S_{v+i} = [S_{v+i-1}\ S_{v+i-2} \ldots S_{i+2}\ S_{i+1}\ S_i] \begin{bmatrix} \sigma_1 \\ \sigma_2 \\ \vdots \\ \sigma_{v-1} \\ \sigma_v \end{bmatrix}. \tag{6.47}$$

For $v = 4$, considering Eq. (6.46), the syndromes $S_1$, $S_2$, $S_3$, and $S_4$ are calculated according to

$$S_1 = [1\ 0\ 0\ 0] \begin{bmatrix} \sigma_1 \\ \sigma_2 \\ \sigma_3 \\ \sigma_4 \end{bmatrix} \tag{6.48}$$

$$S_2 = \left[ S_1\ \underbrace{2}_{=0}\ 0\ 0 \right] \begin{bmatrix} \sigma_1 \\ \sigma_2 \\ \sigma_3 \\ \sigma_4 \end{bmatrix} \rightarrow S_2 = [S_1\ 0\ 0\ 0] \begin{bmatrix} \sigma_1 \\ \sigma_2 \\ \sigma_3 \\ \sigma_4 \end{bmatrix} \tag{6.49}$$

$$S_3 = \left[ S_2\ S_1\ \underbrace{3}_{=1}\ 0 \right] \begin{bmatrix} \sigma_1 \\ \sigma_2 \\ \sigma_3 \\ \sigma_4 \end{bmatrix} \rightarrow S_3 = [S_2\ S_1\ 1\ 0] \begin{bmatrix} \sigma_1 \\ \sigma_2 \\ \sigma_3 \\ \sigma_4 \end{bmatrix} \tag{6.50}$$

$$S_4 = \left[ S_3\ S_2\ S_1\ \underbrace{4}_{=0} \right] \begin{bmatrix} \sigma_1 \\ \sigma_2 \\ \sigma_3 \\ \sigma_4 \end{bmatrix} \rightarrow S_3 = [S_3\ S_2\ S_1\ 0] \begin{bmatrix} \sigma_1 \\ \sigma_2 \\ \sigma_3 \\ \sigma_4 \end{bmatrix} \tag{6.51}$$

and considering Eq. (6.47) the syndromes $S_5$, $S_6$, $S_7$, and $S_8$ are calculated according to

$$S_5 = [S_4 \; S_3 \; S_2 \; S_1]\begin{bmatrix} \sigma_1 \\ \sigma_2 \\ \sigma_3 \\ \sigma_4 \end{bmatrix} \tag{6.52}$$

$$S_6 = [S_5 \; S_4 \; S_3 \; S_2]\begin{bmatrix} \sigma_1 \\ \sigma_2 \\ \sigma_3 \\ \sigma_4 \end{bmatrix} \tag{6.53}$$

$$S_7 = [S_6 \; S_5 \; S_4 \; S_3]\begin{bmatrix} \sigma_1 \\ \sigma_2 \\ \sigma_3 \\ \sigma_4 \end{bmatrix} \tag{6.54}$$

$$S_8 = [S_7 \; S_6 \; S_5 \; S_4]\begin{bmatrix} \sigma_1 \\ \sigma_2 \\ \sigma_3 \\ \sigma_4 \end{bmatrix}. \tag{6.55}$$

Now, using the equations from Eqs. (6.48) to (6.55), we can construct the syndrome matrix equation in Eq. (6.44) as

$$\begin{bmatrix} S_1 \\ S_2 \\ S_3 \\ S_4 \\ S_5 \\ S_6 \\ S_7 \\ S_8 \end{bmatrix} = \begin{bmatrix} 1 & 0 & 0 & 0 \\ S_1 & 0 & 0 & 0 \\ S_2 & S_1 & 1 & 0 \\ S_3 & S_2 & S_1 & 0 \\ S_4 & S_3 & S_2 & S_1 \\ S_5 & S_4 & S_3 & S_2 \\ S_6 & S_5 & S_4 & S_3 \\ S_7 & S_6 & S_5 & S_4 \end{bmatrix}\begin{bmatrix} \sigma_1 \\ \sigma_2 \\ \sigma_3 \\ \sigma_4 \end{bmatrix}. \tag{6.56}$$

Eliminating the rows for $S_2$, $S_4$, $S_8$ from Eq. (6.56), we obtain

$$\begin{bmatrix} S_1 \\ S_3 \\ S_5 \\ S_7 \end{bmatrix} = \begin{bmatrix} 1 & 0 & 0 & 0 \\ S_2 & S_1 & 1 & 0 \\ S_4 & S_3 & S_2 & S_1 \\ S_6 & S_5 & S_4 & S_3 \end{bmatrix}\begin{bmatrix} \sigma_1 \\ \sigma_2 \\ \sigma_3 \\ \sigma_4 \end{bmatrix}. \tag{6.57}$$

**Example 6.23** For any double-error-correcting *BCH* code, find the coefficients of error location polynomial in terms of the syndromes.

**Solution 6.23** For $v = t = 2$, the error location polynomial can be written as

$$\sigma(x) = \sigma_0 + \sigma_1 x + \sigma_2 x^2.$$

Using Eq. (6.45), we can write that

$$\begin{bmatrix} S_1 \\ S_3 \end{bmatrix} = \begin{bmatrix} 1 & 0 \\ S_2 & S_1 \end{bmatrix} \begin{bmatrix} \sigma_1 \\ \sigma_2 \end{bmatrix}$$

from which we obtain that

$$\begin{bmatrix} \sigma_1 \\ \sigma_2 \end{bmatrix} = \begin{bmatrix} 1 & 0 \\ S_2 & S_1 \end{bmatrix}^{-1} \begin{bmatrix} S_1 \\ S_3 \end{bmatrix} \rightarrow \begin{bmatrix} \sigma_1 \\ \sigma_2 \end{bmatrix} = \frac{1}{S_1} \begin{bmatrix} S_1 & 0 \\ S_2 & 1 \end{bmatrix} \begin{bmatrix} S_1 \\ S_3 \end{bmatrix} \rightarrow \begin{bmatrix} \sigma_1 \\ \sigma_2 \end{bmatrix}$$
$$= \begin{bmatrix} S_1 \\ \dfrac{S_1 S_2 + S_3}{S_1} \end{bmatrix}.$$

That is,

$$\sigma_0 = 1 \quad \sigma_1 = S_1 \quad \sigma_2 = \frac{S_1 S_2 + S_3}{S_1} \rightarrow \sigma_2 = \frac{S_1^3 + S_3}{S_1}. \tag{6.58}$$

Note that syndromes are known quantities, and they are calculated from the received word polynomial.

**Example 6.24** For any triple-error-correcting *BCH* code, find the coefficients of the error location polynomial in terms of the syndromes.

**Solution 6.24** For $v = t = 3$, the error location polynomial can be written as

$$\sigma(x) = \sigma_0 + \sigma_1 x + \sigma_2 x^2 + \sigma_3 x^3.$$

Using Eq. (6.45), we can write that

$$\begin{bmatrix} S_1 \\ S_3 \\ S_5 \end{bmatrix} = \begin{bmatrix} 1 & 0 & 0 \\ S_2 & S_1 & 1 \\ S_4 & S_3 & S_2 \end{bmatrix} \begin{bmatrix} \sigma_1 \\ \sigma_2 \\ \sigma_3 \end{bmatrix}$$

from which we can write

$$\sigma_1 = S_1 \tag{6.59}$$

and

$$S_3 = \sigma_1 S_2 + \sigma_2 S_1 + \sigma_3 \rightarrow S_3 = S_1 S_2 + \sigma_2 S_1 + \sigma_3$$
$$S_5 = \sigma_1 S_4 + \sigma_2 S_3 + \sigma_3 S_2 \rightarrow S_5 = S_1 S_4 + \sigma_2 S_3 + \sigma_3 S_2$$

from which we obtain

$$\begin{aligned} \sigma_2 S_1 + \sigma_3 &= S_3 + S_1 S_2 \\ \sigma_2 S_3 + \sigma_3 S_2 &= S_5 + S_1 S_4 \end{aligned} \tag{6.60}$$

where, multiplying the first equation by $S_2$ and adding it to the second one, we get the equation involving only $\sigma_2$ as

$$\sigma_2 S_1 S_2 + \sigma_2 S_3 = S_2 S_3 + S_1 S_2^2 + S_5 + S_1 S_4$$

where, using $S_2 = S_1^2$ and $S_4 = S_2^2$, we obtain

$$\sigma_2 \left( S_1^3 + S_3 \right) = S_1^2 S_3 + S_5 + \underbrace{S_1 S_1^4 + S_1 S_1^4}_{=0}$$

from which we get

$$\sigma_2 = \frac{S_1^2 S_3 + S_5}{S_1^3 + S_3}. \tag{6.61}$$

Finally, substituting Eq. (6.61) into Eq. (6.60), and solving for $\sigma_3$, we obtain

$$\sigma_2 S_1 + \sigma_3 = S_3 + S_1 S_2 \rightarrow \sigma_3 = S_1^3 + S_3 + \frac{S_1^3 S_3 + S_1 S_5}{S_1^3 + S_3}.$$

Hence, the coefficients can be written as

$$\sigma_0 = 1 \quad \sigma_1 = S_1 \quad \sigma_2 = \frac{S_1^2 S_3 + S_5}{S_1^3 + S_3} \quad \sigma_3 = S_1^3 + S_3 + S_1 \sigma_2$$

and the error location polynomial can be expressed as

$$\sigma(x) = \sigma_0 + \sigma_1 x + \sigma_2 x^2 + \sigma_3 x^3 \rightarrow$$

$$\sigma(x) = 1 + S_1 x + \frac{S_1^2 S_3 + S_5}{S_1^3 + S_3} x^2 + \frac{S_1^6 + S_3^2 + S_1^3 S_3 + S_1 S_5}{S_1^3 + S_3} x^3.$$

**Exercise** For a triple-error-correcting *BCH* code, the syndromes $S_1$, $S_3$, and $S_5$ over $GF(2^4)$ are given as

$$S_1 = \alpha^8 \quad S_3 = \alpha^7 \quad S_5 = \alpha^{10}.$$

Assume that the primitive polynomial $p(x) = x^4 + x + 1$ is used for the construction of $GF(2^4)$. Find the coefficients of the error location polynomial.

**Example 6.26** For any four-error-correcting *BCH* code, find the coefficients of error location polynomial in terms of the syndromes.

**Solution 6.26** For $v = t = 4$, the error location polynomial can be written as

$$\sigma(x) = \sigma_0 + \sigma_1 x + \sigma_2 x^2 + \sigma_3 x^3 + \sigma_4 x^4.$$

Using Eq. (6.45), we obtain

$$\begin{bmatrix} S_1 \\ S_3 \\ S_5 \\ S_7 \end{bmatrix} = \begin{bmatrix} 1 & 0 & 0 & 0 \\ S_2 & S_1 & 1 & 0 \\ S_4 & S_3 & S_2 & S_1 \\ S_6 & S_5 & S_4 & S_3 \end{bmatrix} \begin{bmatrix} \sigma_1 \\ \sigma_2 \\ \sigma_3 \\ \sigma_4 \end{bmatrix}$$

from which the coefficients $\sigma_1$, $\sigma_2$, $\sigma_3$, and $\sigma_4$ can be found as

$$\sigma_1 = S_1 \qquad \sigma_2 = \frac{S_1\left(S_7 + S_1^7\right) + S_3\left(S_1^5 + S_5\right)}{S_3\left(S_1^3 + S_3\right) + S_1\left(S_1^5 + S_5\right)}$$

$$\sigma_3 = \left(S_1^3 + S_3\right) + S_1\sigma_1 \qquad \sigma_4 = \frac{\left(S_1^2 S_3 + S_5\right) + \left(S_1^3 + S_3\right)\sigma_2}{S_1}.$$

### 6.4.2 *The Peterson-Gorenstein-Zierler (PGZ) Decoder*

In Eq. (6.47), we obtained that

$$S_{v+i} = [S_{v+i-1} \; S_{v+i-2} \ldots S_{i+2} \; S_{i+1} \; S_i] \begin{bmatrix} \sigma_1 \\ \sigma_2 \\ \vdots \\ \sigma_{v-1} \\ \sigma_v \end{bmatrix}. \tag{6.62}$$

Evaluating Eq. (6.62) for $i = 1, \ldots, v$, and using the results as the rows of a matrix, we get

$$\begin{bmatrix} S_{v+1} \\ S_{v+2} \\ \vdots \\ S_{2v} \end{bmatrix} = \begin{bmatrix} S_1 & S_2 & \ldots & S_{v-1} & S_v \\ S_2 & S_3 & \ldots & S_v & S_{v+1} \\ \vdots & \vdots & \ldots & \vdots & \vdots \\ S_v & S_{v+1} & \ldots & S_{2v-2} & S_{2v-1} \end{bmatrix} \begin{bmatrix} \sigma_v \\ \sigma_{v-1} \\ \vdots \\ \sigma_1 \end{bmatrix}. \tag{6.63}$$

Equation (6.63) can be written as

$$\boldsymbol{S} = \boldsymbol{M\sigma}$$

from which we can write that

$$\boldsymbol{\sigma} = \boldsymbol{M}^{-1}\boldsymbol{S}. \tag{6.64}$$

The inverse of $\boldsymbol{M}$ in Eq. (6.64) exists if there are exactly $v$-bit errors in the received word, and we have $v \leq t$ where $t$ is the maximum number of errors that can be corrected by our code.

However, in practice, we cannot know the exact number of bit errors at the receiver side. For this reason, first, we assume that maximum number of errors occurred, i.e., we take $v = t$, and calculate the inverse of $\boldsymbol{M}$ in Eq. (6.64). If the matrix $\boldsymbol{M}$ is a singular one, then we understand that less number of bit errors occurred. In this case, we construct the $\boldsymbol{M}$ matrix in Eq. (6.64) for $v - 1$ and check its determinant. Now, let's state the Peterson's algorithm to determine the coefficients of error location polynomial.

**PGZ Algorithm**

1. Calculate the syndromes $S_1, S_2, \ldots, S_{2t}$ using the received word polynomial $r(x)$.
2. Construct the syndrome matrix in Eq. (6.64) for $v = t$, i.e., size of $\boldsymbol{M}$ is $t \times t$.
3. Compute the determinant of $\boldsymbol{M}$, and if it is nonzero, go to Step 5.
4. Delete the last row and last column of $\boldsymbol{M}$ and go to Step 3.
5. Computer $\boldsymbol{M}^{-1}$ and determine the coefficients of the error location polynomial using $\boldsymbol{\sigma} = \boldsymbol{M}^{-1}\boldsymbol{S}$.
6. Determine the roots of the error location polynomial $\sigma(x)$. If there are an incorrect number of roots or repeated roots, in this case, we declare a decoding failure.

7. Determine the reciprocals of the roots of $\sigma(x)$ and find the error location numbers.
8. Construct the error word polynomial, and add it to the received word polynomial to correct the transmission errors.

**Example 6.27** For the triple-error-correcting $BCH(15, 5)$ code constructed on the extended field $GF(2^4)$ which is obtained using the primitive polynomial $p(x) = x^4 + x + 1$, the generator polynomial can be formed using the minimal polynomials $m_1(x)$, $m_3(x)$, and $m_5(x)$ as

$$g(x) = \left(x^4 + x + 1\right)\left(x^4 + x^3 + x^2 + x + 1\right)\left(x^2 + x + 1\right)$$

which can be simplified as

$$g(x) = x^{10} + x^8 + x^5 + x^4 + x^2 + x + 1.$$

Assume that a data-word is encoded, and the generated code-word is transmitted. The received word at the receiver side is expressed in polynomial form as

$$r(x) = x^8 + x^5 + x^2 + x + 1.$$

Determine the transmitted code-word using the PGZ algorithm.

**Solution 6.27** The syndromes $S_i = r(\alpha^i)$, $i = 1, \ldots, 2t$ can be calculated using the $GF(2^4)$ field elements in Eq. (6.22) as

$$\begin{aligned}
&S_1 = r(\alpha) \rightarrow S_1 = \alpha^2 \\
&S_2 = r\left(\alpha^2\right) = S_1^2 \rightarrow S_2 = \alpha^4 \\
&S_3 = r\left(\alpha^3\right) \rightarrow S_3 = \alpha^{11} \\
&S_4 = r\left(\alpha^4\right) = S_2^2 \rightarrow S_4 = \alpha^8 \\
&S_5 = r\left(\alpha^5\right) \rightarrow S_5 = 0 \\
&S_6 = r\left(\alpha^6\right) = S_3^2 \rightarrow S_6 = \alpha^7.
\end{aligned}$$

The PGZ decoding algorithm can be outlined as follows:

Step 1: Using the calculated syndromes, we construct the syndrome matrix $\boldsymbol{M}$ of size $3 \times 3$ as in

$$\boldsymbol{M} = \begin{bmatrix} S_1 & S_2 & S_3 \\ S_2 & S_3 & S_4 \\ S_3 & S_4 & S_5 \end{bmatrix} \rightarrow \boldsymbol{M} = \begin{bmatrix} \alpha^2 & \alpha^4 & \alpha^{11} \\ \alpha^4 & \alpha^{11} & \alpha^8 \\ \alpha^{11} & \alpha^8 & 0 \end{bmatrix}.$$

Step 2: The determinant of the matrix is evaluated using the first column as

$$|\boldsymbol{M}| = \begin{vmatrix} \alpha^2 & \alpha^4 & \alpha^{11} \\ \alpha^4 & \alpha^{11} & \alpha^8 \\ \alpha^{11} & \alpha^8 & 0 \end{vmatrix} \rightarrow$$

$$|\boldsymbol{M}| = \alpha^2 \begin{vmatrix} \alpha^{11} & \alpha^8 \\ \alpha^8 & 0 \end{vmatrix} + \alpha^4 \begin{vmatrix} \alpha^4 & \alpha^{11} \\ \alpha^8 & 0 \end{vmatrix} + \alpha^{11} \begin{vmatrix} \alpha^4 & \alpha^{11} \\ \alpha^{11} & \alpha^8 \end{vmatrix} \rightarrow$$

$$|\boldsymbol{M}| = 0.$$

Step 3: Since the determinant of the matrix is 0, i.e., the matrix is a singular matrix, i.e., its inverse does not exist, we decrease the row and column size of the matrix by 1. For this purpose, we remove the last row and last column of the matrix, and we get

$$\boldsymbol{M} = \begin{bmatrix} \alpha^2 & \alpha^4 \\ \alpha^4 & \alpha^{11} \end{bmatrix}.$$

Step 4: The determinant of the modified matrix of Step 3 can be calculated as

$$|\boldsymbol{M}| = \begin{vmatrix} \alpha^2 & \alpha^4 \\ \alpha^4 & \alpha^{11} \end{vmatrix} \rightarrow |\boldsymbol{M}| = \alpha^2\alpha^{11} + \alpha^4\alpha^4 \rightarrow |\boldsymbol{M}| = \alpha^3.$$

Step 5: Since the determinant of the matrix in Step 4 is not zero, we can proceed to find the coefficients of error location polynomial as

$$\boldsymbol{\sigma} = \boldsymbol{M}^{-1}\boldsymbol{S} \rightarrow \begin{bmatrix} \sigma_2 \\ \sigma_1 \end{bmatrix} = \begin{bmatrix} \alpha^2 & \alpha^4 \\ \alpha^4 & \alpha^{11} \end{bmatrix}^{-1} \begin{bmatrix} \alpha^{11} \\ \alpha^8 \end{bmatrix} \rightarrow \begin{bmatrix} \sigma_2 \\ \sigma_1 \end{bmatrix} = \begin{bmatrix} \alpha^{14} \\ \alpha^2 \end{bmatrix}.$$

Note that the determinant of the matrix $\boldsymbol{M}$ is calculated using

$$\boldsymbol{M}^{-1} = \frac{adj(\boldsymbol{M})}{|\boldsymbol{M}|}$$

where the element of the adjoint matrix $adj(\boldsymbol{M})$ at location $(i,j)$ is calculated as

$$adj(\boldsymbol{M}_{ij}) = \text{Remove } i\text{th row and } j\text{th colum of } \boldsymbol{M} \text{ and calculate the determinant of remaining matrix.}$$

Step 6: Using the coefficients found in Step 5, the error location polynomial $\sigma(x)$ is formed as

$$\sigma(x) = 1 + \sigma_1 x + \sigma_2 x^2 \rightarrow \sigma(x) = 1 + \alpha^2 x + \alpha^{14} x^2.$$

Step 7: The roots of the error location polynomial in Step 6 can be determined trivially trying all field elements one by one, i.e., equating $\sigma(x)$ to zero, we get

$$\sigma(x) = 0 \rightarrow 1 + \alpha^2 x + \alpha^{14} x^2 = 0.$$

The roots can be found as $\alpha^5$ and $\alpha^{11}$ trivially.

Step 8: Calculate the reciprocals of the roots as

$$X_1 = \frac{1}{\alpha^5} \rightarrow X_1 = \frac{\alpha^{15}}{\alpha^5} \rightarrow X_1 = \alpha^{10}$$

$$X_2 = \frac{1}{\alpha^{11}} \rightarrow X_2 = \frac{\alpha^{15}}{\alpha^{11}} \rightarrow X_2 = \alpha^4.$$

Step 9: Considering the powers of $X_1$ and $X_2$, the error polynomial is formed as

$$e(x) = x^{10} + x^4.$$

Step 10: The transmitted code-word is decided as

$$\hat{c}(x) = r(x) + e(x) \rightarrow \hat{c}(x) = x^{10} + x^8 + x^5 + x^4 + x^2 + x + 1.$$

**Exercise** For the double-error-correcting $BCH(31, 21)$ code, constructed on the extended field $GF(2^5)$ which is obtained using the primitive polynomial $p(x) = x^5 + x^2 + 1$, the generator polynomial can be formed using the minimal polynomials $m_1(x)$ and $m_3(x)$ as

$$g(x) = \left(x^5 + x^2 + 1\right)\left(x^5 + x^4 + x^3 + x^2 + 1\right)$$

which can be simplified as

$$g(x) = x^{10} + x^9 + x^8 + x^6 + x^5 + x^3 + 1.$$

Assume that a data-word is encoded and the generated code-word is transmitted. The received word at the receiver side is expressed in polynomial form as

$$r(x) = x^{12} + x^{11} + x^8 + x^7 + x^2.$$

Determine the transmitted code-word.

## Problems

1. Obtain the generator polynomial of the single-error-correcting *BCH* code. Use $GF(2^3)$ which is constructed using the primitive polynomial $p(x) = x^3 + x^2 + 1$.
2. Obtain the generator polynomial of the double-error-correcting *BCH* code. Use $GF(2^4)$ which is constructed using the primitive polynomial $p(x) = x^4 + x^3 + 1$.
3. Obtain the generator polynomial of the triple-error-correcting *BCH* code. Use $GF(2^5)$ which is constructed using the primitive polynomial $p(x) = x^5 + x^3 + 1$.
4. Find the generator and parity check matrices of the double-error-correcting *BCH* code obtained using $GF(2^4)$. Use $p(x) = x^4 + x^3 + 1$ for the construction of $GF(2^4)$.
5. Find the generator and parity check matrices of the double-error-correcting *BCH* code. Use $GF(2^5)$ constructed using the primitive polynomial $p(x) = x^5 + x^3 + 1$.
6. The extended field $GF(2^4)$ is constructed using the primitive polynomial $p(x) = x^4 + x + 1$. The generator polynomial of the triple-error-correcting *BCH* (15, 5) code over $GF(2^4)$ is evaluated as

$$g(x) = x^{10} + x^8 + x^5 + x^4 + x^2 + x + 1.$$

Find the parity check and generator matrices of the dual code.

7. The generator polynomial of the double-error-correcting $BCH(15, 7)$ cyclic code is given as

$$g(x) = x^8 + x^7 + x^6 + x^4 + 1.$$

The data-word polynomial is $d(x) = x^6 + x^3 + x + 1$. Obtain the systematic and non-systematic code-words for $d(x)$.

8. For any triple-error-correcting *BCH* code, find the coefficients of error location polynomial in terms of the syndromes.
9. The triple-error-correcting $BCH(15, 5)$ code is constructed on the extended field $GF(2^4)$. The extended field $GF(2^4)$ is obtained using the primitive polynomial $p(x) = x^4 + x^3 + 1$.

The data-word polynomial is $d(x) = x^6 + x^3 + x + 1$. Assume that a data-word is encoded and the generated code-word $c(x)$ is transmitted. The received word at the receiver side is expressed in polynomial form as

$$r(x) = c(x) + e(x)$$

where $e(x) = x^4 + x^2$. Using $r(x)$, determine the transmitted code-word using the PGZ algorithm.

# Chapter 7
# Reed-Solomon Codes

## 7.1 Reed-Solomon Codes and Generator Polynomials of Reed-Solomon Codes

For a $t$-error-correcting Reed-Solomon code, the generator polynomial is constructed using

$$g(x) = (x+\beta)(x+\beta^2)\dots(x+\beta^{2t}) \tag{7.1}$$

where $\beta$ is an extended field element, i.e., $\beta \in GF(2^m)$. If the degree of the generator polynomial is $r$, the parameters of the Reed-Solomon code $RS(n,k)$ are calculated as

$$n = 2^m - 1 \quad k = n - r. \tag{7.2}$$

The error correction capability of Reed-Solomon code $RS(n,k)$ is computed using

$$t_c = \left\lfloor \frac{n-k}{2} \right\rfloor. \tag{7.3}$$

The minimum distance of $RS(n,k)$ is $d_{\min} = n - k + 1$. Reed-Solomon codes satisfy the Singleton bound with equality, i.e., we have

$$d_{\min} = n - k + 1. \tag{7.4}$$

For this reason, Reed-Solomon codes are maximum distance separable codes.

**Example 7.1** The extended field $GF(2^3)$ is generated using the primitive polynomial $p(x) = x^3 + x + 1$. Find the generator polynomial of the single-error-correcting Reed-Solomon code over $GF(2^3)$.


O. Gazi, *Forward Error Correction via Channel Coding*,
https://doi.org/10.1007/978-3-030-33380-5_7

**Solution 7.1** Let $\alpha$ be the root of the equation $p(x) = 0$. Choosing $\beta = \alpha$, the generator polynomial of the single-error-correcting Reed-Solomon code can be calculated using

$$g(x) = (x+\alpha)\left(x+\alpha^2\right)\ldots\left(x+\alpha^{2t}\right) \rightarrow g(x) = (x+\alpha)\left(x+\alpha^2\right) \rightarrow$$
$$g(x) = x^2 + \left(\alpha^2+\alpha\right)x + \alpha^3$$

where, using the recursive expression $\alpha^3 = \alpha + 1$, we can write the coefficients as powers of $\alpha$ and obtain the polynomial expression as

$$g(x) = x^2 + \alpha^4 x + \alpha^3.$$

The degree of the generator polynomial is $r = 2$. The parameters of the Reed-Solomon code can be calculated as

$$n = 2^m - 1 \rightarrow n = 2^3 - 1 \rightarrow n = 7$$
$$k = n - r \rightarrow k = 7 - 2 \rightarrow k = 5.$$

Hence, we can refer to the Reed-Solomon code as $RS(7,5)$.

Note that in $GF(2^m)$ the polynomials are represented by $m$-bit binary vectors. In $GF(2^3)$, we can represent the polynomials using 3 bits. The generator polynomial of the $RS(7,5)$ code can be represented by a bit vector as in

$$g(x) = x^2 + \alpha^4 x + \alpha^3 \rightarrow \boldsymbol{g} = [001 \quad 110 \quad 100]. \tag{7.5}$$

Note that in Eq. (7.5), the bits are not concatenated, but they are grouped, and each group includes three bits.

**Example 7.2** Express the data vector $\boldsymbol{d} = [1 \; 0 \; \alpha \; \alpha^5 \; \alpha^2]$ in bit vector form. The elements of the data vector are chosen from $GF(2^3)$ which is constructed using the primitive polynomial $p(x) = x^3 + x + 1$.

**Solution 7.2** Any polynomial in $GF(2^3)$ can be represented using a bit vector consisting of three bits. Accordingly, the data vector given in the question can be represented as

$$\boldsymbol{d} = \left[1\,0\,\alpha\,\alpha^5\,\alpha^2\right] \rightarrow \boldsymbol{d} = \left[1 \quad 0 \quad \alpha \quad \alpha^2+\alpha+1 \quad \alpha^2\right] \rightarrow$$
$$\boldsymbol{d} = [001 \;\; 000 \;\; 010 \;\; 111 \;\; 100].$$

**Example 7.3** Encode the data vector $\boldsymbol{d} = [1 \; 0 \; \alpha \; \alpha^5 \; \alpha^2]$ using the generator polynomial of the Reed-Solomon code $RS(7,5)$.

**Solution 7.3** In the previous example, we calculated the generator polynomial of the Reed-Solomon code as

$$g(x) = x^2 + \alpha^4 x + \alpha^3.$$

The data vector given in the question can be expressed in polynomial form as

$$d(x) = x^4 + \alpha x^2 + \alpha^5 x + \alpha^2.$$

Using the generator polynomial and data-word polynomial, we can compute the code-word polynomial as

$$c(x) = d(x)g(x) \rightarrow c(x) = \left(x^4 + \alpha x^2 + \alpha^5 x + \alpha^2\right)\left(x^2 + \alpha^4 x + \alpha^3\right)$$

which is simplified as

$$c(x) = x^6 + \alpha^6 x^5 + x^4 + \alpha^4 x^2 + \alpha^5 x + \alpha^5. \tag{7.6}$$

The code-word obtained in Eq. (7.6) is in non-systematic form. Since Reed-Solomon codes are nonbinary cyclic codes, to obtain the code-words in systematic form, we can follow the steps explained in Chap. 5 for the systematic encoding of cyclic codes. The systematic code-word for Reed-Solomon code for the given data-word can be obtained as follows.

Step (1) We calculate $x^{n-k}d(x)$ as

$$x^{n-k}d(x) \rightarrow x^2 d(x) = x^2\left(x^4 + \alpha x^2 + \alpha^5 x + \alpha^2\right) \rightarrow$$
$$x^2 d(x) = x^6 + \alpha x^4 + \alpha^5 x^3 + \alpha^2 x^2.$$

Step (2) We divide $x^2 d(x)$ by $g(x)$ and obtain the remainder polynomial as

$$r(x) = R_{g(x)}\left(x^2 d(x)\right) \rightarrow r(x) = x + \alpha^2.$$

Step (3) We form the code-word polynomial as

$$c(x) = x^{n-k}d(x) + r(x) \rightarrow c(x) = x^2 d(x) + r(x)$$

leading to

$$c(x) = x^6 + \alpha x^4 + \alpha^5 x^3 + \alpha^2 x^2 + x + \alpha^2.$$

**Exercise** Construct the generator polynomial of three-error-correcting Reed-Solomon code over $GF(2^4)$ which is obtained using the primitive polynomial $p(x) = x^4 + x + 1$.

## 7.2 Decoding of Reed-Solomon Codes

For binary codes, an error pattern including $v$ errors can be expressed in polynomial form as

$$e(x) = x^{p_1} + x^{p_2} + \ldots + x^{p_v} \tag{7.7}$$

where the exponentials $p_1, p_2, \ldots, p_v$ indicate the error locations. On the other hand, for Reed-Solomon codes, the error polynomial is expressed in the form

$$e(x) = y_{p_1}x^{p_1} + y_{p_2}x^{p_2} + \ldots + y_{p_v}x^{p_v} \tag{7.8}$$

where the exponentials $p_1, p_2, \ldots, p_v$ indicate the error locations and $y_{p_1}, y_{p_2}, \ldots, y_{p_v}$ denote the error magnitudes at the corresponding error locations.

**Example 7.4** The generator polynomial of the double-error-correcting Reed-Solomon code $RS(7,3)$ over $GF(2^3)$, constructed using the primitive polynomial $p(x) = x^3 + x + 1$, can be calculated as

$$g(x) = x^4 + \alpha^3 x^3 + x^2 + \alpha x + \alpha^3.$$

When the data-word polynomial $d(x) = \alpha^5 x^2 + \alpha^3 x + \alpha$ is systematically encoded, we get the code-word

$$c(x) = \alpha^5 x^6 + \alpha^3 x^5 + \alpha x^4 + \alpha^6 x^3 + \alpha^4 x^2 + \alpha^2 x + 1.$$

The elements of the $GF(2^3)$ are given in Eq. (7.9) for the reminder

$$\begin{aligned} 0 &\to 0 \\ 1 &\to 1 \\ \alpha &\to \alpha \\ \alpha^2 &\to \alpha^2 \\ \alpha^3 &\to \alpha + 1 \\ \alpha^4 &\to \alpha^2 + \alpha \\ \alpha^5 &\to \alpha^2 + \alpha + 1 \\ \alpha^6 &\to \alpha^2 + 1. \end{aligned} \tag{7.9}$$

The code-word polynomial can be represented by a vector as

$$\boldsymbol{c} = \begin{bmatrix}\alpha^5 & \alpha^3 & \alpha & \alpha^6 & \alpha^4 & \alpha^2 & 1\end{bmatrix} \rightarrow$$

$$\boldsymbol{c} = \begin{bmatrix}\alpha^2+\alpha+1 & \alpha+1 & \alpha & \alpha^2+1 & \alpha^2+\alpha & \alpha^2 & 1\end{bmatrix}.$$

where, expressing the polynomials by 3-bit words, we get

$$\boldsymbol{c} = [111 \quad 011 \quad 010 \quad 101 \quad 110 \quad 100 \quad 001].$$

Let the error-word polynomial be

$$e(x) = \alpha^3 x^4 + \alpha^6 x^2. \tag{7.10}$$

Equation (7.10) can be expressed by a vector as

$$\boldsymbol{e} = \begin{bmatrix}0 & 0 & \alpha^3 & 0 & \alpha^6 & 0 & 0\end{bmatrix} \rightarrow \boldsymbol{e} = \begin{bmatrix}0 & 0 & \alpha+1 & 0 & \alpha^2+1 & 0 & 0\end{bmatrix}$$

where, expressing the polynomials by 3-bit words, we get

$$\boldsymbol{e} = [000 \quad 000 \quad 011 \quad 000 \quad 101 \quad 000 \quad 000].$$

The received word can be expressed in polynomial form as

$$r(x) = c(x) + e(x) \rightarrow$$

$$r(x) = \underbrace{\alpha^5 x^6 + \alpha^3 x^5 + \alpha x^4 + \alpha^6 x^3 + \alpha^4 x^2 + \alpha^2 x + 1}_{c(x)} + \underbrace{\alpha^3 x^4 + \alpha^6 x^2}_{e(x)}$$

which can be simplified as

$$r(x) = \alpha^5 x^6 + \alpha^3 x^5 + x^4 + \alpha^6 x^3 + \alpha^3 x^2 + \alpha^2 x + 1. \tag{7.11}$$

Equation (7.11) can be represented by a vector as

$$\boldsymbol{r} = \begin{bmatrix}\alpha^5 & \alpha^3 & 1 & \alpha^6 & \alpha^3 & \alpha^2 & 1\end{bmatrix} \rightarrow$$

$$\boldsymbol{r} = \begin{bmatrix}\alpha^2+\alpha+1 & \alpha+1 & 1 & \alpha^2+1 & \alpha+1 & \alpha^2 & 1\end{bmatrix}$$

where, expressing the polynomials by 3-bit words, we get

$$\boldsymbol{r} = [111 \quad 011 \quad 001 \quad 101 \quad 011 \quad 100 \quad 001].$$

If we compare the binary vectors $\boldsymbol{c}$, $\boldsymbol{e}$, and $\boldsymbol{r}$, we see that

$$\boldsymbol{r} = \boldsymbol{c} + \boldsymbol{e} \rightarrow$$

$$[111 \quad 011 \quad \mathbf{001} \quad 101 \quad \mathbf{011} \quad 100 \quad 001] =$$

$$[111 \quad 011 \quad \mathbf{010} \quad 101 \quad \mathbf{110} \quad 100 \quad 001] + [000 \quad 000 \quad \mathbf{011} \quad 000 \quad \mathbf{101} \quad 000 \quad 000]$$

where we see that **001** = **010** + **011** and **011** = **110** + **101**.

If we inspect the error vector

$$\boldsymbol{e} = \left[ \underbrace{000}_{6} \; \underbrace{000}_{5} \; \underbrace{\mathbf{011}}_{\mathbf{4}} \; \underbrace{000}_{3} \; \underbrace{\mathbf{101}}_{\mathbf{2}} \; \underbrace{000}_{1} \; \underbrace{000}_{0} \right]$$

we see that the errors occur at positions **2** and **4**, and the magnitudes of the errors are **101** and **011** which can be expressed using integers as

$$\mathbf{101} \rightarrow \mathbf{5} \quad \mathbf{011} \rightarrow \mathbf{3}.$$

We can express the error-word in polynomial form as in

$$e(x) = (101)x^2 + (011)x^4 \rightarrow e(x) = 5x^2 + 3x^4.$$

The decoding operation of the Reed-Solomon codes involves determination of error locations and error magnitudes. For the determination of error locations, we can use the approach used for BCH codes. Once we determine the error locations, we can proceed to determine the error magnitudes.

### 7.2.1 Syndrome Decoding of Reed-Solomon Codes

The received word polynomial $r(x)$ can be written as

$$r(x) = c(x) + e(x)$$

where $c(x)$ is the code-word polynomial and $e(x)$ is the error-word polynomial defined as

$$e(x) = y_{p_1}x^{p_1} + y_{p_2}x^{p_2} + \ldots + y_{p_v}x^{p_v}. \tag{7.12}$$

The code-word polynomial satisfies $c(\alpha^i) = 0$, $i = 1, \ldots, 2t$. The syndromes for Reed-Solomon codes can be calculated using

$$S_i = r\left(\alpha^i\right) \rightarrow S_i = \underbrace{c\left(\alpha^i\right)}_{=0} + e\left(\alpha^i\right) \rightarrow S_i = e\left(\alpha^i\right)\ i = 1, \ldots, 2t. \tag{7.13}$$

Calculating the syndromes for $i = 1, \ldots, 2t$, we get the set of equations

$$S_1 = y_{p_1}\alpha^{p_1} + y_{p_2}\alpha^{p_2} + \ldots + y_{p_v}\alpha^{p_v}$$
$$S_2 = y_{p_1}\alpha^{2p_1} + y_{p_2}\alpha^{2p_2} + \ldots + y_{p_v}\alpha^{2p_v}$$
$$\vdots$$
$$S_{2t} = y_{p_1}\alpha^{2tp_1} + y_{p_2}\alpha^{2tp_2} + \ldots + y_{p_v}\alpha^{2tp_v}$$

in which, making use of the parameter change $X_i = \alpha^{p_i}$ and $Y_i = y_{p_i}$, we obtain the equation set

$$\begin{aligned} S_1 &= Y_1X_1 + Y_2X_2 + \ldots + Y_vX_v \\ S_2 &= Y_1X_1^2 + Y_2X_2^2 + \ldots + Y_vX_v^2 \\ &\vdots \\ S_{2t} &= Y_1X_1^{2t} + Y_2X_2^{2t} + \ldots Y_vX_v^{2t}. \end{aligned} \tag{7.14}$$

In general, we can express a syndrome as

$$S_m = \sum_{l=1}^{v} Y_lX_l^m. \tag{7.15}$$

The error location polynomial is expressed as

$$\sigma(x) = \prod_{l=1}^{v}(1 + X_lx) \rightarrow \sigma(x) = \sigma_vx^v + \sigma_{v-1}x^{v-1} + \ldots + \sigma_1x + \sigma_0 \tag{7.16}$$

where $\sigma_v = \sigma_0 = 1$. The error location gets the value of 0 when $x = X_l^{-1}$, i.e., we have

$$\sigma\left(X_l^{-1}\right) = 0 \rightarrow$$
$$\sigma_vX_l^{-v} + \sigma_{v-1}X_l^{-v+1} + \ldots + \sigma_1X_l^{-1} + \sigma_0 = 0. \tag{7.17}$$

Multiplying both sides of Eq. (7.17) by $Y_lX_l^m$, we obtain

$$Y_l\left(\sigma_vX_l^{m-v} + \sigma_{v-1}X_l^{m-v+1} + \ldots + \sigma_1X_l^{m-1} + \sigma_0X_l^m\right) = 0. \tag{7.18}$$

Summing Eq. (7.18) over all indices $l$, we obtain

$$\sigma_v \sum_{l=1}^{v} Y_l X_l^{m-v} + \sigma_{v-1} \sum_{l=1}^{v} Y_l X_l^{m-v+1} + \ldots + \sigma_1 \sum_{l=1}^{v} Y_l X_l^{m-1} + \sigma_0 \sum_{l=1}^{v} Y_l X_l^{m} = 0$$

in which using

$$S_m = \sum_{l=1}^{v} Y_l X_l^m \tag{7.19}$$

we obtain

$$\sigma_v S_{m-v} + \sigma_{v-1} S_{m-v+1} + \ldots + \sigma_1 S_{m-1} + \sigma_0 S_m = 0 \tag{7.20}$$

where using $\sigma_0 = 1$, we get

$$\sigma_v S_{m-v} + \sigma_{v-1} S_{m-v+1} + \ldots + \sigma_1 S_{m-1} = S_m. \tag{7.21}$$

Assuming $v = t$, we can express Eq. (7.21) in matrix form as

$$\begin{bmatrix} S_1 & S_2 & \cdots & S_t \\ S_2 & S_3 & \cdots & S_{t+1} \\ \vdots & \vdots & \cdots & \vdots \\ S_{t-1} & S_t & \cdots & S_{2t-2} \\ S_t & S_{t+1} & \cdots & S_{2t-1} \end{bmatrix} \begin{bmatrix} \sigma_t \\ \sigma_{t-1} \\ \vdots \\ \sigma_2 \\ \sigma_1 \end{bmatrix} = \begin{bmatrix} S_{t+1} \\ S_{t+2} \\ \vdots \\ S_{2t-1} \\ S_{2t} \end{bmatrix}. \tag{7.22}$$

Using Eq. (7.22), we determine the coefficients of the error location polynomial $\sigma(x)$. The reciprocals of the roots of $\sigma(x)$ carry information about the error locations. The equation set in Eq. (7.14) can be written as

$$\begin{bmatrix} X_1 & X_2 & \cdots & X_v \\ X_1^2 & X_2^2 & \cdots & X_v^2 \\ \vdots & \vdots & \cdots & \vdots \\ X_1^{2t} & X_2^{2t} & \cdots & X_v^{2t} \end{bmatrix} \begin{bmatrix} Y_1 \\ Y_2 \\ \vdots \\ Y_v \end{bmatrix} = \begin{bmatrix} S_1 \\ S_2 \\ \vdots \\ S_{2t} \end{bmatrix} \tag{7.23}$$

which can be expressed in short as

$$\boldsymbol{XY} = \boldsymbol{S}$$

from which the error magnitude vector can be obtained via

$$Y = X^{-1}S. \tag{7.24}$$

### 7.2.2 The Error Evaluator Polynomial

The syndrome polynomial $S(x)$ is defined as

$$S(x) = S_1 + S_2x + S_3x^2 + \ldots + S_{2t}x^{2t-1}. \tag{7.25}$$

We define the error evaluator polynomial as

$$B(x) = R_{x^{2t}}\{S(x)\sigma(x)\} \tag{7.26}$$

which is the remainder polynomial obtained from the division of $S(x)\sigma(x)$ by $x^{2t}$. $B(x)$ is a polynomial in the form

$$B(x) = b_1 + b_2x + b_3x^2 + \ldots + b_{v+1}x^v \tag{7.27}$$

where the polynomial coefficients are calculated according to

$$\begin{aligned} b_1 &= 1 \\ b_2 &= S_1 + \sigma_1 \\ b_3 &= S_2 + S_1\sigma_1 + \sigma_2 \\ &\vdots \\ b_v &= S_{v-1} + S_{v-2}\sigma_1 + \ldots + \sigma_{v-1}. \end{aligned} \tag{7.28}$$

In general the coefficient $b_i$ can be calculated from the product of two vectors as in

$$b_i = [1\ S_1 \ldots S_{i-1}] \begin{bmatrix} \sigma_{i-1} \\ \sigma_{i-2} \\ \vdots \\ \sigma_1 \\ 1 \end{bmatrix}.$$

The error magnitudes can be calculated using the error evaluator polynomial $A(x)$ and derivative of the error location polynomial as

$$Y_l = \frac{B\left(X_l^{-1}\right)}{\sigma'\left(X_l^{-1}\right)} \tag{7.29}$$

where the derivative polynomial $\sigma'\left(X_l^{-1}\right)$ is evaluated using

$$\sigma'\left(X_l^{-1}\right) = X_l \prod_{i=1,\, i \neq l}^{v} \left(1 + X_i X_l^{-1}\right). \tag{7.30}$$

**Example 7.5** Using the elements of $GF(2^4)$, the generator polynomial of the triple-error-correcting Reed-Solomon code, $RS(15,9)$, is obtained. For the construction of $GF(2^4)$, the primitive polynomial $p(x) = x^4 + x + 1$ is used. Assume that a data-word is encoded and transmitted. The received word is given as

$$r(x) = \alpha^3 x^{12} + x^8 + \alpha^{10} x^7 + \alpha^2 x^5 + \alpha^8 x^4 + \alpha^{14} x^3 + \alpha^6.$$

The syndromes for the received word are calculated using $S_i = r(\alpha^i)$, $i = 1, \ldots, 6$ as

$$S_1 = \alpha^6 \quad S_2 = 0 \quad S_3 = \alpha^{14} \quad S_4 = \alpha^{11} \quad S_5 = \alpha^{14} \quad S_6 = \alpha^9.$$

Using PGZ method, we can calculate the error location polynomial as

$$\sigma(x) = 1 + x + \alpha^{11} x^2 + \alpha^4 x^3$$

from which the error location numbers can be calculated as

$$X_1 = \alpha^{12} \quad X_2 = \alpha^5 \quad X_3 = \alpha.$$

For $v = 3$, the error evaluator polynomial can be written as

$$A(x) = b_1 + b_2 x + b_3 x^2 + b_4 x^3$$

for which the coefficients $b_1$, $b_2$, $b_3$, and $b_4$ are calculated as

$$b_1 = 1 \rightarrow b_0 = 1$$

$$b_2 = S_1 + \sigma_1 \rightarrow \sigma_2 = \alpha^6 + 1 \rightarrow \sigma_2 = \alpha^3 + \alpha^2 + 1 \rightarrow b_2 = \alpha^{13}$$

$$b_3 = S_2 + S_1\sigma_1 + \sigma_2 \rightarrow b_3 = 0 + \alpha^6 + \alpha^{11} \rightarrow b_3 = \alpha$$

$$b_4 = S_3 + S_2\sigma_1 + S_1\sigma_2 + \sigma_3 \rightarrow b_4 = \alpha^{14} + 0 + \alpha^6\alpha^{11} + \alpha^4 \rightarrow b_4 = \alpha^{11}.$$

Using the calculated coefficients, we can write the error evaluator polynomial as in

$$B(x) = 1 + \alpha^{13}x + \alpha x^2 + \alpha^{11}x^3.$$

We can calculate the error magnitudes using

$$Y_l = \frac{B\left(X_l^{-1}\right)}{\sigma'\left(X_l^{-1}\right)}$$

where

$$\sigma'\left(X_l^{-1}\right) = X_l \prod_{i=1,\, i \neq l}^{v} \left(1 + X_i X_l^{-1}\right)$$

as in

$$Y_1 = \frac{B\left(X_1^{-1}\right)}{\sigma'\left(X_1^{-1}\right)} \quad Y_2 = \frac{B\left(X_2^{-1}\right)}{\sigma'\left(X_2^{-1}\right)} \quad Y_3 = \frac{B\left(X_3^{-1}\right)}{\sigma'\left(X_3^{-1}\right)}$$

which can be written as

$$Y_1 = \frac{B\left(X_1^{-1}\right)}{X_1\left(1 + X_1^{-1}X_2\right)\left(1 + X_1^{-1}X_3\right)} \quad Y_2 = \frac{B\left(X_2^{-1}\right)}{X_2\left(1 + X_2^{-1}X_1\right)\left(1 + X_2^{-1}X_3\right)}$$

$$Y_3 = \frac{B\left(X_3^{-1}\right)}{X_3\left(1 + X_3^{-1}X_1\right)\left(1 + X_3^{-1}X_2\right)}$$

leading to

$$Y_1 = \alpha^3 \quad Y_2 = \alpha^3 \quad Y_3 = \alpha^9.$$

Thus the error polynomial can be written as

$$e(x) = \alpha^3 x^{12} + \alpha^3 x^5 + \alpha^9 x^1.$$

### *7.2.3 Berlekamp Algorithm*

The PGZ algorithm can be used to find the error locations for BCH and RS codes. The PGZ algorithm involves inverse matrix calculation. Hence, as the error correction capability of a code increases, the use of PGZ algorithm becomes infeasible due to heavy computational requirement needed for the calculation of the matrix inverse. For this reason, a less computationally burdensome algorithm called Berlekamp algorithm is proposed in the literature. The Berlekamp algorithm determines the coefficients of the error location polynomial $\sigma(x)$ in an iterative manner. Whenever a set of coefficients are calculated using the Berlekamp algorithm, they are checked employing them for the calculation of a syndrome. If the check fails, in the next iteration, a correction factor is used for the calculation of coefficients of the error location polynomial. Now let's provide the Berlekamp algorithm.

**Berlekamp Algorithm**

Before giving the algorithm, let's introduce some definitions. The error location polynomial for the $i^{th}$ iteration is given as

$$\sigma_i(x) = 1 + \sigma_1^i x + \sigma_2^i x^2 + \ldots + \sigma_{r_i}^i x^{r_i} \tag{7.31}$$

where $r_i$ is the degree of the polynomial $\sigma_i(x)$. To check whether the coefficients are correct or not, we use Newton's equation

$$\begin{bmatrix} S_{v+1} \\ S_{v+2} \\ \vdots \\ S_{2v} \end{bmatrix} = \begin{bmatrix} S_v & S_{v-1} & \cdots & S_1 \\ S_{v+1} & S_v & \cdots & S_2 \\ \vdots & \vdots & \vdots & \vdots \\ S_{2v-1} & S_{2v-2} & \cdots & S_v \end{bmatrix} \begin{bmatrix} \sigma_1 \\ \sigma_2 \\ \vdots \\ \sigma_v \end{bmatrix} \tag{7.32}$$

from which we can write

$$\widehat{S}_{i+1} = \sigma_1^i S_i + \sigma_1^i S_{i-1} + \cdots + \sigma_{r_i}^i S_{i+1-r_i} \tag{7.33}$$

where $\widehat{S}_{i+1}$ is the estimated syndrome obtained using the estimated coefficients $\sigma_1^i, \sigma_2^i, \ldots, \sigma_{r_i}^i$.

Let's define the discrepancy factor as

$$d_i = \widehat{S}_{i+1} + S_{i+1}. \tag{7.34}$$

If the estimated coefficients are correct, then the estimated syndrome $\widehat{S}_{i+1}$ becomes equal to the calculated syndrome $S_{i+1}$, and in this case, we get

$$d_i = \widehat{S}_{i+1} + S_{i+1} \rightarrow d_i = 0.$$

Otherwise, we have

$$d_i = \widehat{S}_{i+1} + S_{i+1} \rightarrow d_i \neq 0.$$

Let's define the parameter $n_k$ as

$$n_k = k - r_k \tag{7.35}$$

where $r_k$ is the degree of $\sigma_k(x)$ and $k$ is the iteration index.

Now, let's explain the Berlekamp algorithm in steps.

1. The iteration index $i$ is initialized to $-1$, i.e., $i = -1$, and for $i = -1$, we have

$$\sigma_{-1}(x) = 1 \quad r_{-1} = 0 \quad d_{-1} = 0 \quad n_{-1} = -1$$

and for $i = 0$, we have

$$\sigma_0(x) = 1 \;\; r_0 = 0 \;\; d_0 = S_1 \quad n_0 = 0.$$

2. The iterations are performed for $i = 1, \ldots, 2t$.
3. For iteration index $i$, calculate the estimated syndrome using

$$\widehat{S}_{i+1} = \sigma'_1 S_i + \sigma'_2 S_{i-1} + \cdots + \sigma'_{r_i} S_{i+1-r_i} \tag{7.36}$$

where $r_i$ is the degree of $\sigma_i(x)$. Using the estimated syndrome and calculated syndrome, compute the discrepancy factor as

$$d_i = \widehat{S}_{i+1} + S_{i+1}. \tag{7.37}$$

4. If $d_i = 0$, then we have

$$\sigma_{i+1}(x) = \sigma_i(x) \qquad r_{i+1} = r_i \tag{7.38}$$

where $r_{i+1}$ denotes the degree of $\sigma_{i+1}(x)$ and $r_i$ is the degree of $\sigma_i(x)$.

If $d_i \neq 0$, then we have

$$\sigma_{i+1}(x) = \sigma_i(x) + e_i(x) \tag{7.39}$$

where $e(x)$ is calculated as

$$e_i(x) = \frac{x^i d_i}{x^k d_k} \sigma_k(x) \qquad k < i \tag{7.40}$$

in which $\sigma_k(x)$ denotes one of the previous polynomials such that

$$n_k = k - r_k \qquad k < i \tag{7.41}$$

has the largest value considering all the previously generated $\sigma_k(x)$. The degree of $e(x)$ can be calculated as

$$p_k = i - k + r_k \rightarrow p_k = i - n_k \tag{7.42}$$

where $r_k$ is the degree of $\sigma_k(x)$. The degree of $\sigma_{i+1}(x)$ can be calculated as

$$r_{i+1} = \max(r_k, p_k). \tag{7.43}$$

5. When $i > 2t$, terminate the loop.

**Example 7.6** The triple-error-correcting $BCH(15,5)$ code is designed in $GF(2^4)$ which is constructed using the primitive polynomial $p(x) = x^4 + x + 1$. The generator polynomial of BCH(15,5) can be calculated as

$$g(x) = x^{10} + x^8 + x^5 + x^4 + x^2 + x + 1.$$

Let the data-word be given as

$$\boldsymbol{m} = [10110].$$

The polynomial form of data-word can be formed as

$$m(x) = x^4 + x^2 + x.$$

If we encode the data-word, we get the code-word polynomial as

$$c(x) = m(x)g(x) \rightarrow c(x) = \left(x^4 + x^2 + x\right)\left(x^{10} + x^8 + x^5 + x^4 + x^2 + x + 1\right) \rightarrow$$
$$c(x) = x^{14} + x^9 + x^7 + x^4 + x^3 + x + 1.$$

Assume that $c(x)$ is transmitted, and during transmission, some bit errors occur. The error pattern is represented in polynomial form as

$$e(x) = x^{13} + x^5 + x^2.$$

The received word polynomial happens to be as in

$$r(x) = c(x) + e(x) \rightarrow r(x) = x^{14} + x^{13} + x^9 + x^7 + x^5 + x^4 + x^3 + x^2 + x + 1.$$

Using $r(x)$,calculate the error location polynomial using Berlekamp algorithm, and using the error location polynomial, determine the error pattern, and find the transmitted code-word.

**Solution 7.6** The syndromes for the received word polynomial can be calculated using

$$S_i = r(\alpha^i) \qquad i = 1, 2, 3, 4, 5, 6$$

as

$$S_1 = \alpha^{12} \quad S_2 = \alpha^9 \quad S_3 = \alpha^{10} \quad S_4 = \alpha^3 \quad S_5 = \alpha^5 \quad S_6 = \alpha^5.$$

Using the initial conditions for the Berlekamp algorithm, we form the decoding table as in Table 7.1.

For $\boldsymbol{i} = \mathbf{0}$, we calculate $\sigma_{i+1}(x)$ as follows:

Since $d_0 = S_1 \neq 0$, then $\sigma_1(x)$ is calculated using the formula

$$\sigma_{i+1}(x) = \sigma_i(x) + e_i(x) \qquad e_i(x) = \frac{x^i d_i}{x^k d_k} \sigma_k(x) \qquad k < i$$

as

$$\sigma_1(x) = \sigma_0(x) + e_0(x) \qquad e_0(x) = \frac{x^0 d_0}{x^k d_k} \sigma_k(x) \;\; k < 0 \tag{7.44}$$

where we have only one choice for $k$, and it is $-1$, i.e., $k = -1$. Then, Eq. (7.44) is calculated as

$$\sigma_0(x) = 1 \quad e_0(x) = \frac{x^0 d_0}{x^{-1} d_{-1}} \sigma_{-1}(x) \rightarrow e_0(x) = \frac{\overbrace{x^0}^{=1} \overbrace{d_0}^{=S_1}}{x^{-1} \underbrace{d_{-1}}_{=1}} \underbrace{\sigma_{-1}(x)}_{=1} \rightarrow e_0(x) = \underbrace{S_1}_{\alpha^{12}} x$$

$\rightarrow$

$$\sigma_1(x) = \sigma_0(x) + e_0(x) \rightarrow \sigma_1(x) = 1 + \alpha^{12} x. \tag{7.45}$$

For $\boldsymbol{i} = \mathbf{1}$, first, we calculate $r_1$ and $n_1$ using $\sigma_1(x)$ found in the previous step as

$$r_1 = \text{degree of } \sigma_1(x) \rightarrow r_1 = 1 \quad n_1 = 1 - r_1 \rightarrow n_1 = 1 - 1 \rightarrow n_1 = 0.$$

**Table 7.1** Calculation of $\sigma(x)$ using Berlekamp algorithm

| $i$ | $\sigma_i(x)$ | $r_i$: degree of $\sigma_i(x)$ | $n_i = i - r_i$ | $d_i = \widehat{S}_{i+1} + S_{i+1}$ |
|---|---|---|---|---|
| $-1$ | $\sigma_{-1}(x) = 1$ | $r_{-1} = 0$ | $n_{-1} = -1$ | $d_{-1} = 1$ |
| 0 | $\sigma_0(x) = 1$ | $r_0 = 0$ | $n_0 = 0$ | $d_0 = S_1$ |

Using the error location polynomial in Eq. (7.45), we calculate the estimated syndrome $\widehat{S}_2$ using the formula

$$\widehat{S}_{i+1} = \sigma'_1 S_i + \sigma'_2 S_{i-1} + \cdots + \sigma'_{r_i} S_{i+1-r_i}$$

for $i = 1$ as

$$\widehat{S}_2 = \sigma'_1 S_1 \rightarrow \widehat{S}_2 = \alpha^{12} S_1 \rightarrow \widehat{S}_2 = \alpha^{12}\alpha^{12} \rightarrow \widehat{S}_2 = \alpha^9. \quad (7.46)$$

The discrepancy factor $d_1$ is calculated as

$$d_1 = S_2 + \widehat{S}_2$$

leading to

$$d_1 = \alpha^9 + \alpha^9 \rightarrow d_1 = 0.$$

Since $d_1 = 0$, we can write that

$$\sigma_2(x) = \sigma_1(x) \rightarrow \sigma_2(x) = 1 + \alpha^{12}x. \quad (7.47)$$

The coefficients of the polynomial in Eq. (7.47) can be written as

$$\sigma'_0 = 1 \qquad \sigma'_1 = \alpha^{12} \qquad \sigma'_2 = 0.$$

With the calculated values, Table 7.1 can be updated as Table 7.2.

***Note*** The estimated error location polynomial $\sigma_i(x)$ has the general form

$$\sigma_i(x) = \sigma'_0 + \sigma'_1 x^1 + \sigma'_2 x^2 + \ldots$$

For $\boldsymbol{i} = \mathbf{2}$, first, we calculate $r_2$ and $n_2$ using $\sigma_2(x)$ found in the previous step as

$$r_2 = \text{degree of } \sigma_2(x) \rightarrow r_2 = 1 \quad n_2 = 2 - r_2 \rightarrow n_2 = 2 - 1 \rightarrow n_2 = 1.$$

Using the error location polynomial $\sigma_2(x)$ in Eq. (7.47), we calculate the estimated syndrome $\widehat{S}_3$ using the formula

$$\widehat{S}_{i+1} = \sigma'_1 S_i + \sigma'_2 S_{i-1} + \cdots + \sigma'_{r_i} S_{i+1-r_i}$$

for $i = 2$ as

**Table 7.2** Calculation of $\sigma(x)$ using Berlekamp algorithm

| $i$ | $\sigma_i(x)$ | $r_i$: degree of $\sigma_i(x)$ | $n_i = i - r_i$ | $d_i = \widehat{S}_{i+1} + S_{i+1}$ |
|---|---|---|---|---|
| −1 | $\sigma_{-1}(x) = 1$ | $r_{-1} = 0$ | $n_{-1} = -1$ | $d_{-1} = 1$ |
| 0 | $\sigma_0(x) = 1$ | $r_0 = 0$ | $n_0 = 0$ | $d_0 = S_1$ |
| 1 | $\sigma_1(x) = 1 + \alpha^{12}x$ | $r_1 = 1$ | $n_1 = 0$ | $d_1 = 0$ |

$$\widehat{S}_3 = \sigma_1' S_2 + \sigma_2' S_1 \rightarrow \widehat{S}_3 = \alpha^{12} S_2 + 0 S_1 \rightarrow \widehat{S}_3 = \alpha^{12}\alpha^9 \rightarrow \widehat{S}_3 = \alpha^6. \tag{7.48}$$

The discrepancy factor $d_2$ is calculated as

$$d_2 = S_3 + \widehat{S}_3$$

leading to

$$d_2 = \alpha^{10} + \alpha^7 \rightarrow d_2 = \alpha^7 \rightarrow d_2 \neq 0.$$

Since $d_2 \neq 0$, then $\sigma_3(x)$ is calculated using the formula

$$\sigma_{i+1}(x) = \sigma_i(x) + e_i(x) \qquad e_i(x) = \frac{x^i d_i}{x^k d_k}\sigma_k(x) \qquad k < i$$

as

$$\sigma_3(x) = \sigma_2(x) + e_2(x) \qquad e_2(x) = \frac{x^2 d_2}{x^k d_k}\sigma_k(x) \;\; k < 2. \tag{7.49}$$

To determine the value of $k$ in Eq. (7.49), we inspect Table 7.2 and choose $\sigma_k(x)$ such that $n_k$ is maximum considering all the previously generated $\sigma_k(x)$ polynomials. In Table 7.2, we have

$$n_{-1} = -1 \quad n_0 = 0 \quad n_1 = 0$$

from which we can conclude that $k$ can be chosen as 0 or 1. For $k = 1$, we have $d_1 = 0$, and this makes the denominator of $e_2(x)$ in Eq. (7.49) zero. For this reason, we choose $k = 0$, and Eq. (7.49) is calculated as

$$\sigma_2(x) = 1 + \alpha^{12}x \quad e_2(x) = \frac{x^2 d_2}{x^k d_k}\sigma_k(x) \overset{k=0}{\rightarrow} e_2(x) = \frac{x^2 \underbrace{d_2}_{\alpha^7}}{x^0 \underbrace{d_0}_{\alpha^{12}}}\underbrace{\sigma_0(x)}_{=1} \rightarrow$$

$$e_2(x) = \alpha^{10}x^2$$

$$\sigma_3(x) = \sigma_2(x) + e_2(x) \rightarrow \sigma_3(x) = 1 + \alpha^{12}x + \alpha^{10}x^2. \tag{7.50}$$

The coefficients of the polynomial in Eq. (7.50) can be written as

$$\sigma'_0 = 1 \quad \sigma'_1 = \alpha^{12} \quad \sigma'_2 = \alpha^{10} \quad \sigma'_3 = 0.$$

With the calculated values, Table 7.2 can be updated as Table 7.3.

For $\boldsymbol{i} = \mathbf{3}$, first, we calculate $r_3$ and $n_3$ using $\sigma_3(x)$ found in the previous step as

$$r_3 = \text{degree of } \sigma_3(x) \rightarrow r_3 = 2 \quad n_3 = 3 - r_3 \rightarrow n_3 = 3 - 2 \rightarrow n_3 = 1.$$

Using the error location polynomial $\sigma_3(x)$ in Eq. (7.50), we calculate the estimated syndrome $\widehat{S}_4$ using the formula

$$\widehat{S}_{i+1} = \sigma'_1 S_i + \sigma'_2 S_{i-1} + \cdots + \sigma'_{r_i} S_{i+1-r_i}$$

for $i = 3$ as

$$\widehat{S}_4 = \sigma'_1 S_3 + \sigma'_2 S_2 + \sigma'_3 S_1 \rightarrow \widehat{S}_4 = \alpha^{12} S_3 + \alpha^{10} S_2 + 0S_1 \rightarrow \widehat{S}_4 = \alpha^{12}\alpha^{10} + \alpha^{10}\alpha^9 \rightarrow$$

$$\widehat{S}_4 = \alpha^7 + \alpha^4 \rightarrow \widehat{S}_3 = \alpha^3. \tag{7.51}$$

The discrepancy factor $d_3$ is calculated as

$$d_3 = S_4 + \widehat{S}_4 \rightarrow d_3 = \alpha^3 + \alpha^3 \rightarrow d_3 = 0.$$

Since $d_3 = 0$, then $\sigma_4(x)$ is calculated using the formula

$$\sigma_{i+1}(x) = \sigma_i(x)$$

as

**Table 7.3** The calculation of $\sigma(x)$ using Berlekamp algorithm

| $i$ | $\sigma_i(x)$ | $r_i$: degree of $\sigma_i(x)$ | $n_i = i - r_i$ | $d_i = \widehat{S}_{i+1} + S_{i+1}$ |
|---|---|---|---|---|
| $-1$ | $\sigma_{-1}(x) = 1$ | $r_{-1} = 0$ | $n_{-1} = -1$ | $d_{-1} = 1$ |
| 0 | $\sigma_0(x) = 1$ | $r_0 = 0$ | $n_0 = 0$ | $d_0 = S_1$ |
| 1 | $\sigma_1(x) = 1 + \alpha^{12}x$ | $r_1 = 1$ | $n_1 = 0$ | $d_1 = 0$ |
| 2 | $\sigma_2(x) = 1 + \alpha^{12}x$ | $r_2 = 1$ | $n_2 = 1$ | $d_2 = \alpha^7$ |

$$\sigma_4(x) = \sigma_3(x) \rightarrow \sigma_4(x) = 1 + \alpha^{12}x + \alpha^{10}x^2. \tag{7.52}$$

The coefficients of the polynomial in Eq. (7.52) can be written as

$$\sigma_0' = 1 \quad \sigma_1' = \alpha^{12} \quad \sigma_2' = \alpha^{10} \quad \sigma_3' = \sigma_4' = 0.$$

With the calculated values, Table 7.3 can be updated as Table 7.4.

For $\boldsymbol{i} = \mathbf{4}$, first, we calculate $r_4$ and $n_4$ using $\sigma_4(x)$ found in the previous step as

$$r_4 = \text{degree of } \sigma_4(x) \rightarrow r_4 = 2 \quad n_4 = 4 - r_4 \rightarrow n_4 = 4 - 2 \rightarrow n_4 = 2.$$

Using the error location polynomial $\sigma_4(x)$ in Eq. (7.52), we calculate the estimated syndrome $\widehat{S}_5$ using the formula

$$\widehat{S}_{i+1} = \sigma_1' S_i + \sigma_2' S_{i-1} + \cdots + \sigma_{r_i}' S_{i+1-r_i}$$

for $i = 4$ as

$$\widehat{S}_5 = \sigma_1' S_4 + \sigma_2' S_3 + \sigma_3' S_2 + \sigma_4' S_1 \rightarrow \widehat{S}_5 = \alpha^{12} S_4 + \alpha^{10} S_3 + 0 S_2 + 0 S_1 \rightarrow$$
$$\widehat{S}_5 = \alpha^{12}\alpha^3 + \alpha^{10}\alpha^{10} \rightarrow \widehat{S}_5 = 1 + \alpha^5 \rightarrow \widehat{S}_5 = \alpha^{10}.$$

The discrepancy factor $d_4$ is calculated as

$$d_4 = S_5 + \widehat{S}_5 \rightarrow d_4 = \alpha^5 + \alpha^{10} \rightarrow d_4 = 1.$$

Since $d_4 \neq 0$, then $\sigma_5(x)$ is calculated using the formula

$$\sigma_{i+1}(x) = \sigma_i(x) + e_i(x) \quad e_i(x) = \frac{x^i d_i}{x^k d_k} \sigma_k(x) \quad k < i$$

as

**Table 7.4** Calculation of $\sigma(x)$ using Berlekamp algorithm

| $i$ | $\sigma_i(x)$ | $r_i$: degree of $\sigma_i(x)$ | $n_i = i - r_i$ | $d_i = \widehat{S}_{i+1} + S_{i+1}$ |
|---|---|---|---|---|
| $-1$ | $\sigma_{-1}(x) = 1$ | $r_{-1} = 0$ | $n_{-1} = -1$ | $d_{-1} = 1$ |
| 0 | $\sigma_0(x) = 1$ | $r_0 = 0$ | $n_0 = 0$ | $d_0 = S_1$ |
| 1 | $\sigma_1(x) = 1 + \alpha^{12}x$ | $r_1 = 1$ | $n_1 = 0$ | $d_1 = 0$ |
| 2 | $\sigma_2(x) = 1 + \alpha^{12}x$ | $r_2 = 1$ | $n_2 = 1$ | $d_2 = \alpha^7$ |
| 3 | $\sigma_3(x) = 1 + \alpha^{12}x + \alpha^{10}x^2$ | $r_3 = 2$ | $n_3 = 1$ | $d_3 = 0$ |

$$\sigma_5(x) = \sigma_4(x) + e_4(x) \qquad e_4(x) = \frac{x^4 d_4}{x^k d_k}\sigma_k(x) \;\; k < 4. \tag{7.53}$$

To determine the value of $k$ in Eq. (7.53), we inspect Table 7.4 and choose $\sigma_k(x)$ such that $n_k$ is maximum considering all the previously generated $\sigma_k(x)$ polynomials. In Table 7.4, we have

$$n_{-1} = -1 \quad n_0 = 0 \quad n_1 = 0 \quad n_2 = 1 \quad n_3 = 0$$

from which we can conclude that $k$ should be chosen as 2; note that $k = \{l \,|\, \max(n_l)\}$. For $k = 2$, we calculate Eq. (7.53) as

$$\sigma_4(x) = 1 + \alpha^{12}x + \alpha^{10}x^2 \quad e_4(x) = \frac{x^4 d_4}{x^k d_k}\sigma_k(x) \overset{k=2}{\rightarrow} e_4(x) = \frac{\overbrace{x^4 \; \underbrace{1}_{\alpha^7}}}{\underbrace{x^2 \; d_2}_{\alpha^7}} \underbrace{\sigma_2(x)}_{=1+\alpha^{12}x} \rightarrow$$

$$e_4(x) = x^2\alpha^8\left(1 + \alpha^{12}x\right) \rightarrow e_4(x) = \alpha^8x^2 + \alpha^5x^3$$

$$\sigma_5(x) = \sigma_4(x) + e_4(x) \rightarrow \sigma_5(x) = 1 + \alpha^{12}x + \alpha^{10}x^2 + \alpha^8x^2 + \alpha^5x^3 \rightarrow$$

$$\sigma_5(x) = 1 + \alpha^{12}x + \left(\alpha^{10} + \alpha^8\right)x^2 + \alpha^5x^3 \rightarrow \sigma_5(x) = 1 + \alpha^{12}x + \alpha x^2 + \alpha^5x^3. \tag{7.54}$$

The coefficients of the polynomial in Eq. (7.54) can be written as

$$\sigma_0' = 1 \qquad \sigma_1' = \alpha^{12} \qquad \sigma_2' = \alpha \qquad \sigma_3' = \alpha^5 \qquad \sigma_4' = \sigma_5' = 0.$$

With the calculated values, Table 7.4 can be updated as Table 7.5.
For $\boldsymbol{i} = \mathbf{5}$, first, we calculate $r_5$ and $n_5$ using $\sigma_5(x)$ found in the previous step as

$$r_5 = \text{degree of } \sigma_5(x) \rightarrow r_5 = 3 \quad n_5 = 5 - r_5 \rightarrow n_5 = 5 - 3 \rightarrow n_5 = 2.$$

Using the error location polynomial $\sigma_5(x)$ in Eq. (7.54), we calculate the estimated syndrome $\widehat{S}_6$ using the formula

$$\widehat{S}_{i+1} = \sigma_1'S_i + \sigma_2'S_{i-1} + \cdots + \sigma_{r_i}'S_{i+1-r_i}$$

for $i = 5$ as

$$\widehat{S}_6 = \sigma_1'S_5 + \sigma_2'S_4 + \sigma_3'S_3 + \sigma_4'S_2 + \sigma_5'S_1 \rightarrow \widehat{S}_6 = \alpha^{12}S_5 + \alpha S_4 + \alpha^5S_3 + 0S_2 + 0S_2$$

**Table 7.5** Calculation of $\sigma(x)$ using Berlekamp algorithm

| $i$ | $\sigma_i(x)$ | $r_i$: degree of $\sigma_i(x)$ | $n_i = i - r_i$ | $d_i = \widehat{S}_{i+1} + S_{i+1}$ |
|---|---|---|---|---|
| $-1$ | $\sigma_{-1}(x) = 1$ | $r_{-1} = 0$ | $n_{-1} = -1$ | $d_{-1} = 1$ |
| 0 | $\sigma_0(x) = 1$ | $r_0 = 0$ | $n_0 = 0$ | $d_0 = S_1$ |
| 1 | $\sigma_1(x) = 1 + \alpha^{12}x$ | $r_1 = 1$ | $n_1 = 0$ | $d_1 = 0$ |
| 2 | $\sigma_2(x) = 1 + \alpha^{12}x$ | $r_2 = 1$ | $n_2 = 1$ | $d_2 = \alpha^7$ |
| 3 | $\sigma_3(x) = 1 + \alpha^{12}x + \alpha^{10}x^2$ | $r_3 = 2$ | $n_3 = 1$ | $d_3 = 0$ |
| 4 | $\sigma_4(x) = 1 + \alpha^{12}x + \alpha^{10}x^2$ | $r_4 = 2$ | $n_4 = 2$ | $d_4 = 1$ |

$$\rightarrow \widehat{S}_6 = \alpha^{12}\alpha^5 + \alpha\alpha^3 + \alpha^5\alpha^{10} \rightarrow \widehat{S}_6 = \alpha^2 + \alpha^4 + 1 \rightarrow \widehat{S}_6 = \alpha^5.$$

The discrepancy factor $d_5$ is calculated as

$$d_5 = S_6 + \widehat{S}_6 \rightarrow d_5 = \alpha^5 + \alpha^5 \rightarrow d_5 = 0.$$

Since $d_5 = 0$, then $\sigma_6(x)$ is calculated using the formula

$$\sigma_{i+1}(x) = \sigma_i(x)$$

as

$$\sigma_6(x) = 1 + \alpha^{12}x + \alpha x^2 + \alpha^5 x^3. \tag{7.55}$$

Since $6 = 2 \times 3$, we can stop the iteration. Thus, the error location polynomial is found as in Eq. (7.55).

If we search the roots of Eq. (7.55), i.e., the roots of

$$\sigma_6(x) = 0 \rightarrow 1 + \alpha^{12}x + \alpha x^2 + \alpha^5 x^3 = 0$$

we can find them as

$$\alpha^2 \quad \alpha^{10} \quad \alpha^{13}.$$

Then, the inverse of the roots can be calculated as

$$X_1 = \alpha^{13} \quad X_2 = \alpha^5 \quad X_3 = \alpha^2$$

from which the error polynomial can be determined as

$$e(x) = x^{13} + x^5 + x^2.$$

Using error polynomial, we can determine the code-word polynomial as

$$c(x) = r(x) + e(x) \rightarrow$$
$$c(x) = x^{14} + x^{13} + x^9 + x^7 + x^5 + x^4 + x^3 + x^2 + x + 1 + x^{13} + x^5 + x^2 \rightarrow$$
$$c(x) = x^{14} + x^9 + x^7 + x^4 + x^3 + x + 1.$$

We can find the data-word polynomial as

$$m(x) = \frac{c(x)}{g(x)} \rightarrow m(x) = x^4 + x^2 + x$$

if $c(x)$ is not in systematic form, which is the case for this example; otherwise, we can just choose the most significant $k$-bits of the code-word as the data bits.

## 7.3 Generator Matrices of Reed-Solomon Codes

Reed-Solomon codes are cyclic codes, and a cyclic shift of a code-word is another code-word. The generator polynomial of a Reed-Solomon code can be used to construct the generator matrix of the corresponding Reed-Solomon code, or as an alternative approach, we can find the generator matrix of a Reed-Solomon code using

$$\boldsymbol{G} = \begin{bmatrix} g_1 \\ g_2 \\ g_3 \\ \vdots \\ g_k \end{bmatrix} = \begin{bmatrix} 1 & 1 & 1 & \cdots & 1 \\ 1 & \alpha & \alpha^2 & \cdots & \alpha^{n-1} \\ 1 & (\alpha)^2 & (\alpha^2)^2 & \cdots & (\alpha^{n-1})^2 \\ \vdots & \vdots & \vdots & \vdots & \vdots \\ 1 & (\alpha)^{k-1} & (\alpha^2)^{k-1} & \cdots & (\alpha^{n-1})^{k-1} \end{bmatrix} \quad (7.56)$$

where each $\alpha^i$ term can be represented using a column vector including $m$-bits, where $m$ is the number appearing in $GF(2^m)$.

### 7.3.1 *Encoding of Reed-Solomon Codes*

We can achieve encoding operation following different approaches. Let's explain these approaches.

**First Encoding Method**
If the data-word polynomial is expressed in vector form using the extended field elements as

$$\boldsymbol{d} = [d_1\ d_2 \ldots d_k],$$

then the encoding operation can be achieved using

$$\boldsymbol{c} = \boldsymbol{dG}$$

leading to

$$\boldsymbol{c} = d_1\boldsymbol{g_1} + d_2\boldsymbol{g_2} + \ldots d_k\boldsymbol{g_k}.$$

The encoding operation can also be achieved using

$$c(x) = d(x)g(x)$$

where $g(x)$ is the generator polynomial of the $RS(n,k)$ and $d(x)$ is the polynomial form of $\boldsymbol{d}$.

**Second Encoding Method**

In the second approach, we first express the reversed polynomial form of

$$\boldsymbol{d} = [d_k\ d_{k-1} \ldots d_1] \tag{7.57}$$

as

$$d_r(x) = d_k x + d_{k-1}x^2 + \ldots + d_1 x^k. \tag{7.58}$$

Next, we calculate the code-word $\boldsymbol{c}$ using field elements as

$$\boldsymbol{c} = \left[d_r(1)\ \ d_r(\alpha)\ \ d_r\left(\alpha^2\right) \ldots d_r\left(\alpha^{n-1}\right)\right] \tag{7.59}$$

which can be illustrated in polynomial form by means of

$$c(x) = d_r(1)x^{n-1} + d_r(\alpha)x^{n-2} + \ldots + d_r\left(\alpha^{n-2}\right)x + d_r\left(\alpha^{n-1}\right). \tag{7.60}$$

*Note that* the polynomial form of Eq. (7.57) given in Eq. (7.58) is different than the conventional form used through the book. However, such different form is necessary for the correct calculation of the code-word.

**Example 7.7** The generator polynomial of the double-error-correcting $RS(7,3)$ using the extended field elements $GF(2^3)$ generated using the primitive polynomial $p(x) = x^3 + x + 1$ can be calculated as

$$g(x) = x^4 + \alpha^3 x^3 + x^2 + \alpha x + \alpha^3.$$

We want to encode the data-word $\boldsymbol{d} = [110111010]$ using $RS(7,3)$. The encoding of the data-word can be achieved using different methods. Now, let's consider these methods separately.

**Method 1** First, let's give the field elements of $GF(2^3)$ for the reminder

$$0 \rightarrow 0$$
$$1 \rightarrow 1$$
$$\alpha \rightarrow \alpha$$
$$\alpha^2 \rightarrow \alpha^2$$
$$\alpha^3 \rightarrow \alpha + 1$$
$$\alpha^4 \rightarrow \alpha^2 + \alpha$$
$$\alpha^5 \rightarrow \alpha^2 + \alpha + 1$$
$$\alpha^6 \rightarrow \alpha^2 + 1.$$

The generator matrix of the code can be formed using Eq. (7.56) as

$$\boldsymbol{G} = \begin{bmatrix} 1 & 1 & 1 & 1 & 1 & 1 & 1 \\ 1 & \alpha & \alpha^2 & \alpha^3 & \alpha^4 & \alpha^5 & \alpha^6 \\ 1 & \alpha^2 & \alpha^4 & \alpha^6 & \alpha^8 & \alpha^{10} & \alpha^{12} \end{bmatrix}$$

which can be simplified using $\alpha^7 = 1$ as

$$\boldsymbol{G} = \begin{bmatrix} 1 & 1 & 1 & 1 & 1 & 1 & 1 \\ 1 & \alpha & \alpha^2 & \alpha^3 & \alpha^4 & \alpha^5 & \alpha^6 \\ 1 & \alpha^2 & \alpha^4 & \alpha^6 & \alpha & \alpha^3 & \alpha^5 \end{bmatrix}. \quad (7.61)$$

The data-word given in the example can be expressed in polynomial form as

$$\boldsymbol{d} = \left[ \underbrace{110}_{\alpha^2+\alpha} \underbrace{111}_{\alpha^2+\alpha+1} \underbrace{010}_{\alpha} \right] \rightarrow \boldsymbol{d} = \left[ \alpha^4 \ \alpha^5 \ \alpha \right].$$

The encoding operation can be performed as

$$c = dG \rightarrow c = \alpha^4[1\ 1\ 1\ 1\ 1\ 1\ 1] + \alpha^5\left[1\ \alpha\ \alpha^2\ \alpha^3\ \alpha^4\ \alpha^5\ \alpha^6\right] + \alpha\left[1\ \alpha^2\ \alpha^4\ \alpha^6\ \alpha\ \alpha^3\ \alpha^5\right] \rightarrow$$

$$c = \left[\alpha^4\ \alpha^4\ \alpha^4\ \alpha^4\ \alpha^4\ \alpha^4\ \alpha^4\right] + \left[\alpha^5\ \alpha^6\ \alpha^7\ \alpha^8\ \alpha^9\ \alpha^{10}\ \alpha^{11}\right] + \left[\alpha\ \alpha^3\ \alpha^5\ \alpha^7\ \alpha^2\ \alpha^4\ \alpha^6\right]$$

$$c = \left[\alpha^4 + \alpha^5 + \alpha \quad \alpha^4 + \alpha^6 + \alpha^3 \quad \alpha^4 + \alpha^7 + \alpha^5 \quad \alpha^4 + \alpha^8 + \alpha^7 \quad \alpha^4 + \alpha^9 + \alpha^2 \quad \alpha^4 + \alpha^{10} + \alpha^4 \quad \alpha^4 + \alpha^{11} + \alpha^6\right] \rightarrow$$

$$c = \left[\alpha^3\ \ 0\ \ 0\ \ \alpha^6\ \ \alpha^4\ \ \alpha^3\ \ \alpha^6\right] \tag{7.62}$$

which can be expressed in binary form as

$$c = \left[\underbrace{\alpha^3}_{\alpha+1}\ \ 0\ \ 0\ \ \underbrace{\alpha^6}_{\alpha^2+1}\ \ \underbrace{\alpha^4}_{\alpha^2+\alpha}\ \ \underbrace{\alpha^3}_{\alpha+1}\ \ \underbrace{\alpha^6}_{\alpha^2+1}\right] \rightarrow$$

$$c = [011\ \ 000\ \ 000\ \ 101\ \ 110\ \ 011\ \ 101]$$

which can be expressed using concatenated bits as

$$c = [011000000101110011101].$$

**Method 2** Expressing each element of $G$ in Eq. (7.61) using three bits, we get the binary equivalent of Eq. (7.61) as

$$G = \begin{bmatrix} 001 & 001 & 001 & 001 & 001 & 001 & 001 \\ 001 & 010 & 100 & 011 & 110 & 111 & 101 \\ 001 & 100 & 110 & 101 & 010 & 011 & 111 \end{bmatrix} \tag{7.63}$$

The encoding operation using the data-word $d = [110111010]$ can be achieved using the convolution

$$c = d \star G \rightarrow c$$

$$= [110\ 111\ 010] \star \begin{bmatrix} 001 & 001 & 001 & 001 & 001 & 001 & 001 \\ 001 & 010 & 100 & 011 & 110 & 111 & 101 \\ 001 & 100 & 110 & 101 & 010 & 011 & 111 \end{bmatrix} \rightarrow$$

$$c = [(110 \star 001 + 111 \star 001 + 010 \star 001), \ldots, (110 \star 001 + 111 \star 101 + 010 \star 111)] \rightarrow \tag{7.64}$$

$$c = [011\ \ 000\ \ 000\ \ 101\ \ 110\ \ 011\ \ 101].$$

Note that the result of the convolutions in Eq. (7.64) is a bit stream consisting of 5 bits. After summing the convolution results, we take the three least significant bits of the result. For instance,

$$[110\ 111\ 010] \star \begin{bmatrix} 001 \\ 011 \\ 101 \end{bmatrix} \rightarrow \underbrace{110 \star 001}_{=00110} + \underbrace{111 \star 011}_{=01001} + \underbrace{010 \star 101}_{01001} \rightarrow 00\mathbf{101}$$
$$\rightarrow 101.$$

**Method 3** We first express the data bit stream $\boldsymbol{d} = [110111010]$ using bit groups containing 3 bits as

$$\boldsymbol{d} = [110\ \ 111\ \ 010]$$

which can be written in vector form as

$$\boldsymbol{d} = \left[\alpha^2 + \alpha \quad \alpha^2 + \alpha + 1 \quad \alpha\right] \rightarrow \boldsymbol{d} = \left[\alpha^4\ \ \alpha^5\ \ \alpha\right].$$

The reversed polynomial form for $\boldsymbol{d} = [\alpha^4\ \ \alpha^5\ \ \alpha]$ can be written as

$$d_r(x) = \alpha^4 + \alpha^5 x + \alpha x^2.$$

The code-word vector can be calculated using $d(x)$ as

$$\boldsymbol{c} = \left[d_r(1)\ \ d_r(\alpha)\ \ d_r(\alpha^2)\ \ d_r(\alpha^3)\ \ d_r(\alpha^4)\ \ d_r(\alpha^5)\ \ d_r(\alpha^6)\right]$$

where we have

$$d(1) = \alpha^4 + \alpha^5 + \alpha \rightarrow d(1) = \alpha^3$$
$$d(\alpha) = \alpha^4 + \alpha^6 + \alpha^3 \rightarrow d(\alpha) = 0$$
$$d(\alpha^2) = \alpha^4 + \alpha^7 + \alpha^5 \rightarrow d(\alpha^2) = 0$$
$$d(\alpha^3) = \alpha^4 + \alpha^8 + \alpha^7 \rightarrow d(\alpha^3) = \alpha^6$$
$$d(\alpha^4) = \alpha^4 + \alpha^9 + \alpha^9 \rightarrow d(\alpha^4) = \alpha^4$$
$$d(\alpha^5) = \alpha^4 + \alpha^{10} + \alpha^{11} \rightarrow d(\alpha^5) = \alpha^3$$

$$d(\alpha^6) = \alpha^4 + \alpha^{11} + \alpha^{13} \rightarrow d(\alpha^6) = \alpha^6$$

from which we obtain

$$\boldsymbol{c} = \begin{bmatrix}\alpha^3 & 0 & 0 & \alpha^6 & \alpha^4 & \alpha^3 & \alpha^6\end{bmatrix}$$

which is the same as that of Eq. (7.62).

### 7.3.2 Systematic Encoding Using Generator Polynomial

For a $t$-error-correcting Reed-Solomon code, the code-word polynomial can be obtained in systematic form as

$$c(x) = d(x)x^{2t} + r(x) \quad r(x) = R_{g(x)}\left[d(x)x^{2t}\right] \tag{7.65}$$

where $r(x)$ is the remainder polynomial obtained after the division of $d(x)x^{2t}$ by $g(x)$.

**Example 7.8** The generator matrix of the systematic Reed-Solomon code $RS(7,3)$ can be put into the form $\boldsymbol{G}_s = [\boldsymbol{P} \mid \boldsymbol{I}]$ using elementary row operations as

$$\boldsymbol{G}_s = \begin{bmatrix} \alpha^3 & \alpha & 1 & \alpha^3 & 1 & 0 & 0 \\ \alpha^6 & \alpha^6 & 1 & \alpha^2 & 0 & 1 & 0 \\ \alpha^5 & \alpha^4 & 1 & \alpha^4 & 0 & 0 & 1 \end{bmatrix}.$$

Besides, the generator polynomial of $RS(7,3)$ can be calculated as

$$g(x) = x^4 + \alpha^3 x^3 + x^2 + \alpha x + \alpha^3.$$

Let the data vector be $\boldsymbol{d} = [\alpha \ \alpha^2 \ 1]$. The encoding operation for the given data vector can be achieved using the systematic generator matrix as

$$\boldsymbol{c} = \boldsymbol{dG} \rightarrow \boldsymbol{c} = \begin{bmatrix}\alpha & \alpha^2 & 1\end{bmatrix} \begin{bmatrix} \alpha^3 & \alpha & 1 & \alpha^3 & 1 & 0 & 0 \\ \alpha^6 & \alpha^6 & 1 & \alpha^2 & 0 & 1 & 0 \\ \alpha^5 & \alpha^4 & 1 & \alpha^4 & 0 & 0 & 1 \end{bmatrix} \rightarrow$$

$$\boldsymbol{c} = \alpha \times \left[\alpha^3\ \alpha\ 1\ \alpha^3\ 1\ 0\ 0\right] + \alpha^2 \times \left[\alpha^6\ \alpha^6\ 1\ \alpha^2\ 0\ 1\ 0\right] + 1 \times \left[\alpha^5\ \alpha^4\ 1\ \alpha^2\ 0\ 0\ 1\right] \rightarrow$$

$$\boldsymbol{c} = \left[\alpha^3\ 0\ \alpha^5\ \alpha^4\ \alpha\ \alpha^2\ 1\right].$$

On the other hand, if we employ the generator polynomial for the encoding operation, we can obtain the systematic code-word in polynomial form as follows. For

$$t = 2 \text{ and } d(x) = \alpha x^2 + \alpha^2 x + 1$$

the formulas

$$c(x) = d(x)x^{2t} + r(x) \quad r(x) = R_{g(x)}\left[d(x)x^{2t}\right]$$

happen to be

$$c(x) = \left(\alpha x^2 + \alpha^2 x + 1\right)x^4 + r(x) \quad r(x) = R_{g(x)}\left[\left(\alpha x^2 + \alpha^2 x + 1\right)x^4\right]$$

from which $c(x)$ can be calculated as

$$c(x) = \alpha^3 x^6 + \alpha^5 x^4 + \alpha^4 x^3 + \alpha x^2 + \alpha^2 x + 1.$$

## 7.4 Reed-Solomon Code as a Cyclic Code

The generator matrix of a cyclic code can be obtained as

$$G = \begin{bmatrix} g(x) \\ xg(x) \\ x^2 g(x) \\ \vdots \\ x^{k-1} g(x) \end{bmatrix} = \begin{bmatrix} g_{2t} & g_{2t-1} & \cdots & g_1 & g_0 & 0 & 0 & 0 & \cdots & 0 \\ 0 & g_{2t} & g_{2t-1} & \cdots & g_1 & g_0 & 0 & 0 & \cdots & 0 \\ 0 & 0 & g_{2t} & g_{2t-1} & \cdots & g_1 & g_0 & 0 & \cdots & 0 \\ \vdots & \vdots & \vdots & \vdots & \vdots & \vdots & \vdots & \vdots & \vdots & \vdots \\ 0 & \cdots & 0 & 0 & 0 & g_{2t} & g_{2t-1} & \cdots & g_1 & g_0 \end{bmatrix} \tag{7.66}$$

**Example 7.9** The generator polynomial of the double-error-correcting $RS(7,3)$ is given as

$$g(x) = x^4 + \alpha^3 x^3 + x^2 + \alpha x + \alpha^3.$$

The generator matrix can be formed using the generator polynomial as

$$G = \begin{bmatrix} 1 & \alpha^3 & 1 & \alpha & \alpha^3 & 0 & 0 \\ 0 & 1 & \alpha^3 & 1 & \alpha & \alpha^3 & 0 \\ 0 & 0 & 1 & \alpha^3 & 1 & \alpha & \alpha^3 \end{bmatrix}.$$

The data-word $\boldsymbol{d} = [001 \ \ 100 \ \ 010] \rightarrow \boldsymbol{d} = [1 \ \alpha^2 \ \alpha]$ can be encoded as

$$\boldsymbol{c} = \boldsymbol{dG} \rightarrow c = \begin{bmatrix} 1 \; \alpha^2 \; \alpha \end{bmatrix} \begin{bmatrix} 1 & \alpha^3 & 1 & \alpha & \alpha^3 & 0 & 0 \\ 0 & 1 & \alpha^3 & 1 & \alpha & \alpha^3 & 0 \\ 0 & 0 & 1 & \alpha^3 & 1 & \alpha & \alpha^3 \end{bmatrix}$$

leading to

$$\boldsymbol{c} = 1 \times \begin{bmatrix} 1 \; \alpha^3 \; 1 \; \alpha \; \alpha^3 \; 0 \; 0 \end{bmatrix} + \alpha^2 \times \begin{bmatrix} 0 \; 1 \; \alpha^3 \; 1 \; \alpha \; \alpha^3 \; 0 \end{bmatrix}$$
$$+\alpha \begin{bmatrix} 0 \; 0 \; 1 \; \alpha^3 \; 1 \; \alpha \; \alpha^3 \end{bmatrix} \rightarrow$$

$$\boldsymbol{c} = \begin{bmatrix} 1 \quad \alpha^3 + \alpha^2 \quad 1 + \alpha + \alpha^5 \quad \alpha + \alpha^2 + \alpha^4 \quad \alpha + \alpha^3 + \alpha^3 \quad \alpha^2 + \alpha^5 \quad \alpha^4 \end{bmatrix}$$

which can be simplified as

$$\boldsymbol{c} = \begin{bmatrix} 1 \; \alpha^5 \; \alpha^2 \; 0 \; \alpha \; \alpha^3 \; \alpha^4 \end{bmatrix} \tag{7.67}$$

which can be expressed in binary form as

$$\boldsymbol{c} = \begin{bmatrix} 1 \; \alpha^5 \; \alpha^2 \; 0 \; \alpha \; \alpha^3 \; \alpha^4 \end{bmatrix} \rightarrow$$
$$\boldsymbol{c} = \begin{bmatrix} 1 \quad \alpha^2 + \alpha + 1 \quad \alpha^2 \quad 0 \quad \alpha \quad \alpha + 1 \quad \alpha^2 + \alpha \end{bmatrix} \rightarrow$$
$$\boldsymbol{c} = [001 \quad 111 \quad 100 \quad 000 \quad 010 \quad 011 \quad 110]$$

which can be written using concatenated bits as

$$\boldsymbol{c} = [001111100000010011110].$$

The encoding operation can be achieved using the polynomial multiplication

$$c(x) = d(x)g(x) \rightarrow c(x) = \left(x^2 + \alpha^2 x + \alpha\right)\left(x^4 + \alpha^3 x^3 + x^2 + \alpha x + \alpha^3\right) \rightarrow$$
$$c(x) = x^6 + \alpha^3 x^5 + x^4 + \alpha x^3 + \alpha^3 x^2$$
$$+\alpha^2 x^5 + \alpha^5 x^4 + \alpha^2 x^3 + \alpha^3 x^2 + \alpha^5 x$$
$$+\alpha x^4 + \alpha^4 x^3 + \alpha x^2 + \alpha^2 x + \alpha^4 \rightarrow$$
$$c(x) = x^6 + (\alpha^3 + \alpha^2)x^5 + (1 + \alpha + \alpha^5)x^4 + (\alpha + \alpha^2 + \alpha^4)x^3 +$$
$$(\alpha^3 + \alpha^3 + \alpha)x^2 + (\alpha^5 + \alpha^2)x + \alpha^4 \rightarrow$$
$$c(x) = x^6 + \alpha^5 x^5 + \alpha^2 x^4 + 0x^3 + \alpha x^2 + \alpha^3 x + \alpha^4$$

which can be written in vector form as

$$\mathbf{c} = \left[1\ \alpha^5\ \alpha^2\ 0\ \alpha\ \alpha^3\ \alpha^4\right]. \tag{7.68}$$

We see that Eqs. (7.67) and (7.68) are the same.

**Exercise**
Find the systematic form of the generator matrix

$$\mathbf{G} = \begin{bmatrix} 1 & \alpha^3 & 1 & \alpha & \alpha^3 & 0 & 0 \\ 0 & 1 & \alpha^3 & 1 & \alpha & \alpha^3 & 0 \\ 0 & 0 & 1 & \alpha^3 & 1 & \alpha & \alpha^3 \end{bmatrix}.$$

### *7.4.1 Parity Check Matrix of Reed-Solomon Code and Syndrome Calculation*

The parity check matrix of a $t$-error-correcting $RS(n, k)$ can be calculated as

$$\mathbf{H} = \begin{bmatrix} 1 & \alpha & \alpha^2 & \cdots & \alpha^{n-1} \\ 1 & (\alpha)^2 & (\alpha^2)^2 & \cdots & (\alpha^{n-1})^2 \\ 1 & (\alpha)^3 & (\alpha^2)^3 & \cdots & (\alpha^{n-1})^3 \\ \vdots & \vdots & \vdots & \vdots & \vdots \\ 1 & (\alpha)^{2t} & (\alpha^2)^{2t} & \cdots & (\alpha^{n-1})^{2t} \end{bmatrix} \tag{7.69}$$

For a received-word

$$\mathbf{r} = \left[r_{n-1}\ r_{n-2} \ldots r_1\ r_0\right]$$

which can also be expressed in polynomial form as

$$r(x) = r_{n-1}x^{n-1} + r_{n-2}x^{n-2} + \ldots r_1 x + r_0$$

the syndrome can be calculated using

$$\mathbf{s} = \mathbf{r}\mathbf{H}^T \rightarrow \mathbf{s} = \mathbf{e}\mathbf{H}^T \tag{7.70}$$

as

$$s = [r_{n-1}\ r_{n-2} \ldots r_1\ r_0] \begin{bmatrix} 1 & \alpha & \alpha^2 & \cdots & \alpha^{n-1} \\ 1 & (\alpha)^2 & (\alpha^2)^2 & \cdots & (\alpha^{n-1})^2 \\ 1 & (\alpha)^3 & (\alpha^2)^3 & \cdots & (\alpha^{n-1})^3 \\ \vdots & \vdots & \vdots & \vdots & \vdots \\ 1 & (\alpha)^{2t} & (\alpha^2)^{2t} & \cdots & (\alpha^{n-1})^{2t} \end{bmatrix}^T \tag{7.71}$$

leading to

$$\begin{aligned} s = {} & r_{n-1}[1\ 1\ 1 \ldots 1] + r_{n-2}\left[(\alpha)\ (\alpha)^2\ (\alpha)^3 \ldots (\alpha)^{2t}\right] \\ & + r_{n-3}\left[\alpha^2\ (\alpha^2)^2\ (\alpha^2)^3 \ldots (\alpha^2)^{2t}\right] + \ldots \\ & + r_0\left[\alpha^{n-1}\ (\alpha^{n-1})^2\ (\alpha^{n-1})^3 \ldots (\alpha^{n-1})^{2t}\right] \end{aligned}$$

which can be written as

$$s = \left[\sum_{i=0}^{n-1} r_i \alpha^{n-i-1}\quad \sum_{i=0}^{n-1} r_i (\alpha^2)^{n-i-1}\quad \sum_{i=0}^{n-1} r_i (\alpha^3)^{n-i-1} \ldots \sum_{i=0}^{n-1} r_i (\alpha^{2t})^{n-i-1}\right] \tag{7.72}$$

leading to

$$s = \left[r(\alpha)\ r(\alpha^2)\ r(\alpha^3) \ldots r(\alpha^{2t})\right]. \tag{7.73}$$

The elements of the syndrome in Eq. (7.73) can be separately written as

$$\begin{aligned} S_1 &= r(\alpha) \\ S_2 &= r(\alpha^2) \\ S_3 &= r(\alpha^3) \\ &\vdots \\ S_{2t} &= r(\alpha^{2t}). \end{aligned}$$

Since $s = eH^T$, we can construct syndrome table for a $RS(n, k)$ code considering all possible error patterns, and such a table contains two columns and $2^{m(n-k)}$ rows assuming that $RS(n, k)$ is constructed using $GF(2^m)$. However, even for $RS(7, 3)$ code constructed on $GF(2^3)$, we have $2^{3 \times (7-3)} = 2^{12} \rightarrow 4 \times 1024$ rows, and this is a large number. For this reason, syndrome tables are not preferred for Reed-Solomon codes.

**Table 7.6** Syndrome table for $RS(7,5)$

| Error patterns | Syndrome vector |
|---|---|
| [0 0 0 0 0 0 1] | |
| [0 0 0 0 0 0 $\alpha$] | |
| [0 0 0 0 0 0 $\alpha^2$] | |
| [0 0 0 0 0 0 $\alpha^3$] | |
| [0 0 0 0 0 0 $\alpha^4$] | |
| [0 0 0 0 0 0 $\alpha^5$] | |
| [0 0 0 0 0 0 $\alpha^6$] | |
| [0 0 0 0 0 1 0] | |
| [0 0 0 0 0 $\alpha$ 0] | |
| [0 0 0 0 0 $\alpha^2$ 0] | |
| [0 0 0 0 0 $\alpha^3$ 0] | |
| [0 0 0 0 0 $\alpha^4$ 0] | |
| [0 0 0 0 0 $\alpha^5$ 0] | |
| [0 0 0 0 0 $\alpha^6$ 0] | |
| $\vdots$ | |
| [1 0 0 0 0 0 0] | |
| [$\alpha$ 0 0 0 0 0 0] | |
| [$\alpha^2$ 0 0 0 0 0 0] | |
| [$\alpha^3$ 0 0 0 0 0 0] | |
| [$\alpha^4$ 0 0 0 0 0 0] | |
| [$\alpha^5$ 0 0 0 0 0 0] | |
| [$\alpha^6$ 0 0 0 0 0 0] | |

**Example 7.10** Construct the syndrome table for single-error-correcting $RS(7,5)$ code constructed on $GF(2^4)$ which is formed using the primitive polynomial $p(x) = x^3 + x + 1$.

The syndrome table includes $2^{3 \times (7-5)} \rightarrow 64$ rows. The error patterns can be tabulated as in Table 7.6.

The syndromes corresponding to the error patterns can be calculated using either

$$S = eH^T \tag{7.74}$$

where $H^T$ is the parity check matrix of the $RS(7,5)$ code, or using the polynomial equivalents of the error patterns, i.e., $e(x)$, the syndromes can be calculated using

$$S_i = e(\alpha^i), i = 1, 2, \ldots, 2t. \tag{7.75}$$

For instance, for the error pattern [0 0 0 0 0 $\alpha^3$ 0], we have the polynomial representation

$$e(x) = \alpha^3 x$$

from which the syndromes are calculated as

$$S_1 = e(\alpha) \rightarrow S_1 = \alpha^3 \alpha \rightarrow S_1 = \alpha^4$$
$$S_2 = e(\alpha^2) \rightarrow S_2 = \alpha^3 \alpha^2 \rightarrow S_2 = \alpha^5$$

and the syndrome vector corresponding to the error pattern $[0\ 0\ 0\ 0\ 0\ \alpha^3\ 0]$ can be written as

$$[\alpha^4\ \alpha^5].$$

Following a similar approach for the other error patterns, we obtain the syndrome table as in Table 7.7.

The parity check matrix of $RS(7,5)$ can be calculated using Eq. (7.10) as

$$\boldsymbol{H} = \begin{bmatrix} 1 & \alpha & \alpha^2 & \alpha^3 & \alpha^4 & \alpha^5 & \alpha^6 \\ 1 & \alpha^2 & \alpha^4 & \alpha^6 & \alpha^8 & \alpha^{10} & \alpha^{12} \end{bmatrix}$$

which can be simplified using $\alpha^7 = 1$ as

$$\boldsymbol{H} = \begin{bmatrix} 1 & \alpha & \alpha^2 & \alpha^3 & \alpha^4 & \alpha^5 & \alpha^6 \\ 1 & \alpha^2 & \alpha^4 & \alpha^6 & \alpha & \alpha^3 & \alpha^5 \end{bmatrix}.$$

For the error pattern $[0\ 0\ 0\ 0\ 0\ \alpha^3\ 0]$, we can calculate the syndrome using

$$\boldsymbol{s} = \boldsymbol{e}\boldsymbol{H}^{\boldsymbol{T}}$$

as

$$\boldsymbol{s} = [0\,0\,0\,0\,0\ \alpha^3\ 0] \begin{bmatrix} 1 & 1 \\ \alpha & \alpha^2 \\ \alpha^2 & \alpha^4 \\ \alpha^3 & \alpha^6 \\ \alpha^4 & \alpha \\ \alpha^5 & \alpha^3 \\ \alpha^6 & \alpha^5 \end{bmatrix}$$

leading to

**Table 7.7** Syndrome table for $RS(7,5)$

| Error patterns | Syndrome vector |
|---|---|
| [0 0 0 0 0 0 1] | [1 1] |
| $[0\ 0\ 0\ 0\ 0\ 0\ \alpha]$ | $[\alpha\ \alpha]$ |
| $[0\ 0\ 0\ 0\ 0\ 0\ \alpha^2]$ | $[\alpha^2\ \alpha^2]$ |
| $[0\ 0\ 0\ 0\ 0\ 0\ \alpha^3]$ | $[\alpha^3\ \alpha^3]$ |
| $[0\ 0\ 0\ 0\ 0\ 0\ \alpha^4]$ | $[\alpha^4\ \alpha^4]$ |
| $[0\ 0\ 0\ 0\ 0\ 0\ \alpha^5]$ | $[\alpha^5\ \alpha^5]$ |
| $[0\ 0\ 0\ 0\ 0\ 0\ \alpha^6]$ | $[\alpha^6\ \alpha^6]$ |
| [0 0 0 0 0 1 0] | $[\alpha\ \alpha^2]$ |
| $[0\ 0\ 0\ 0\ 0\ \alpha\ 0]$ | $[\alpha^2\ \alpha^3]$ |
| $[0\ 0\ 0\ 0\ 0\ \alpha^2\ 0]$ | $[\alpha^3\ \alpha^4]$ |
| $[0\ 0\ 0\ 0\ 0\ \alpha^3\ 0]$ | $[\alpha^4\ \alpha^5]$ |
| $[0\ 0\ 0\ 0\ 0\ \alpha^4\ 0]$ | $[\alpha^5\ \alpha^6]$ |
| $[0\ 0\ 0\ 0\ 0\ \alpha^5\ 0]$ | $[\alpha^6\ \alpha^7]$ |
| $[0\ 0\ 0\ 0\ 0\ \alpha^6\ 0]$ | $[\alpha^7\ \alpha^8]$ |
| ⋮ ⋮⋮ ⋮ | ⋮ ⋮⋮ ⋮ |
| [1 0 0 0 0 0 0] | $[\alpha^6\ \alpha^{12}]$ |
| $[\alpha\ 0\ 0\ 0\ 0\ 0\ 0]$ | $[\alpha^7\ \alpha^{13}]$ |
| $[\alpha^2\ 0\ 0\ 0\ 0\ 0\ 0]$ | $[\alpha^8\ \alpha^{14}]$ |
| $[\alpha^3\ 0\ 0\ 0\ 0\ 0\ 0]$ | $[\alpha^9\ \alpha^{15}]$ |
| $[\alpha^4\ 0\ 0\ 0\ 0\ 0\ 0]$ | $[\alpha^{10}\ \alpha^{16}]$ |
| $[\alpha^5\ 0\ 0\ 0\ 0\ 0\ 0]$ | $[\alpha^{11}\ \alpha^{17}]$ |
| $[\alpha^6\ 0\ 0\ 0\ 0\ 0\ 0]$ | $[\alpha^{12}\ \alpha^{18}]$ |

$$\mathbf{s} = \left[\alpha^8\ \ \alpha^6\right] \rightarrow \mathbf{s} = \left[\alpha\ \ \alpha^6\right].$$

**Example 7.11** In Example 7.9, the generator polynomial of the double-error-correcting $RS(7,3)$ is found as

$$\mathbf{G} = \begin{bmatrix} 1 & \alpha^3 & 1 & \alpha & \alpha^3 & 0 & 0 \\ 0 & 1 & \alpha^3 & 1 & \alpha & \alpha^3 & 0 \\ 0 & 0 & 1 & \alpha^3 & 1 & \alpha & \alpha^3 \end{bmatrix}. \tag{7.76}$$

The systematic form of Eq. (7.76) can be obtained using $\mathbf{G} = [\mathbf{I} \mid \mathbf{P}]$ form as

$$\begin{bmatrix} 1 & \alpha^3 & 1 & \alpha & \alpha^3 & 0 & 0 \\ 0 & 1 & \alpha^3 & 1 & \alpha & \alpha^3 & 0 \\ 0 & 0 & 1 & \alpha^3 & 1 & \alpha & \alpha^3 \end{bmatrix} : R_1 + \alpha^3 R_2 \rightarrow R_1$$

$$\rightarrow \begin{bmatrix} 1 & 0 & 1+\alpha^6 & \alpha+\alpha^3 & \alpha^3+\alpha^4 & \alpha^6 & 0 \\ 0 & 1 & \alpha^3 & 1 & \alpha & \alpha^3 & 0 \\ 0 & 0 & 1 & \alpha^3 & 1 & \alpha & \alpha^3 \end{bmatrix}$$

$$\rightarrow \begin{bmatrix} 1 & 0 & \alpha^2 & 1 & \alpha^6 & \alpha^6 & 0 \\ 0 & 1 & \alpha^3 & 1 & \alpha & \alpha^3 & 0 \\ 0 & 0 & 1 & \alpha^3 & 1 & \alpha & \alpha^3 \end{bmatrix} : R_1 + \alpha^2 R_3 \rightarrow R_1$$

$$\rightarrow \begin{bmatrix} 1 & 0 & 0 & 1+\alpha^5 & \alpha^6+\alpha^2 & \alpha^6+\alpha^3 & \alpha^5 \\ 0 & 1 & \alpha^3 & 1 & \alpha & \alpha^3 & 0 \\ 0 & 0 & 1 & \alpha^3 & 1 & \alpha & \alpha^3 \end{bmatrix}$$

$$\rightarrow \begin{bmatrix} 1 & 0 & 0 & \alpha^4 & 1 & \alpha^4 & \alpha^5 \\ 0 & 1 & \alpha^3 & 1 & \alpha & \alpha^3 & 0 \\ 0 & 0 & 1 & \alpha^3 & 1 & \alpha & \alpha^3 \end{bmatrix} : R_2 + \alpha^3 R_3 \rightarrow R_2$$

$$\rightarrow \begin{bmatrix} 1 & 0 & 0 & \alpha^4 & 1 & \alpha^4 & \alpha^5 \\ 0 & 1 & 0 & 1+\alpha^6 & \alpha+\alpha^3 & \alpha^3+\alpha^4 & \alpha^6 \\ 0 & 0 & 1 & \alpha^3 & 1 & \alpha & \alpha^3 \end{bmatrix}$$

$$\rightarrow \begin{bmatrix} 1 & 0 & 0 & \alpha^4 & 1 & \alpha^4 & \alpha^5 \\ 0 & 1 & 0 & \alpha^2 & 1 & \alpha^6 & \alpha^6 \\ 0 & 0 & 1 & \alpha^3 & 1 & \alpha & \alpha^3 \end{bmatrix}.$$

Hence, the systematic $\boldsymbol{G}$ in $\boldsymbol{G} = [\boldsymbol{I} \mid \boldsymbol{P}]$ form is

$$\boldsymbol{G} = \begin{bmatrix} 1 & 0 & 0 & \alpha^4 & 1 & \alpha^4 & \alpha^5 \\ 0 & 1 & 0 & \alpha^2 & 1 & \alpha^6 & \alpha^6 \\ 0 & 0 & 1 & \alpha^3 & 1 & \alpha & \alpha^3 \end{bmatrix}.$$

The systematic form of Eq. (7.76) can be obtained using $\boldsymbol{G} = [\boldsymbol{P} \mid \boldsymbol{I}]$ as

$$\boldsymbol{G} = \begin{bmatrix} \alpha^3 & \alpha & 1 & \alpha^3 & 1 & 0 & 0 \\ \alpha^6 & \alpha^6 & 1 & \alpha^2 & 0 & 1 & 0 \\ \alpha^5 & \alpha^4 & 1 & \alpha^4 & 0 & 0 & 1 \end{bmatrix}. \tag{7.77}$$

**Example 7.12** The $RS(7,3)$ code is constructed using the elements of $GF(2^3)$ which is constructed using the primitive polynomial $p(x) = x^3 + x + 1$.For a given data-word $\boldsymbol{d} = [101110001]$, how many different code-word can be formed?

**Solution 7.12** In the first approach, using the generator polynomial of $RS(7,3)$, we can construct the code-word polynomial via

$$c(x) = d(x)g(x).$$

In the second approach, first, we obtain the generator matrix using Eq. (7.56), or using the construction in Eq. (7.66), and then obtain the code-word polynomial as

$$\boldsymbol{c} = \boldsymbol{d}\boldsymbol{G}.$$

In the third approach, we put the generator matrices obtained using Eqs. (7.56) and (7.66) into systematic form and obtain the code-word as

$$\boldsymbol{c} = \boldsymbol{d}\boldsymbol{G}_s$$

where $\boldsymbol{G}_s$ is the systematic form of the generator matrix. In total, we can have three different code-words for the same data-word.

**Exercise** The parity check matrix of $RS(7,3)$ can be calculated using Eq. (7.69) as

$$\boldsymbol{H} = \begin{bmatrix} 1 & \alpha & \alpha^2 & \alpha^3 & \alpha^4 & \alpha^5 & \alpha^6 \\ 1 & \alpha^2 & \alpha^4 & \alpha^6 & \alpha & \alpha^3 & \alpha^5 \\ 1 & \alpha^3 & \alpha^6 & \alpha^2 & \alpha^5 & \alpha & \alpha^4 \\ 1 & \alpha^4 & \alpha & \alpha^5 & \alpha^2 & \alpha^6 & \alpha^3 \end{bmatrix}. \tag{7.78}$$

Put the parity check matrix in Eq. (7.78) into systematic form.

### 7.4.2 Parity Check Polynomials for RS Codes

Let $g(x)$ be the generator polynomial of a $RS(n,k)$ code. The parity check polynomial of the $RS(n,k)$ code can be calculated as

$$h(x) = \frac{x^n + 1}{g(x)}.$$

If $h(x)$ is in the form

$$h(x) = h_k x^k + h_{k-1}x^{k-1} + \ldots + h_1 x + h_0$$

then the parity check matrix of the $RS(n,k)$ can be constructed as

$$H = \begin{bmatrix} h_0 & h_1 & \ldots & h_{k-1} & h_k & 0 & 0 & 0 & \ldots & 0 \\ 0 & h_0 & h_1 & \ldots & h_{k-1} & h_k & 0 & 0 & \ldots & 0 \\ 0 & 0 & h_0 & h_1 & \ldots & h_{k-1} & h_k & 0 & \ldots & 0 \\ \vdots & \vdots & \vdots & \vdots & \vdots & \vdots & \vdots & \vdots & \vdots & \vdots \\ 0 & \ldots & 0 & 0 & 0 & h_0 & h_1 & \ldots & h_{k-1} & h_k \end{bmatrix}. \tag{7.79}$$

### 7.4.3 Chien Search

To determine the roots of a polynomial $g(x)$, we try every field element in $g(x) = 0$ and check whether the results are zero or not, i.e., we check

$$g(\alpha^i) = 0 \quad i = 1, \ldots, n \text{ where } n = 2^m - 1$$

and determine those $\alpha^i$ satisfying $g(\alpha^i) = 0$.

To make the implementation of the roots search in hardware or software, a systematic search method is proposed, and this search method is called Chien search. Let's explain the Chien search for error location polynomial given as

$$\sigma(x) = \sigma_v x^v + \sigma_{v-1} x^{v-1} + \ldots + \sigma_2 x^2 + \sigma_1 x + \sigma_0 \tag{7.80}$$

where $\sigma_0 = 1$. When Eq. (7.20) is evaluated for $\alpha^i$, we obtain

$$\sigma(\alpha^i) = \sigma_v \alpha^{iv} + \sigma_{v-1}\alpha^{i(v-1)} + \ldots + \sigma_2 \alpha^{2i} + \sigma_1 \alpha^i + \sigma_0. \tag{7.81}$$

Let's indicate each term of Eq. (7.81) as a state, i.e., define

$$q_v = \sigma_v \alpha^{iv} \qquad q_{v-1} = \sigma_{v-1}\alpha^{i(v-1)} \ldots \quad q_2 = \sigma_2 \alpha^{2i} \qquad q_1 = \sigma_1 \alpha^i \qquad q_0 = \sigma_0. \tag{7.82}$$

For $\alpha^{i+1}$, using Eq. (7.80), we obtain

$$\sigma(\alpha^{i+1}) = \sigma_v \alpha^{(i+1)v} + \sigma_{v-1}\alpha^{(i+1)(v-1)} + \ldots + \sigma_2 \alpha^{2(i+1)} + \sigma_1 \alpha^{i+1} + q_0$$

for which we can define the states,

$$q'_v = \sigma_v \alpha^{(i+1)v} \quad q'_{v-1} = \sigma_{v-1}\alpha^{(i+1)(v-1)} \ldots q'_2 = \sigma_2 \alpha^{2(i+1)} \quad q'_1 = \sigma_1 \alpha^{i+1} \quad q'_0 = q_0. \tag{7.83}$$

Among states $q_l$ and $q'_l$, $l > 0$, we have the relationship

$$q'_v = q_v\alpha^v \qquad q'_{v-1} = q_{v-1}\alpha^{v-1}\ldots\; q'_2 = q_2\alpha^2 \qquad q'_1 = q_1\alpha^1 \qquad q'_0 = q_0. \tag{7.84}$$

In Chien search algorithm, we first calculate Eq. (7.82) for $i = 1$, and check whether we have

$$\sum_{j=0}^{v} q_j = 0 \tag{7.85}$$

or not. If Eq. (7.85) is satisfied, then $\alpha$ is a root. For $i = 2$, first, we update the states using Eq. (7.84), and check whether we have

$$\sum_{j=0}^{v} q'_j = 0 \tag{7.86}$$

or not. If Eq. (7.86) is satisfied, then $\alpha^2$ is a root.

For $i = 3$, the state values calculated for $i = 2$ in the previous step are considered as $q_v, q_{v-1}, \ldots, q_2, q_1$, and they are updated using Eq. (7.84), and we check whether we have

$$\sum_{j=0}^{v} q'_j = 0 \tag{7.87}$$

or not. If Eq. (7.86) is satisfied, $\alpha^3$ is a root. The same process is repeated for $i = 4, \ldots, n = 2^m - 2$, i.e., for all the other field elements.

**Example 7.13** Considering the extended field $GF(2^4)$ constructed using the primitive polynomial $p(x) = x^4 + x + 1$, find the roots of the polynomial

$$r(x) = x^2 + \alpha^3 x + \alpha^6$$

using Chien search algorithm.

**Solution 7.13** Let's sort the extended field $GF(2^4)$ elements generated by the primitive polynomial $p(x) = x^4 + x + 1$ for the reminder as in

$$
\begin{aligned}
&0 \rightarrow 0\\
&1 \rightarrow 1\\
&\alpha \rightarrow \alpha\\
&\alpha^2 \rightarrow \alpha^2\\
&\alpha^3 \rightarrow \alpha^3\\
&\alpha^4 \rightarrow \alpha + 1\\
&\alpha^5 \rightarrow \alpha^2 + \alpha\\
&\alpha^6 \rightarrow \alpha^3 + \alpha^2\\
&\alpha^7 \rightarrow \alpha^3 + \alpha + 1\\
&\alpha^8 \rightarrow \alpha^2 + 1\\
&\alpha^9 \rightarrow \alpha^3 + \alpha\\
&\alpha^{10} \rightarrow \alpha^2 + \alpha + 1\\
&\alpha^{11} \rightarrow \alpha^3 + \alpha^2 + \alpha\\
&\alpha^{12} \rightarrow \alpha^3 + \alpha^2 + \alpha + 1\\
&\alpha^{13} \rightarrow \alpha^3 + \alpha^2 + 1\\
&\alpha^{14} \rightarrow \alpha^3 + 1.
\end{aligned}
\tag{7.88}
$$

If we try the field elements in $x^2 + \alpha^3 x + \alpha^6 = 0$ one by one, we see that the first root is $\alpha^8$. To find the second root, we first form the states as

$$q_2 = \alpha^{16} \quad q_1 = \alpha^{11} \quad q_0 = \alpha^6.$$

Next, we update the states as

$$q_2' = \alpha^2 q_2 \quad q_1' = \alpha q_1 \quad q_0' = q_0$$

and check the summation

$$\sum_i q_i' = 0$$

to determine the next root. According to this information, we can construct the Table 7.8.

As we mentioned before, Chien search is useful from the hardware implementation point of view.

Table 7.8 Update of states and sum of the states

| $i$ | $q_2$ | $q_1$ | $q_0$ | $\sum_i q_i$ |
|---|---|---|---|---|
| 8 | $\alpha^{16}$ | $\alpha^{11}$ | $\alpha^{6}$ | $=0$ |
| 9 | $\alpha^{18}$ | $\alpha^{12}$ | $\alpha^{6}$ | $\neq 0$ |
| 10 | $\alpha^{20}$ | $\alpha^{13}$ | $\alpha^{6}$ | $\neq 0$ |
| 11 | $\alpha^{22}$ | $\alpha^{14}$ | $\alpha^{6}$ | $\neq 0$ |
| 12 | $\alpha^{24}$ | $\alpha^{15}$ | $\alpha^{6}$ | $\neq 0$ |
| 13 | $\alpha^{26}$ | $\alpha^{16}$ | $\alpha^{6}$ | $=0$ |

## Problems

1. The extended field $GF(2^3)$ is generated using the primitive polynomial $p(x) = x^3 + x^2 + 1$. Find the generator polynomial of the single-error-correcting Reed-Solomon code over $GF(2^3)$.
2. Express the data vector $\boldsymbol{d} = [\alpha^3\ 0\ \alpha^4\ \alpha^5\ \alpha^2]$ in bit vector form. The elements of the data vector are chosen from $GF(2^3)$ which is constructed using the primitive polynomial $p(x) = x^3 + x^2 + 1$.
3. Encode the data vector $\boldsymbol{d} = [\alpha^3\ \alpha^2\ \alpha\ \alpha^4\ \alpha^2]$ using the generator polynomial of the Reed-Solomon code $RS(7, 5)$ designed in problem 1.
4. Construct the generator polynomial of three-error-correcting Reed-Solomon code over $GF(2^4)$ which is obtained using the primitive polynomial $p(x) = x^4 + x^2 + 1$.
5. The generator polynomial of the double-error-correcting Reed-Solomon code $RS(7, 3)$ over $GF(2^3)$, constructed using the primitive polynomial $p(x) = x^3 + x + 1$, can be calculated as

$$g(x) = x^4 + \alpha^3 x^3 + x^2 + \alpha x + \alpha^3.$$

The data-word polynomial is given as $d(x) = \alpha^3 x^2 + \alpha^5 x + \alpha^4$. Systematically encode $d(x)$.

6. Using the elements of $GF(2^4)$, the generator polynomial of the triple-error-correcting Reed-Solomon code, $RS(15, 9)$, is obtained. For the construction of $GF(2^4)$, the primitive polynomial $p(x) = x^4 + x^2 + 1$ is used. Assume that the data-word to be encoded is given as

$$d(x) = x^8 + \alpha^6 x^5 + \alpha^3 x^2 + x + 1.$$

Encode $d(x)$, and let $c(x)$ be the code-word obtained after encoding operation. The error pattern is given as $e(x) = \alpha^2 x^9 + \alpha^5 x^4$. Decode the received word polynomial $r(x) = c(x) + e(x)$. Use Berlekamp algorithm in your decoding operation.

# Chapter 8
# Convolutional Codes

## 8.1 Convolutional Coding

Before introducing the convolutional encoding operation, let's explain some fundamental concepts.

***Memory Cell***

A memory cell is an electronic circuit capable of holding 1-bit of information. Memory cells are flip-flops. The commercially available flip-flops can be listed as *SR*, *JK*, *D*, and *T*. Memory cells are usually selected from *D*-type flip-flops. A flip-flop is triggered either by a rising or falling pulse at its clock input, and when a flip-flop is triggered, its output may take new value depending on the value at its input.

The *D*-type flip-flop and its characteristic table are depicted in Fig. 8.1.

In Fig. 8.1, $Q(t + 1)$ represents the flip-flop output value after low-to-high clock pulse transition at the clock input. It is seen from the characteristic table of *D*-type flip-flop in Fig. 8.1 that $Q(t + 1) = D$. The flip-flop, i.e., memory cell, shown in Fig. 8.1 can be graphically represented as in Fig. 8.2. In convolutional encoding operations, the graphical representations of the memory cells are used, as it is seen from Fig. 8.2 that clock symbol is not shown on the graphical representation of the memory cell.

***Mod-2 Summer***

Module-2 summer is an electronic logic circuit. This circuit is an XOR gate. Module-2 summers with two inputs and three inputs are depicted in Fig. 8.3.

### *8.1.1 Convolutional Encoder Circuit*

A convolutional encoder circuit is a logic circuit consisting of logic cells and mod-2 summers. For instance, a typical convolutional encoder circuit is depicted in Fig. 8.4.


O. Gazi, *Forward Error Correction via Channel Coding*,
https://doi.org/10.1007/978-3-030-33380-5_8

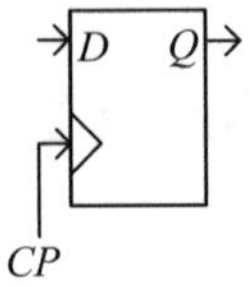


| D | Q(t+1) |
|---|---|
| 0 | 0 |
| 1 | 1 |

**Fig. 8.1** The *D*-type flip-flop and its characteristic table

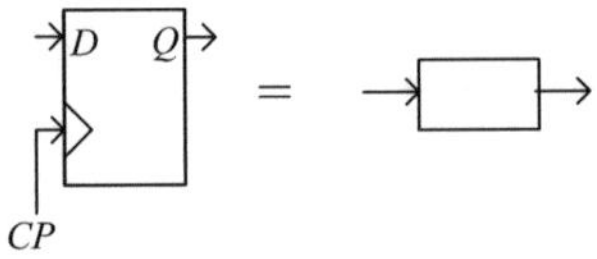


**Fig. 8.2** Memory cell and its graphical representation

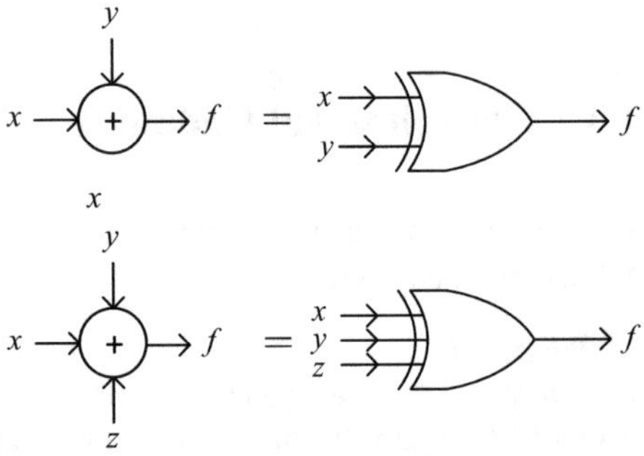


**Fig. 8.3** Module-2 summers are electronic logic circuits

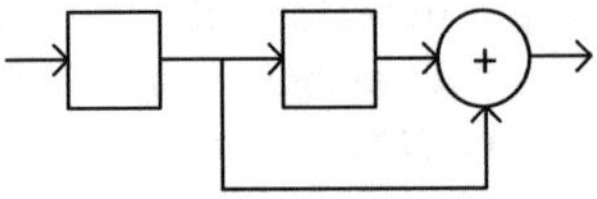


**Fig. 8.4** A convolutional encoder circuit

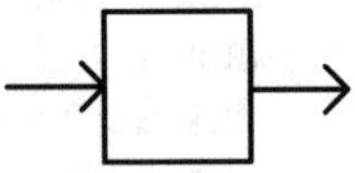

**Fig. 8.5** A single memory cell

While calculating the outputs of the convolutional circuits, we assume that all the cells are connected to a common clock pulse generator, and at the rising or falling edge of each clock pulse, the memory cells are triggered, and this means that the contents of the memory cells are updated regarding their input values. Let's illustrate the concept with an example.

**Example 8.1** A single memory cell is shown in Fig. 8.5, and at the input of the memory cell, we have the input stream $\boldsymbol{d} = [10110]$. Find the output of the memory cell.

**Solution 8.1** We assume the most significant bit of the binary vector enter to the memory cell first. The most significant bit of $\boldsymbol{d} = [10110]$ can be indicated in bold as $\boldsymbol{d} = [10110]$. Hence, at the input of the cell, we have the stream 01101, and this is depicted in Fig. 8.6.

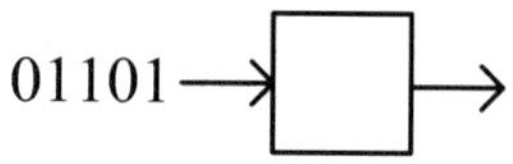


**Fig. 8.6** Memory cell and its input sequence

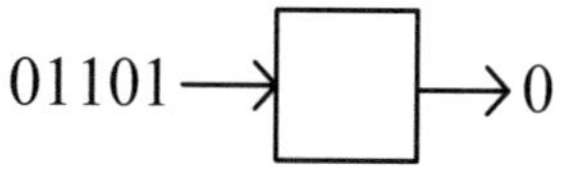


**Fig. 8.7** The first output of the memory cell

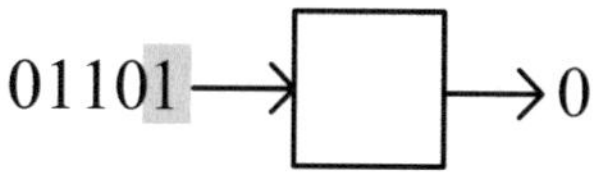


**Fig. 8.8** Logic-1 at the input of the memory cell

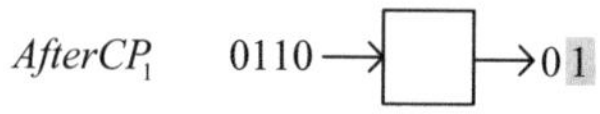


**Fig. 8.9** The second output of the memory cell

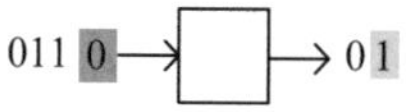


**Fig. 8.10** The bit-0 at the input of the memory cell

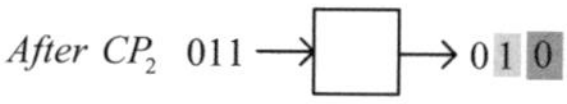


**Fig. 8.11** Third output of the memory cell

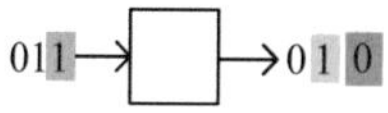


**Fig. 8.12** Bit 1 at the input of the memory cell

Initially the circuit is in reset state and the output of the circuit is 0 as depicted in Fig. 8.7.

At the input of the circuit, logic-1 as shadowed in Fig. 8.8 is available.

Upon the application of first clock pulse, the logic-1 shown shadowed goes to the output of the logic cell and stays there as illustrated in Fig. 8.9.

After the application of first clock pulse, i.e., $CP_1$, we have the output sequence 01, and at the input of the memory cell, we have bit-0 as shown in Fig. 8.10.

Upon the application of second clock pulse, the logic-0 shown in pink color goes to the output of the logic cell and stays there as illustrated in Fig. 8.11.

After the application of second clock pulse, i.e., $CP_2$, we have the output sequence 010, and at the input of the memory cell, we have bit-1as shown in Fig. 8.12.

Upon the application of third clock pulse, the logic-1 shown in green color goes to the output of the logic cell and stays there as illustrated in Fig. 8.13.

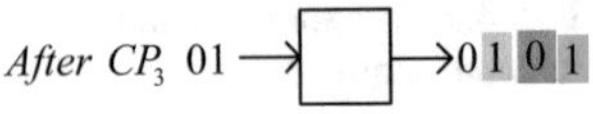


**Fig. 8.13** Fourth cell output

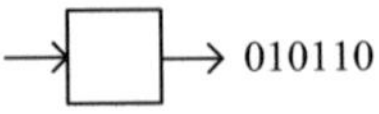


**Fig. 8.14** Memory cell outputs

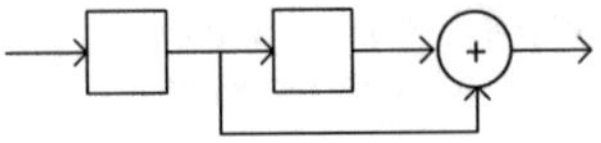


**Fig. 8.15** Convolutional encoder for Example 8.2

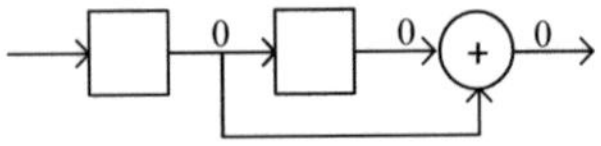


**Fig. 8.16** Initial outputs of the encoder

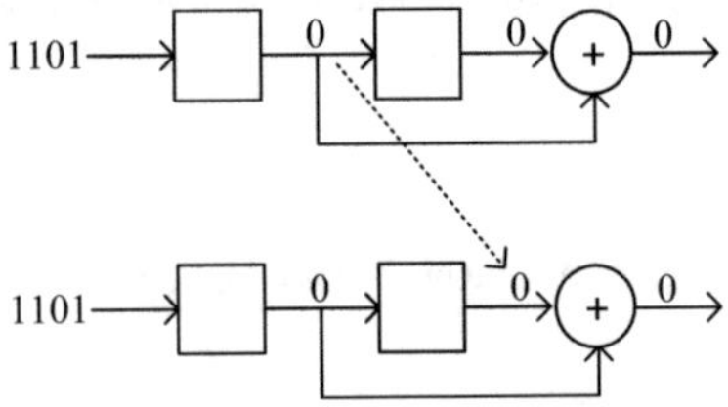


**Fig. 8.17** Calculation of the output of the second cell

In a similar manner after the application of fourth and fifth clock pulse, we get the output stream 011010 at the output of the memory cell as depicted in Fig. 8.14.

Note that the initial output value of the memory cell output is also accounted in the output stream of Fig. 8.14.

**Example 8.2** For the convolutional encoder shown in Fig. 8.15, find the bit stream at the convolutional encoder output for the input bit vector $\boldsymbol{d} = [1011]$.

**Solution 8.2** First, we indicate the initial output values of the cells as shown in Fig. 8.16.

When clocks are applied, memory cell outputs are evaluated starting from the rightmost cell proceeding toward the leftward. Finally, using the cell outputs, circuit output is evaluated.

According to this information, after the application of the first clock pulse, we fist evaluate the output of the second cell, i.e., the output of the rightmost cell, as shown in Fig. 8.17.

Next, we evaluate the output of the first cell as shown in Fig. 8.18.

In the last step, we calculate the circuit output using the cell outputs as shown in Fig. 8.19.

That is, after the application of $CP_1$, we have the cell and circuit outputs as shown in Fig. 8.19.

When second clock pulse, i.e., $CP_2$, is applied, we proceed in a similar manner. That is, we first calculate the output of the second cell and then the first cell,

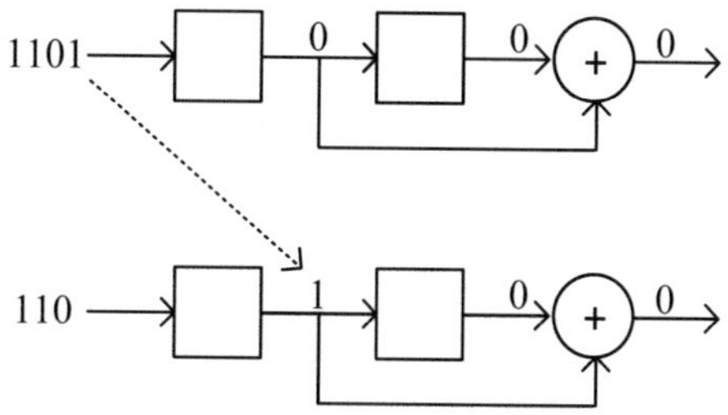


**Fig. 8.18** Calculation of the output of the first cell

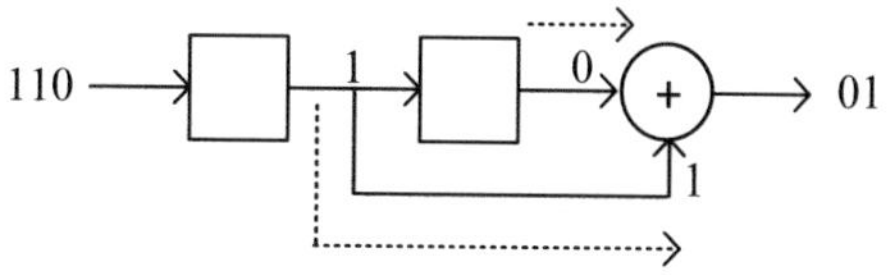


**Fig. 8.19** Calculation of the circuit output

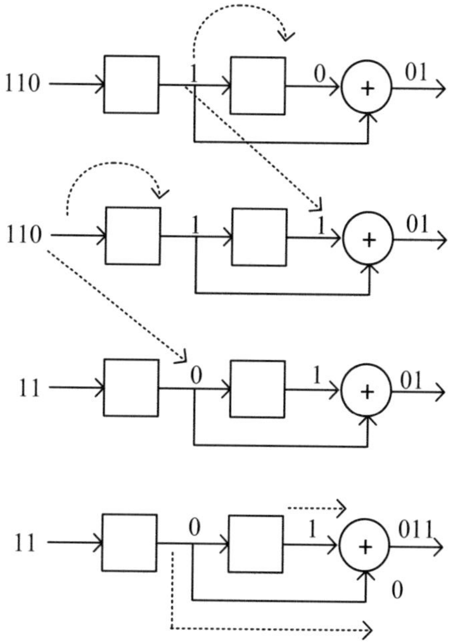


**Fig. 8.20** Calculation of the circuit output after second clock pulse

and finally, using cell outputs, we evaluate the circuit output. These steps are illustrated in Fig. 8.20.

For the third clock pulse, i.e., $CP_3$, the determination of cells and circuit outputs are illustrated in Fig. 8.21.

For fourth clock pulse, i.e., $CP_4$, the determination of the outputs of the cells and circuit outputs are illustrated in Fig. 8.22.

It is seen from Fig. 8.22 that there are more input bits at the input of the convolutional encoder. However, we should continue evaluating the circuit output

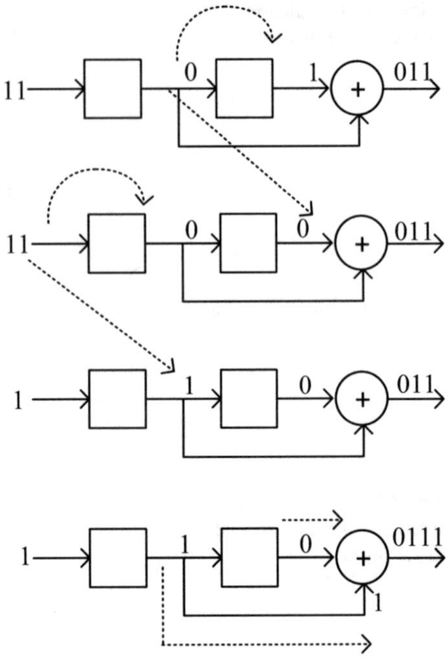


Fig. 8.21 Calculation of the circuit output after third clock pulse

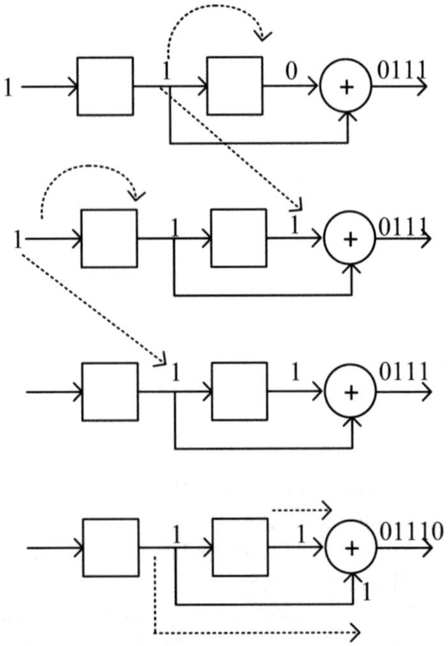


Fig. 8.22 Calculation of the circuit output after fourth clock pulse

until the memory cells contain no information bits. For this reason, we evaluate the circuit output for one more clock pulse and assume the external incoming bits as 0s.

For fifth clock pulse, i.e., $CP_5$, the determination of the outputs of the cells and circuit is illustrated in Fig. 8.23.

It is seen that the circuit output is 011101. Our input sequence was 1011. Hence, for 4-bit input sequence, we obtained 6-bit output sequence. In fact, the convolutional encoder in Fig. 8.23 has the impulse response 011, and the output of the sequence can be written as

$$011101 = 1011 * 011 \tag{8.1}$$

where $*$ denotes the convolutional operation. We will talk about the convolutional operation and impulse response of convolutional encoders in the subsequent sections.

Note that if $\boldsymbol{x}$ is an $N$-bit sequence and $\boldsymbol{h}$ is a $M$-bit sequence, then the convolution of these two sequences produces a $L$-bit sequence $\boldsymbol{y}$ such that $L = N + M - 1$. For the previous example, we have $N = 4$, $M = 3$, and $L = 6$, and it is obvious that $6 = 4 + 3 - 1$.

## 8.2 Impulse Response of a Convolutional Encoder

The output of the convolutional encoder when its input is only a single "1" is called the impulse response of the convolutional encoder. If the convolutional encoder has $m$ memory cells connected in series, then we need to apply $m$ clock pulses to get the encoder output when the input is a single "1." Considering the initial output of the encoder, we can say that the impulse response has $m + 1$ bits.

Let's illustrate the concept by an example.

**Example 8.3** Find the impulse response of the convolutional encoder shown in Fig. 8.24.

**Solution 8.3** The circuit has two memory cells, and we need to apply two clock pulses to get the circuit output for the single input bit "1." Since we need to apply two clock pulses, then we concatenate a single "0" to the input bit "1," and our circuit input becomes "01." This is illustrated in Fig. 8.25.

The first output value is the initial output of the circuit, and the calculation of the initial output value is depicted in Fig. 8.26.

Upon the application of first clock pulse, i.e., $CP_1$, the calculation of the circuit output is illustrated in Fig. 8.27.

Upon the application of second clock pulse, i.e., $CP_2$, the calculation of the circuit output is illustrated in Fig. 8.28.

It is seen from Fig. 8.28 that the impulse response of the convolutional encoder is $\boldsymbol{h} = [011]$.

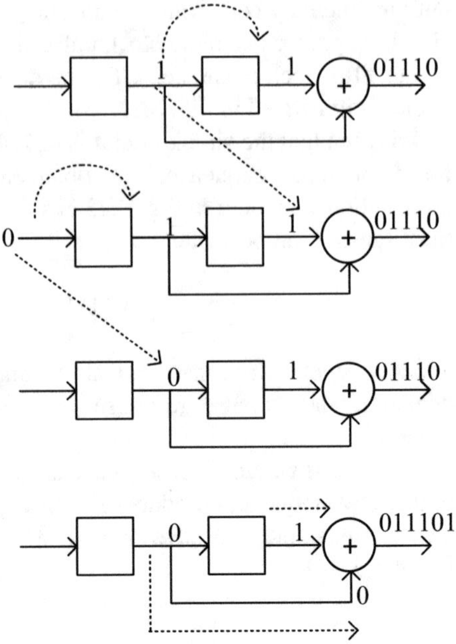


**Fig. 8.23** Calculation of the circuit output fifth second clock pulse

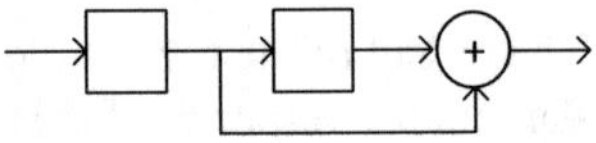

**Fig. 8.24** Convolutional encoder for Example 8.3

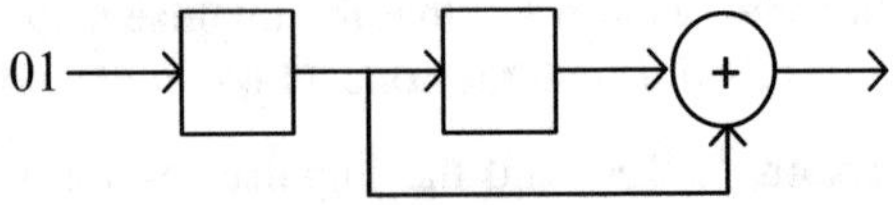


**Fig. 8.25** Determination of the encoder inputs

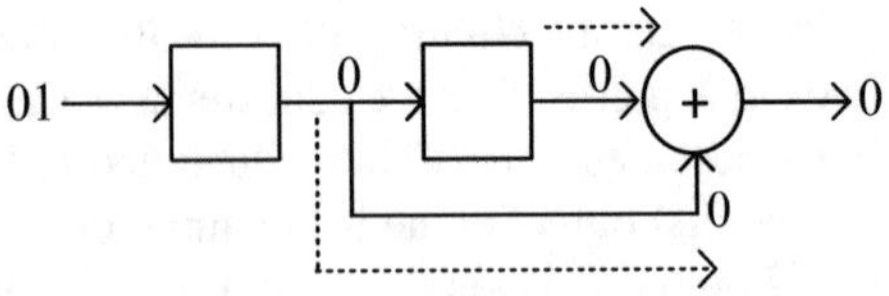


**Fig. 8.26** Initial output of the encoder

If we know the impulse response of a convolutional circuit, then, for any arbitrary binary input sequence, we can find the circuit output by taking the convolution of the arbitrary input with the impulse response of the circuit.

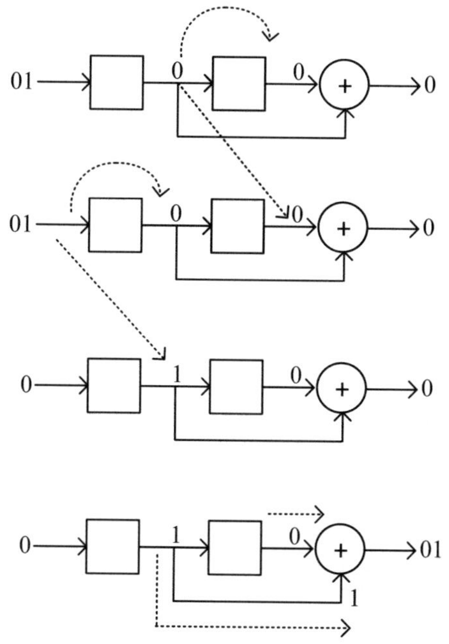


**Fig. 8.27** Calculation of the circuit output after first clock pulse

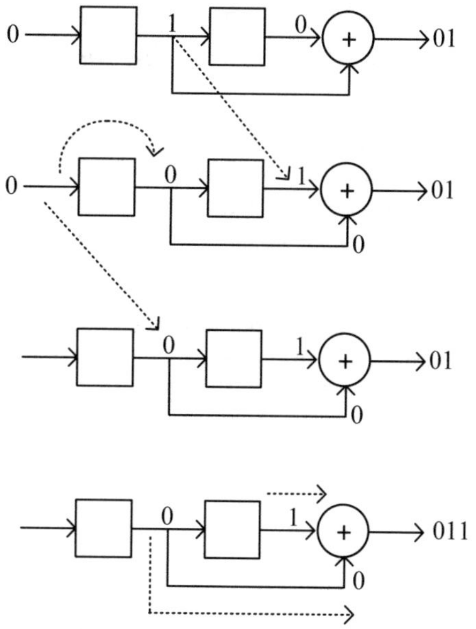


**Fig. 8.28** Calculation of the circuit output after second clock pulse

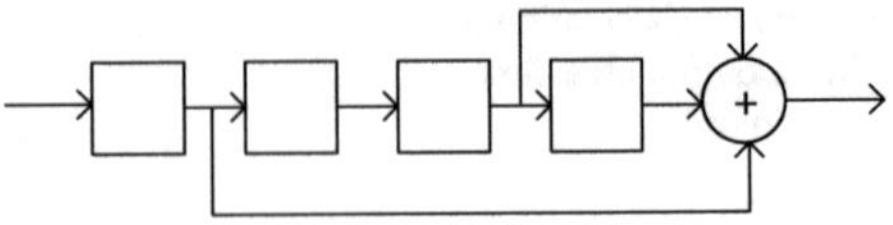


**Fig. 8.29** Convolutional encoder for Example 8.4

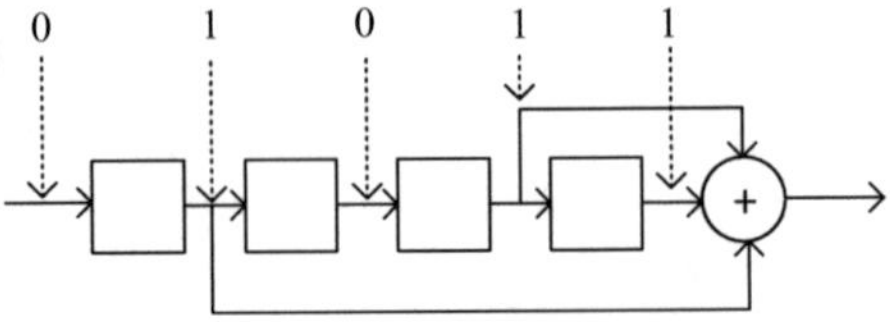


**Fig. 8.30** Determination of the impulse response

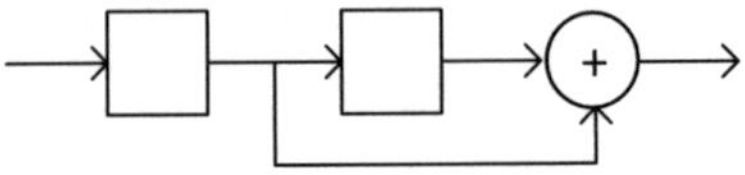


**Fig. 8.31** Convolutional encoder for Example 8.5

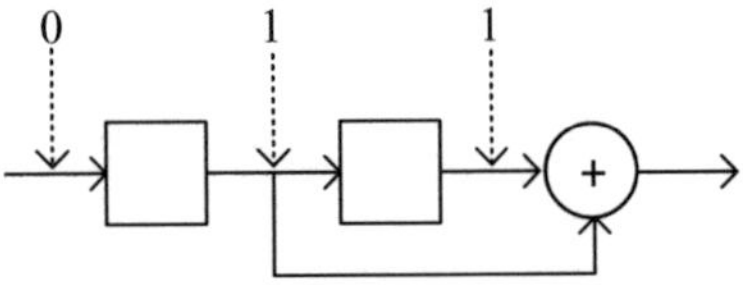


**Fig. 8.32** Determination of the impulse response

## 8.2.1 *Short Method to Find the Impulse Response of a Convolutional Encoder*

The impulse response of a convolutional encoder can be determined by inspecting the memory cell input/output connections to the circuit output. If there is a line connected between circuit output and cell input/output, we label the connection by "1." Otherwise, we label it by "0." Let's illustrate the concept by an example.

**Example 8.4** A convolutional encoder circuit is depicted in Fig. 8.29.

The impulse response of the convolutional encoder depicted in Fig. 8.29 can be determined as in Fig. 8.30. It is seen from Fig. 8.30 that the impulse response of the convolutional encoder can be written as

$$\boldsymbol{h} = [01011].$$

**Example 8.5** The impulse response of the convolutional encoder circuit in Fig. 8.31 is studied in Example 8.3, and the impulse response is found as $\boldsymbol{h} = [011]$.

We can find the impulse response of the convolutional encoder in Fig. 8.31 using the short method as in Fig. 8.32.

From Fig. 8.32, the impulse response of the convolutional circuit is found as 011.

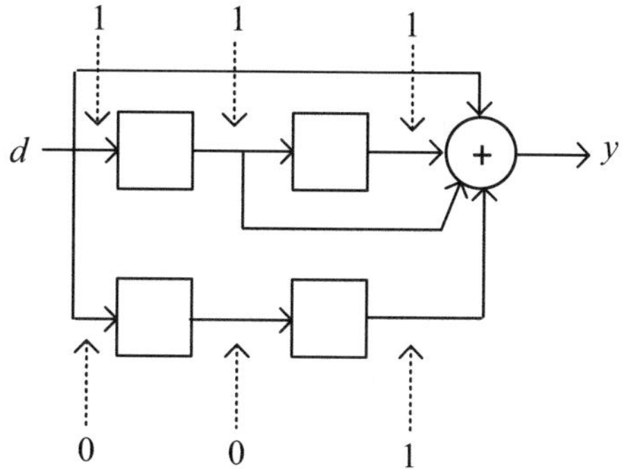


**Fig. 8.33** Determination of two impulse responses

### 8.2.2 *Parallel Path Case*

If there is more than one path between an input and an output of the convolutional encoder, then, to find the impulse response of the convolutional encoder for the given input output pair, we calculate the impulse responses of the parallel paths and sum them to find the required impulse response.

**Example 8.6** In Fig. 8.33, a convolutional encoder is depicted. It is clear from Fig. 8.33 that there are two parallel paths between the input and output of the convolutional encoder, and each path has its impulse response as shown in Fig. 8.33.

The impulse response of the upper path is $\boldsymbol{h}_{\boldsymbol{u}} = [1\ 1\ 1]$, and the impulse response of the lower path is $\boldsymbol{h}_{\boldsymbol{l}} = [0\ 0\ 1]$.

The impulse response of the convolutional encoder is the sum of the impulse responses of upper and lower paths, i.e., we have

$$\boldsymbol{h} = \boldsymbol{h}_u + \boldsymbol{h}_l \rightarrow \boldsymbol{h} = [1\ 1\ 1] + [0\ 0\ 1] \rightarrow \boldsymbol{h} = [1\ 1\ 0].$$

### 8.2.3 *Transfer Function Approach to Find the Impulse Responses of Convolutional Codes*

In Sect. 8.4, we provide detailed information about the polynomial representation of convolutional codes. In this section, we will provide short information to determine the impulse response of convolutional codes using the polynomial approach.

We can consider the input stream of a convolutional encoder as a digital sequence and consider the memory cell as a delay element. Let $x[n]$ represent a digital sequence, i.e., a binary sequence. The operation of the memory cell can be illustrated as in Fig. 8.34.

For a digital sequence $x[n]$, i.e., digital signal $x[n]$, $Z$-transform is defined as

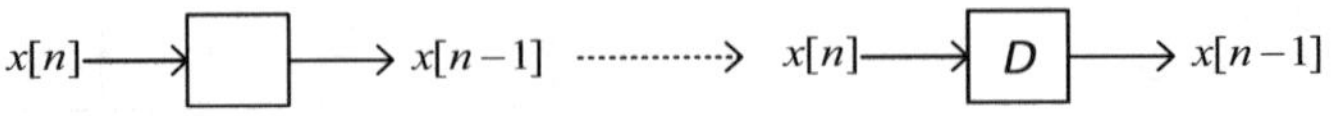


**Fig. 8.34** Delay element

$$X(z) = \sum_{n=-\infty}^{\infty} x[n]z^{-n}. \tag{8.2}$$

The shifted signal $x[n - n_0]$ has the Z-transform $z^{-n_0}X(z)$. If we let $D = z^{-1}$, then the Z-transform can be written as

$$X(z) = \sum_{n=-\infty}^{\infty} x[n]D^n. \tag{8.3}$$

Transfer function of a convolutional encoder between an input port and an output port is defined as

$$H(z) = \frac{Y(z)}{X(z)} \leftrightarrow H(D) = \frac{Y(D)}{X(D)}. \tag{8.4}$$

where $H(z)$ is the Z-transform of the impulse response, $X(z)$ is the Z-transform of the input sequence, and $Y(z)$ is the Z-transform of the output sequence. The inverse Z-transform of $H(z)$ is a bit sequence which is the impulse response of the convolutional encoder.

Note that $H(D)$ is the polynomial representation of the impulse response of the convolutional encoder, and since $x[n] \in \{0, 1\}$, Z-transform of a binary sequence equals to its polynomial representation.

**Example 8.7** Find the polynomial representation and Z-transform of $h = [0\ 1\ 0\ 1\ 1]$.

**Solution 8.7** The polynomial representation of

$$h = \left[ \underbrace{0}_{h[0]}\ \underbrace{1}_{h[1]}\ \underbrace{0}_{h[2]}\ \underbrace{1}_{h[3]}\ \underbrace{1}_{h[4]} \right]$$

is

$$h(D) = D + D^3 + D^4.$$

The Z-transform of $h$ is calculated as

$$H(D) = \sum_{n=-\infty}^{\infty} h[n]D^n \rightarrow H(D) = h[0]D^0 + h[1]D + h[2]D^2 + h[3]D^3 + h[4]D^4$$

leading to

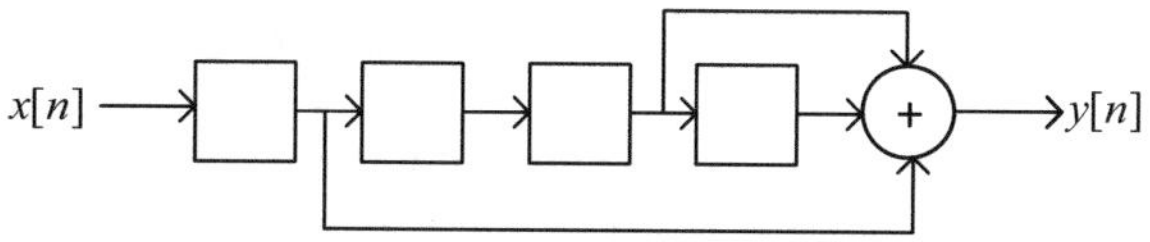


**Fig. 8.35** Convolutional encoder for Example 8.8

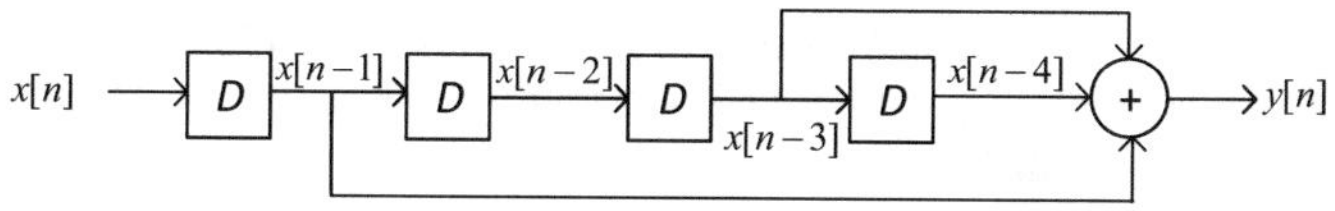


**Fig. 8.36** Determination of the outputs of the delay elements

$$H(D) = D + D^3 + D^4$$

which is the same as polynomial representation of $h$, i.e., same as $h(D)$.

**Example 8.8** Find the impulse response of the convolutional encoder shown in Fig. 8.35 using transfer function approach.

Considering the memory cells as delay elements, we can obtain the signal flow graph of the convolutional encoder as shown in Fig. 8.36.

Using Fig. 8.36, we can write that

$$y[n] = x[n-1] + x[n-3] + x[n-4]$$

where, taking the $Z$-transform of both sides, we get

$$Y(z) = DX(z) + D^3X(z) + D^4X(Z)$$

from which we obtain that

$$\frac{Y(Z)}{X(Z)} = D + D^3 + D^4 \rightarrow H(D) = D + D^3 + D^4. \tag{8.5}$$

The $H(D)$ expression in Eq. (8.5) is the polynomial representation of the impulse response $h$, and then, we can write the impulse response as

$$h = [0\ 1\ 0\ 1\ 1]$$

which is the same result as the one found in Example 8.4.

### 8.2.4 Convolution Operation

Let $\boldsymbol{x}$ and $\boldsymbol{y}$ be two binary sequences of lengths $M$ and $N$, respectively. The convolution of $\boldsymbol{x}$ and $\boldsymbol{y}$ produces another sequence $\boldsymbol{z}$ shown as

$$z = \boldsymbol{x} * \boldsymbol{y}. \tag{8.6}$$

To calculate the convolution of $\boldsymbol{x}$ and $\boldsymbol{y}$, we reverse one of the sequences $\boldsymbol{x}$ and $\boldsymbol{y}$ and shift it toward left/right until they overlap on a single bit. We multiply the overlapping points, and then, we shift the sequence toward right/left so that they overlap on two bits. We multiply the overlapping bits and find the mod-2 summation result. Shifting, overlapping, and mod-2 summation operations are continued till the sequences do not overlap.

**Example 8.9** The convolution of $\boldsymbol{x} = [1011]$ and $\boldsymbol{y} = [110]$ can be performed as follows.

If we reverse $\boldsymbol{y}$, we obtain 011. Using the reversed $\boldsymbol{y}$, and the sequence $\boldsymbol{x} = [1011]$, we evaluate the convolution of $\boldsymbol{x}$ and $\boldsymbol{y}$ as

```
  1011
011    ········> 1×1 → 1········> LSB

 1011
011    ········> 1×1+ 0×1→ 1

1011
011    ········> 1×0+ 0×1+ 1×1→ 1

1011
 011   ········> 0×0+ 1×1+ 1×1→ 0

1011
  011  ········> 1×0+ 1×1→ 1

1011
   011 ········> 0×1→ 0········> MSB
```

leading to the result

$$z = [010111].$$

**Remark** If the impulse response $\boldsymbol{h}$ of a convolutional encoder circuit is known, then, for an arbitrary input binary sequence $\boldsymbol{x}$, the circuit output can be found by evaluating the convolution of $\boldsymbol{h}$ and $\boldsymbol{x}$.

**Example 8.10** In Example 8.2, we calculated the output of the convolutional encoder shown in Fig. 8.37.

For the binary input sequence $\boldsymbol{d} = [1011]$, we found it as 101110. The impulse response of the convolutional encoder circuit shown in Fig. 8.37 is $\boldsymbol{h} = [011]$. The convolution of $\boldsymbol{h}$ and $\boldsymbol{d}$ can be obtained as

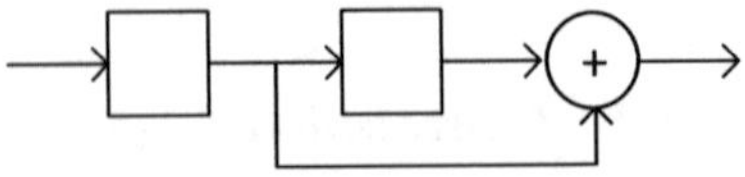


**Fig. 8.37** Convolutional encoder for Example 8.10

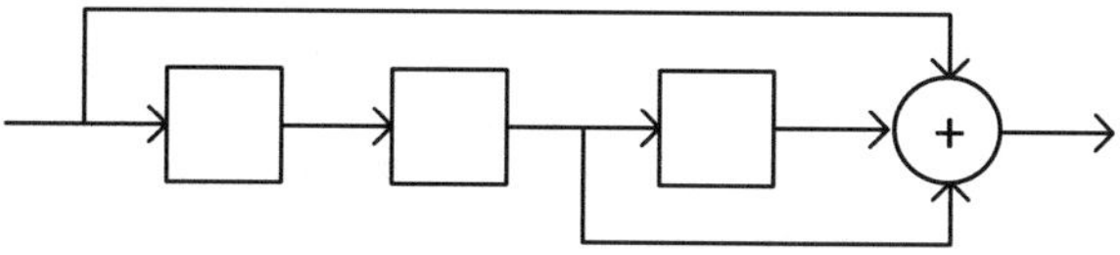


**Fig. 8.38** Convolutional encoder for Example 8.11

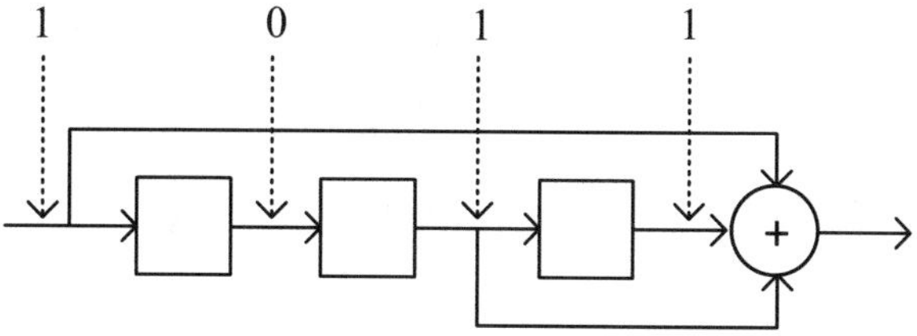


**Fig. 8.39** Determination of the impulse response

1011 / 110 ⋯⋯> $1\times 0 \rightarrow 0$ ⋯⋯> LSB

1011 / 110 ⋯⋯> $1\times 1+0\times 0 \rightarrow 1$

1011 / 110 ⋯⋯> $1\times 1+0\times 1+1\times 0 \rightarrow 1$

1011 / 110 ⋯⋯> $0\times 1+1\times 1+1\times 0 \rightarrow 1$

1011 / 110 ⋯⋯> $1\times 1+1\times 1 \rightarrow 0$

1011 / 110 ⋯⋯> $1\times 1 \rightarrow 1$ ⋯⋯> MSB

leading to the result $z = [101110]$.

**Example 8.11** Find the impulse response of the convolutional circuit in Fig. 8.38, and calculate the output of the circuit for the binary input sequence $\boldsymbol{d} = [11010]$.

**Solution 8.11** We can label the flip-flop inputs and outputs considering their connections to the circuit output as in Fig. 8.39.

Using Fig. 8.39, we can determine the impulse response of the convolutional encoder as

$$\boldsymbol{h} = [1011].$$

For the input sequence $\boldsymbol{d} = [11010]$, the convolutional encoder output can be obtained by taking the convolution of $\boldsymbol{d}$ and $\boldsymbol{h}$ as

```
   11010
1101     ········> 1×1→1 ········> LSB

   11010
  1101    ········> 1×0+1×1→1

   11010
   1101   ········> 1×1+1×0+0×1→1

   11010
   1101   ········> 1×1+1×1+0×0+1×1→1

   11010
    1101  ········> 1×1+0×1+1×0+0×1→1

   11010
     1101 ········> 0×1+1×1+0×0→1

   11010
      1101 ········> 1×1+0×1→1

   11010
       1101 ········> 0×1→0 ········> MSB
```

resulting in

$$\boldsymbol{c} = \boldsymbol{d} * \boldsymbol{h} \rightarrow \boldsymbol{c} = [01111111].$$

A convolutional encoder may have more than one input and one output. In this case, between every input-output pair, there exists an impulse response. If there are $M$ inputs and $N$ outputs, then there are $M \times N$ impulse responses between inputs and outputs.

**Example 8.12** Find the impulse response between every input and output pair of the convolutional encoder shown in Fig. 8.40.

If we denote the impulse response between $x_i$ and $y_j$ by $\boldsymbol{h}_{ij}$, then we have four impulse responses, and these impulse responses are $\boldsymbol{h}_{11}$, $\boldsymbol{h}_{12}$, $\boldsymbol{h}_{21}$, and $\boldsymbol{h}_{22}$. To determine the impulse response $\boldsymbol{h}_{ij}$, we look at the path between $x_i$ and $x_j$. According to this, we can determine the impulse $\boldsymbol{h}_{11}$ response considering the path between $x_1$ and $y_1$ as indicated in Fig. 8.41

where it is seen that $\boldsymbol{h}_{11} = [0101]$. Since there is no signal flow from $x_1$ to $y_2$, the impulse response $\boldsymbol{h}_{12}$ is the all-zero sequence, i.e., $\boldsymbol{h}_{12} = [0000]$.

The impulse response between $x_2$ and $y_1$ can be found considering the path between $x_2$ and $y_1$ as shown in Fig. 8.42

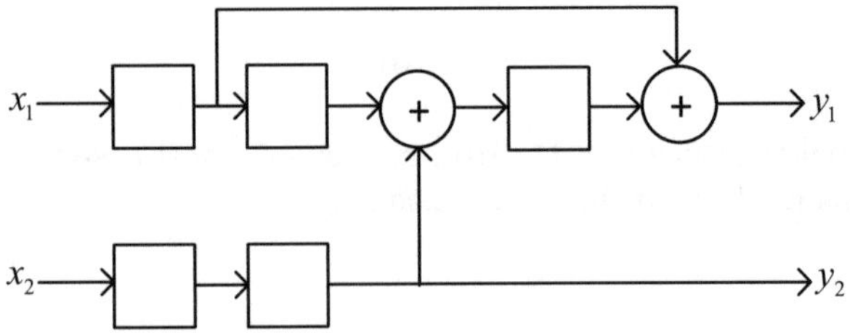


**Fig. 8.40** Convolutional encoder for Example 8.12

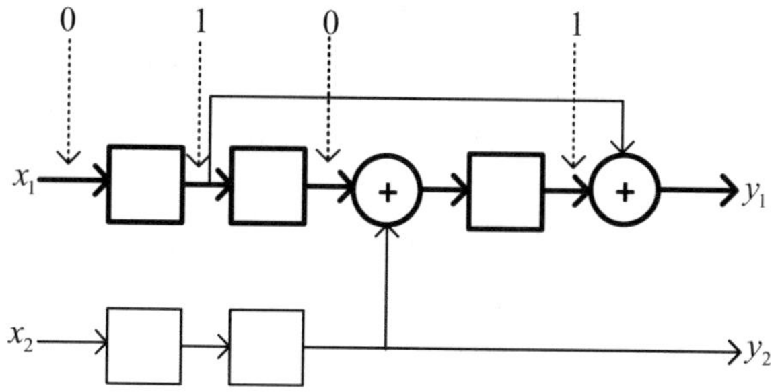


**Fig. 8.41** Determination of impulse response $h_{11}$

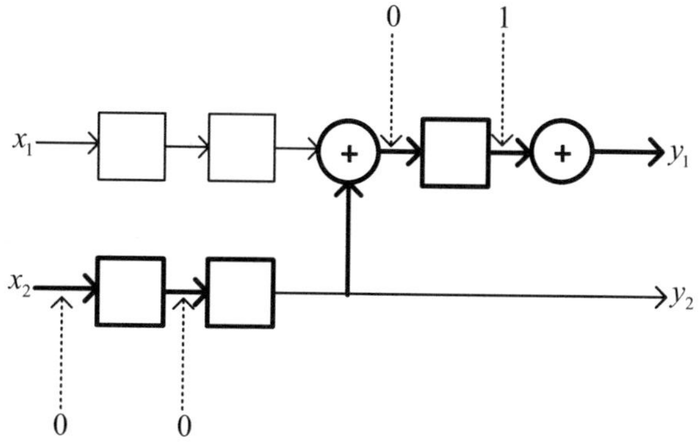


**Fig. 8.42** Determination of the impulse response $h_{21}$

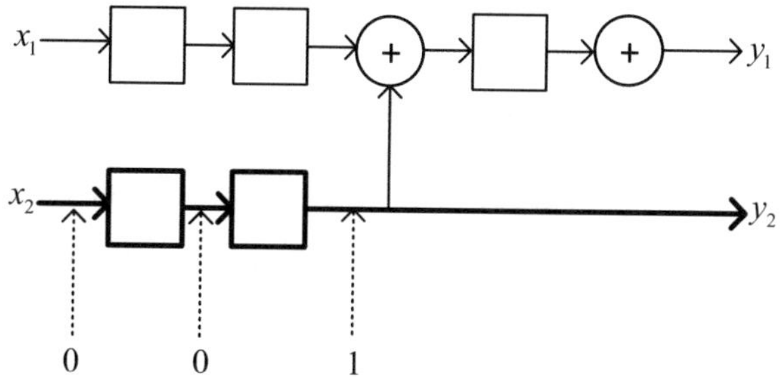


**Fig. 8.43** Determination of the impulse response $h_{22}$

where it is seen that impulse response $\boldsymbol{h_{21}}$ between $x_2$ and $y_1$ is [0001].

In a similar manner, we can determine the impulse response $\boldsymbol{h_{22}}$ considering the path between $x_2$ and $y_2$ as indicated in Fig. 8.43

where it is seen that the impulse response between $x_2$ and $y_2$ equals to 001. However, we prefer that all the impulse responses have equal lengths. For this reason, shorter impulse responses are padded by zeros so as to have the same number of bits as the longest impulse response. Then, the impulse response $\boldsymbol{h_{22}}$ can be formed as $\boldsymbol{h_{22}} = [0010]$.

Hence, the impulse responses of the convolutional encoder in Fig. 8.40 can altogether be written as

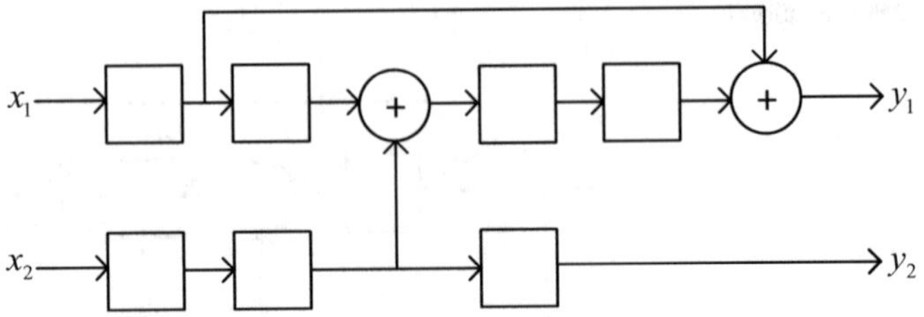


**Fig. 8.44** Convolutional encoder for exercise

**Fig. 8.45** Convolutional encoder for exercise

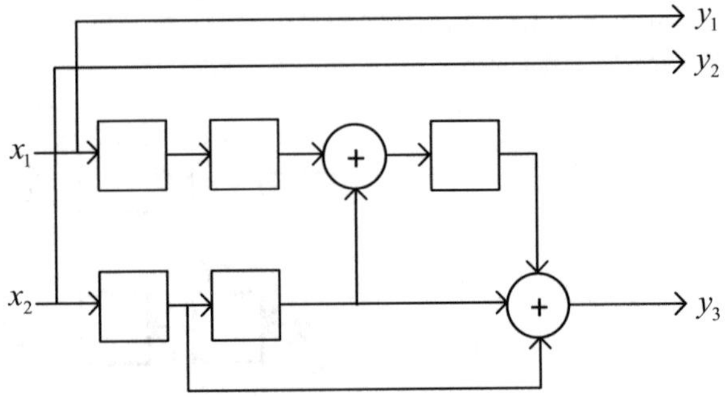


$$\boldsymbol{h}_{11} = [0101] \quad \boldsymbol{h}_{12} = [0000] \quad \boldsymbol{h}_{21} = [0001] \quad \boldsymbol{h}_{22} = [0010].$$

***Note*** If there is a maximum of $m$ memory cells along a path of the convolutional encoder, then all the impulse responses should have $m + 1$ bits. Those shorter ones are padded by zeros so that their lengths equal to $m + 1$.

**Exercise** Find all the impulse responses of the convolutional encoder shown in Fig. 8.44.

**Exercise** Find all the impulse responses of the convolutional encoder shown in Fig. 8.45.

### 8.2.5 *Recursive Convolutional Encoders*

If a convolutional encoder includes a feedback path, then the convolutional encoder is of recursive type. A recursive convolutional encoder may have infinite impulse response, i.e., its impulse response may have infinite number of bits. In other words, when a single "1" is used as the input, the convolutional encoder output never produces all-zero sequence, i.e., always a mixture of zeros and ones is produced till infinity.

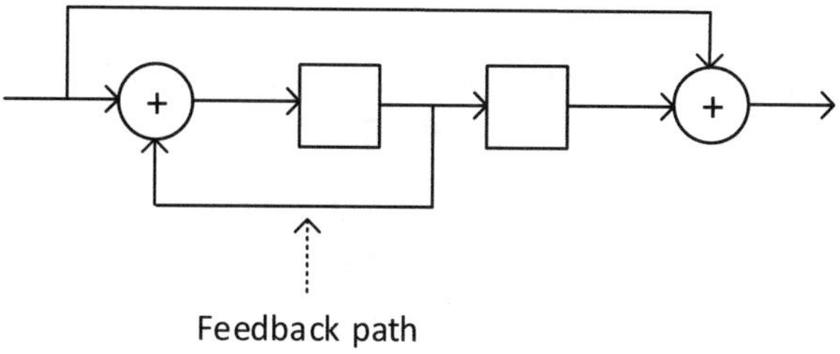


**Fig. 8.46** A recursive convolutional encoder

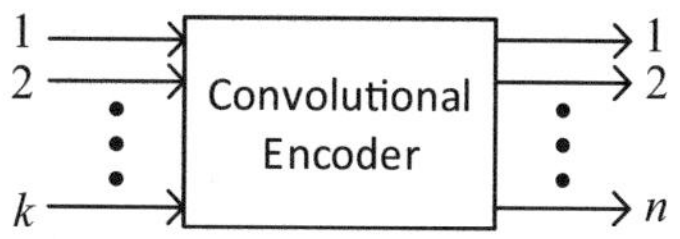


**Fig. 8.47** Black box representation of a convolutional encoder

A recursive convolutional encoder is depicted in Fig. 8.46.

## 8.3 Generator Matrices for Convolutional Codes

In general a convolutional encoder has $k$ parallel inputs and $n$ parallel outputs which can be depicted as in Fig. 8.47.

The convolutional encoder with $k$ inputs and $n$ outputs employing the black box diagram shown in Fig. 8.47 has $k \times n$ impulse responses. Let's denote the impulse response between input $d_i$ and output $y_j$ by $\boldsymbol{h}_{ij}$ such that $i = 1, \ldots, k$ and $j = 1, \ldots, n$. If there are a maximum of $m$ memory cells in any path, then each impulse response contains $m + 1$ bits.

Assume that we have an $l$-bit serial input stream denoted by $\boldsymbol{d}_i$ at the input port $i$. The output of the convolutional encoder at output port $j$ can be calculated by taking the convolution of input stream at input port $i$ and impulse response $\boldsymbol{h}_{ij}$ as

$$\boldsymbol{c}_j = \boldsymbol{d}_i * \boldsymbol{h}_{ij}. \tag{8.7}$$

If we consider all the input streams and the produced output streams, we can write the relation

$$\boldsymbol{c} = \boldsymbol{d}\boldsymbol{G} \tag{8.8}$$

where $\boldsymbol{G}$ is the generator matrix of the convolutional encoder. Now, let's see the calculation of the generator matrix of a convolutional encoder.

Let's denote the $s$th bit of $h_{ij}$ by $h_{ij}^{s}$. To construct the generator matrix of the convolutional encoder, first, we construct the sub-generator matrices $\boldsymbol{G}_r$, $r = 1, \ldots,$ $m$ using the bits of impulse responses as

$$\boldsymbol{G}_r = \begin{bmatrix} h_{11}^r & h_{12}^r & h_{13}^r & \cdots & h_{1n}^r \\ h_{21}^r & h_{22}^r & h_{23}^r & \cdots & h_{2n}^r \\ \vdots & \vdots & \vdots & \cdots & \vdots \\ h_{p1}^r & h_{p2}^r & h_{p3}^r & \cdots & h_{pn}^r \end{bmatrix}. \quad (8.9)$$

Next, using the sub-generator matrices calculated in Eq. (8.9), we construct the generator matrix of the convolutional encoder as

$$\boldsymbol{G} = \begin{bmatrix} \boldsymbol{G}_0 & \boldsymbol{G}_1 & \boldsymbol{G}_2 & \cdots & \boldsymbol{G}_m & \boldsymbol{0} & \boldsymbol{0} & \cdots \\ \boldsymbol{0} & \boldsymbol{G}_0 & \boldsymbol{G}_1 & \cdots & \cdots & \boldsymbol{G}_m & \boldsymbol{0} & \cdots \\ \boldsymbol{0} & \boldsymbol{0} & \boldsymbol{G}_0 & \cdots & \cdots & \cdots & \boldsymbol{G}_m & \cdots \\ \vdots & \vdots & \vdots & \vdots & \vdots & \vdots & \vdots & \cdots \end{bmatrix}. \quad (8.10)$$

**Example 8.13** Assume that the maximum number of memory elements between an input and an output of a convolutional encoder is $m = 2$, and there are $k = 3$ input ports and $n = 4$ output ports. There are $3 \times 4 = 12$ impulse responses each having $m + 1 = 3$ bits, and the impulse responses are given as

$$\begin{array}{llll} \boldsymbol{h}_{11} = [100] & \boldsymbol{h}_{12} = [100] & \boldsymbol{h}_{13} = [100] & \boldsymbol{h}_{14} = [100] \\ \boldsymbol{h}_{21} = [000] & \boldsymbol{h}_{22} = [110] & \boldsymbol{h}_{23} = [010] & \boldsymbol{h}_{24} = [100] \\ \boldsymbol{h}_{31} = [000] & \boldsymbol{h}_{32} = [010] & \boldsymbol{h}_{33} = [101] & \boldsymbol{h}_{34} = [101]. \end{array}$$

Assume that, at each input port, we have binary sequences consisting of 3 bits. That means that we have nine information bits in total at the port inputs to be encoded. This leads to the conclusion that the generator matrix should have nine rows.

We can construct the sub-generator matrix $\boldsymbol{G}_0$ selecting the first bits of impulse responses as

$$\begin{array}{llll} \boldsymbol{h}_{11} = [\mathbf{1}00] & \boldsymbol{h}_{12} = [\mathbf{1}00] & \boldsymbol{h}_{13} = [\mathbf{1}00] & \boldsymbol{h}_{14} = [\mathbf{1}00] \\ \boldsymbol{h}_{21} = [\mathbf{0}00] & \boldsymbol{h}_{22} = [\mathbf{1}10] & \boldsymbol{h}_{23} = [\mathbf{0}10] & \boldsymbol{h}_{24} = [\mathbf{1}00] \\ \boldsymbol{h}_{31} = [\mathbf{0}00] & \boldsymbol{h}_{32} = [\mathbf{0}10] & \boldsymbol{h}_{33} = [\mathbf{1}01] & \boldsymbol{h}_{34} = [\mathbf{1}01] \end{array}$$

and placing them into a matrix as

$$\boldsymbol{G}_0 = \begin{bmatrix} 1111 \\ 0101 \\ 0011 \end{bmatrix}.$$

We can construct the sub-generator matrix $\boldsymbol{G}_1$ selecting the second bits of the impulse responses as

$$\begin{aligned} &\boldsymbol{h}_{11} = [100] \quad \boldsymbol{h}_{12} = [100] \quad \boldsymbol{h}_{13} = [100] \quad \boldsymbol{h}_{14} = [100] \\ &\boldsymbol{h}_{21} = [000] \quad \boldsymbol{h}_{22} = [110] \quad \boldsymbol{h}_{23} = [010] \quad \boldsymbol{h}_{24} = [100] \\ &\boldsymbol{h}_{31} = [000] \quad \boldsymbol{h}_{32} = [010] \quad \boldsymbol{h}_{33} = [101] \quad \boldsymbol{h}_{34} = [101] \end{aligned}$$

and placing them into a matrix as

$$\boldsymbol{G}_1 = \begin{bmatrix} 0000 \\ 0110 \\ 0100 \end{bmatrix}.$$

Lastly, proceeding in a similar manner, we can construct the sub-generator matrix $\boldsymbol{G}_2$ selecting the third bits of the impulse responses as

$$\begin{aligned} &\boldsymbol{h}_{11} = [100] \quad \boldsymbol{h}_{12} = [100] \quad \boldsymbol{h}_{13} = [100] \quad \boldsymbol{h}_{14} = [100] \\ &\boldsymbol{h}_{21} = [000] \quad \boldsymbol{h}_{22} = [110] \quad \boldsymbol{h}_{23} = [010] \quad \boldsymbol{h}_{24} = [100] \\ &\boldsymbol{h}_{31} = [000] \quad \boldsymbol{h}_{32} = [010] \quad \boldsymbol{h}_{33} = [101] \quad \boldsymbol{h}_{34} = [101]. \end{aligned}$$

and placing them into a matrix as

$$\boldsymbol{G}_2 = \begin{bmatrix} 0000 \\ 0000 \\ 0011 \end{bmatrix}.$$

Using the sub-generator matrices, $\boldsymbol{G}_0$, $\boldsymbol{G}_1$, $\boldsymbol{G}_2$, we can construct the generator matrix $\boldsymbol{G}$ using

$$\boldsymbol{G} = \begin{bmatrix} \boldsymbol{G}_0 & \boldsymbol{G}_1 & \boldsymbol{G}_2 & \boldsymbol{0} & \boldsymbol{0} \\ \boldsymbol{0} & \boldsymbol{G}_0 & \boldsymbol{G}_1 & \boldsymbol{G}_2 & \boldsymbol{0} \\ \boldsymbol{0} & \boldsymbol{0} & \boldsymbol{G}_0 & \boldsymbol{G}_1 & \boldsymbol{G}_2 \end{bmatrix}$$

as

$$\boldsymbol{G}=\begin{bmatrix} 1111 & 0000 & 0000 & 0000 & 0000 \\ 0101 & 0110 & 0000 & 0000 & 0000 \\ 0011 & 0100 & 0011 & 0000 & 0000 \\ \\ 0000 & 1111 & 0000 & 0000 & 0000 \\ 0000 & 0101 & 0110 & 0000 & 0000 \\ 0000 & 0011 & 0100 & 0011 & 0000 \\ \\ 0000 & 0000 & 1111 & 0000 & 0000 \\ 0000 & 0000 & 0101 & 0110 & 0000 \\ 0000 & 0000 & 0011 & 0100 & 0011 \end{bmatrix}$$

which can be expressed using concatenated bits as

$$\boldsymbol{G}=\begin{bmatrix} 11110000000000000000 \\ 01010110000000000000 \\ 00001111000000000000 \\ 00000101011000000000 \\ 00000011010000110000 \\ 00000000111100000000 \\ 00000000010101100000 \\ 00000000001101000011 \end{bmatrix}.$$

***Note*** The size of the generator matrix of a convolutional encoder depends on the length of the input streams. For this reason, there is no a fixed size definition for the generator matrices of convolutional codes.

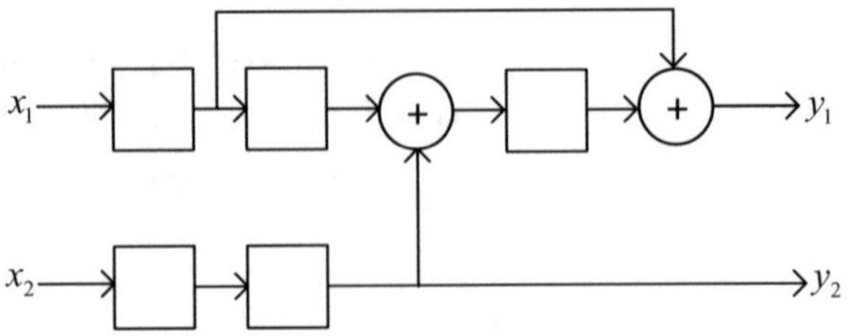


**Fig. 8.48** Convolutional encoder for Example 8.14

**Example 8.14** Find the outputs of the convolutional encoder shown in Fig. 8.48 for the input streams $\boldsymbol{x_1} = [1011]$ and $\boldsymbol{x_2} = [0101]$ using both encoding by generator matrix and encoding via convolution approaches.

**Solution 8.14** The impulse responses between inputs and outputs are found in Example 8.12 as

$$\boldsymbol{h}_{11} = [0101] \quad \boldsymbol{h}_{12} = [0000] \quad \boldsymbol{h}_{21} = [0001] \quad \boldsymbol{h}_{22} = [0010].$$

The bit streams resulting from the convolution of input streams and corresponding impulse responses can be evaluated as

$$\begin{aligned} \boldsymbol{x}_1 * \boldsymbol{h}_{11} &\rightarrow [1011] * [0101] = [0\,1\,0\,0\,1\,1\,1] \\ \boldsymbol{x}_1 * \boldsymbol{h}_{12} &\rightarrow [1011] * [0000] = [0\,0\,0\,0\,0\,0\,0] \\ \boldsymbol{x}_2 * \boldsymbol{h}_{21} &\rightarrow [0101] * [0001] = [0\,0\,0\,0\,1\,0\,1] \\ \boldsymbol{x}_2 * \boldsymbol{h}_{22} &\rightarrow [0101] * [0101] = [0\,0\,0\,1\,0\,1\,0]. \end{aligned}$$

The output streams can be obtained as

$$\begin{aligned} \boldsymbol{y}_1 = \boldsymbol{x}_1 * \boldsymbol{h}_{11} + \boldsymbol{x}_2 * \boldsymbol{h}_{21} \rightarrow \boldsymbol{y}_1 &= [0\,1\,0\,0\,1\,1\,1] + [0\,0\,0\,0\,1\,0\,1] \rightarrow \\ \boldsymbol{y}_1 &= [0\,1\,0\,0\,0\,1\,0] \\ \boldsymbol{y}_2 = \boldsymbol{x}_1 * \boldsymbol{h}_{12} + \boldsymbol{x}_2 * \boldsymbol{h}_{22} \rightarrow \boldsymbol{y}_2 &= [0\,0\,0\,0\,0\,0\,0] + [0\,0\,0\,1\,0\,1\,0] \rightarrow \\ \boldsymbol{y}_2 &= [0\,0\,0\,1\,0\,1\,0] \end{aligned}$$

and multiplexing $\boldsymbol{y_1}$ and $\boldsymbol{y_2}$, we get

$$\boldsymbol{y} = [0\,0\,1\,0\,0\,0\,0\,1\,0\,0\,1\,1\,0\,0]. \tag{8.11}$$

Now, let's use the generator matrix approach to find the output of the convolutional encoder. First, let's calculate the generator matrix of the convolutional encoder. The generator submatrices can be found using the impulse responses

$$\begin{aligned} \boldsymbol{h}_{11} &= [0101] \quad \boldsymbol{h}_{12} = [0000] \\ \boldsymbol{h}_{21} &= [0001] \quad \boldsymbol{h}_{22} = [0010] \end{aligned}$$

and following the approach in Example 8.13 as

$$\boldsymbol{G}_0 = \begin{bmatrix} 00 \\ 00 \end{bmatrix} \quad \boldsymbol{G}_1 = \begin{bmatrix} 10 \\ 00 \end{bmatrix} \quad \boldsymbol{G}_2 = \begin{bmatrix} 00 \\ 01 \end{bmatrix} \quad \boldsymbol{G}_3 = \begin{bmatrix} 10 \\ 10 \end{bmatrix}$$

from which the generator matrix is constructed as

$$G = \begin{bmatrix} G_0 & G_1 & G_2 & G_3 & 0 & 0 & 0 \\ 0 & G_0 & G_1 & G_2 & G_3 & 0 & 0 \\ 0 & 0 & G_0 & G_1 & G_2 & G_3 & 0 \\ 0 & 0 & 0 & G_0 & G_1 & G_2 & G_3 \end{bmatrix}$$

leading to

$$G = \begin{bmatrix} 00 & 10 & 00 & 10 & 00 & 00 & 00 \\ 00 & 00 & 01 & 10 & 00 & 00 & 00 \\ \\ 00 & 00 & 10 & 00 & 10 & 00 & 00 \\ 00 & 00 & 00 & 01 & 10 & 00 & 00 \\ \\ 00 & 00 & 00 & 10 & 00 & 10 & 00 \\ 00 & 00 & 00 & 00 & 01 & 10 & 00 \\ \\ 00 & 00 & 00 & 00 & 10 & 00 & 10 \\ 00 & 00 & 00 & 00 & 00 & 01 & 10 \end{bmatrix}$$

which can be written as

$$G = \begin{bmatrix} 00100010000000 \\ 00000110000000 \\ 00001000100000 \\ 00000001100000 \\ 00000010001000 \\ 00000000011000 \\ 00000000100010 \\ 00000000000110 \end{bmatrix}.$$

Note that row size of the generator matrix is 8, and this number is determined considering the input sequences $\boldsymbol{x_1} = [1011]$ and $\boldsymbol{x_2} = [0101]$. Multiplexing $\boldsymbol{x_1}$ and $\boldsymbol{x_2}$, we obtain

$$\boldsymbol{x} = [10011011].$$

When $x$ is encoded, we get

$$c = xG \rightarrow c = [10011011]\begin{bmatrix} 00100010000000 \\ 00000110000000 \\ 00001000100000 \\ 00000001100000 \\ 00000010001000 \\ 00000000011000 \\ 00000000100010 \\ 00000000000110 \end{bmatrix} \rightarrow c = [00100001001100]$$

which is the same as that of the result in Eq. (8.11).

## 8.4 Polynomial Representation of Convolutional Codes

Let $\boldsymbol{h}$ be the impulse response between an input and an output port of a convolutional encoder. For an input stream $\boldsymbol{d}$, the output stream is calculated using the convolutional operation as

$$\boldsymbol{y} = \boldsymbol{d} * \boldsymbol{h}.$$

If $y(D)$, $d(D)$, and $h(D)$ are the polynomial representations of the bit vectors $\boldsymbol{y}$, $\boldsymbol{h}$, and $\boldsymbol{d}$, then we have

$$y(D) = d(D)h(D). \tag{8.12}$$

That is, convolution operation can be achieved by the polynomial multiplication. Let's illustrate the concept by an example.

**Example 8.15** The impulse response of the convolutional encoder shown in Fig. 8.49 can be evaluated as $\boldsymbol{h} = [0101]$. Assume that we have the input sequence $\boldsymbol{d} = [1011]$.

The output of the convolutional encoder can be found as

$$\boldsymbol{c} = \boldsymbol{d} * \boldsymbol{h} \rightarrow \boldsymbol{c} = [1011] * [0101] \rightarrow \boldsymbol{c} = [0100111].$$

The polynomial representation of $\boldsymbol{d}$ is

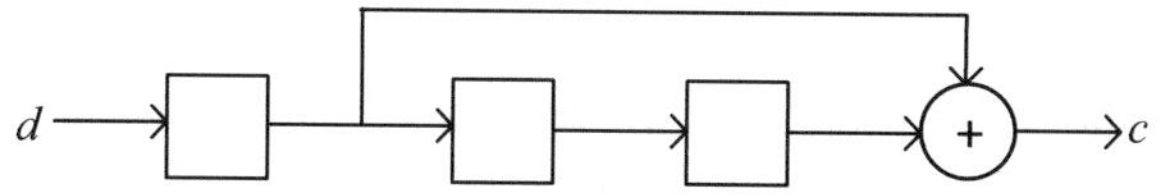


**Fig. 8.49** Convolutional encoder for Example 8.15

$$d(D) = 1 + D^2 + D^3$$

and the polynomial representation of $\boldsymbol{h}$ is

$$h(D) = D + D^3.$$

The output of the convolutional encoder can be found using polynomial multiplication

$$c(D) = d(D)h(D) \rightarrow c(D) = \left(1 + D^2 + D^3\right)\left(D + D^3\right) \rightarrow$$
$$c(D) = D + D^4 + D^5 + D^6$$

which is the polynomial representation of $\boldsymbol{c} = [0100111]$.

***Note*** If you consider your bit vector as $\boldsymbol{b} = [b_0\ b_1\ b_2 \ldots]$, then its polynomial representation becomes as

$$b(D) = b_0 + b_1 D + b_2 D^2 + \ldots$$

On the other hand, if you consider your bit vector $\boldsymbol{b} = \left[\ldots b_2'\ b_1'\ b_0'\right]$, it has the polynomial representation

$$b(D) = b_0' + b_1' D + b_2' D^2 + \ldots$$

Whichever form you adapt is not a matter; however, you should always use the same form for every bit vector. For instance, if we represent the bit vectors in our example as

$$d(D) = 1 + D + D^3 \quad h(D) = 1 + D^2$$

then the output of the convolutional encoder happens to be

$$c(D) = d(D)h(D) \rightarrow c(D) = \left(1 + D + D^3\right)\left(1 + D^2\right) \rightarrow c(D)$$
$$= 1 + D + D^2 + D^5$$

which is still the polynomial representation of $\boldsymbol{c} = [0100111]$.

### 8.4.1 Generator Matrix in Polynomial Form

The polynomial form of the generator matrix can be obtained in two different approaches. In the first approach, the sub-generator matrices are used to obtain the polynomial form of the generator matrix as

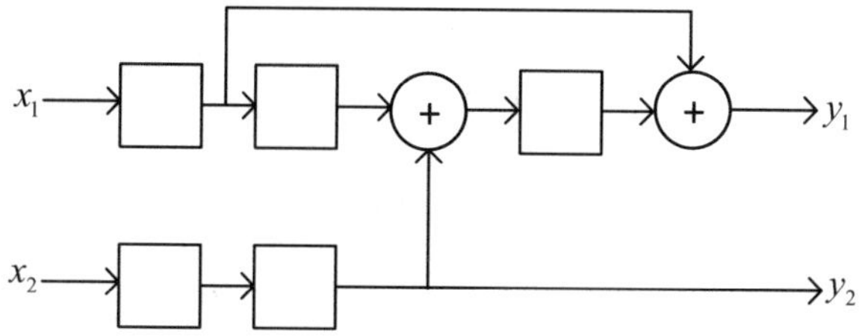


**Fig. 8.50** Convolutional encoder for Example 8.16

$$G(D) = G_0 + G_1 D + \ldots + G_m D^m. \tag{8.13}$$

In the second approach, we first obtain polynomial forms of all the impulse responses. The element of the generator matrix at the location $(i, j)$ is $h_{ij}(D)$ which is the polynomial representation of the impulse response $h_{ij}$. The generator polynomial can be expressed as

$$G(D) = \left[h_{ij}(D)\right]. \tag{8.14}$$

**Example 8.16** For the convolutional encoder shown in Fig. 8.50, the impulse responses can be obtained as

$$\boldsymbol{h}_{11} = [0101] \quad \boldsymbol{h}_{12} = [0000] \quad \boldsymbol{h}_{21} = [0001] \quad \boldsymbol{h}_{22} = [0010].$$

The impulse responses are represented in polynomial form

$$h_{11}(D) = D + D^3 \quad h_{12}(D) = 0 \quad h_{21}(D) = D^3 \quad h_{22}(D) = D^2.$$

Polynomial forms of the impulse responses are also called generator polynomials of the convolutional encoder.

The matrix form of the generator matrix can be obtained using the polynomial form of the impulse responses as

$$G(D) = \begin{bmatrix} h_{11}(D) & h_{12}(D) \\ h_{21}(D) & h_{22}(D) \end{bmatrix}$$

leading to

$$G(D) = \begin{bmatrix} D + D^3 & 0 \\ D^3 & D^2 \end{bmatrix}. \tag{8.15}$$

The sub-generator matrices of the convolutional encoder shown in Fig. 8.50 can be obtained as

$$\boldsymbol{G_0} = \begin{bmatrix} 00 \\ 00 \end{bmatrix} \quad \boldsymbol{G_1} = \begin{bmatrix} 10 \\ 00 \end{bmatrix} \quad \boldsymbol{G_2} = \begin{bmatrix} 00 \\ 01 \end{bmatrix} \quad \boldsymbol{G_3} = \begin{bmatrix} 10 \\ 10 \end{bmatrix}$$

from which the generator matrix in polynomial form is obtained as

$$G(D) = \boldsymbol{G_0} + \boldsymbol{G_1}D + \boldsymbol{G_2}D^2 + \boldsymbol{G_3}D^3$$

in which, substituting the sub-generator matrices, we obtain

$$G(D) = \begin{bmatrix} 00 \\ 00 \end{bmatrix} + \begin{bmatrix} 10 \\ 00 \end{bmatrix} D + \begin{bmatrix} 00 \\ 01 \end{bmatrix} D^2 + \begin{bmatrix} 10 \\ 10 \end{bmatrix} D^3$$

which is simplified as

$$G(D) = \begin{bmatrix} D + D^3 & 0 \\ D^3 & D^2 \end{bmatrix}. \tag{8.16}$$

When we compare Eqs. (8.15) and (8.16), we see that they are the same of each other.

The polynomial form of the output $y_1$ can be expressed as

$$y_1(D) = x_1(D)h_{11}(D) + x_2(D)h_{21}(D).$$

In a similar manner, the polynomial form of the output $y_2$ can be expressed as

$$y_2(D) = x_1(D)h_{12}(D) + x_2(D)h_{22}(D).$$

**Exercise** Find the generator matrices of the convolutional encoders shown in Figs. 8.51 and 8.52 in polynomial form.

## 8.5 Generator Matrices of the Recursive Convolutional Encoders

When we try to find the generator matrices of the convolutional encoders, we pay attention to the impulse responses of the forward and backward paths. The path between an input port and an output port of a convolutional encoder can be shown as in Fig. 8.53.

In Fig. 8.53, the symbols $f_i$, $b_j$ are used to indicate the connection of the corresponding node to the circuit output. If $f_i = 1$, then the corresponding node has a direct connection to the circuit output. The forward impulse response can be written as

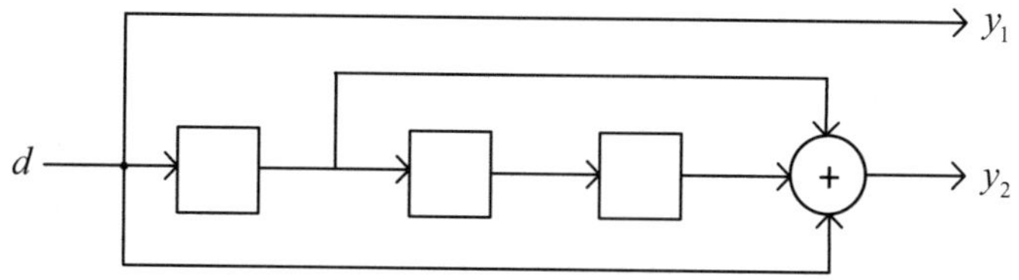


**Fig. 8.51** Convolutional encoder for exercise

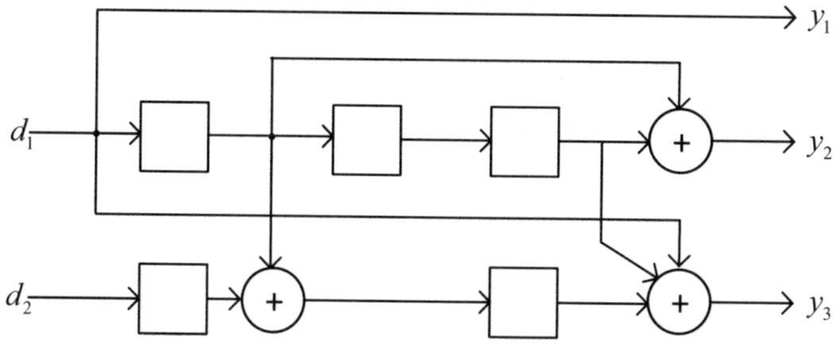


**Fig. 8.52** Convolutional encoder for exercise

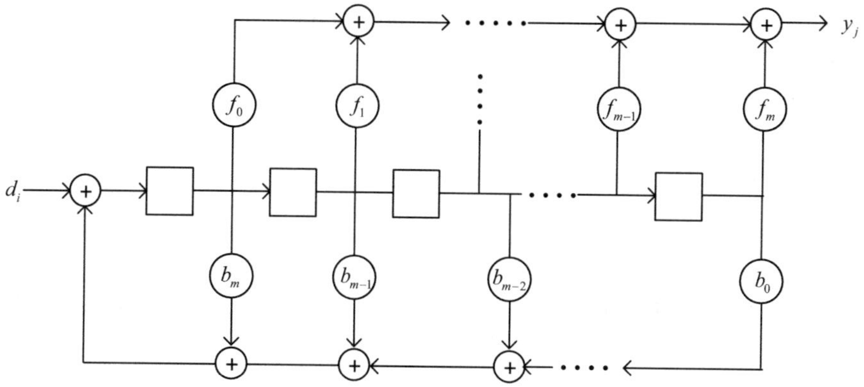


**Fig. 8.53** Recursive convolutional encoder in controllable canonical form

$$h_f = [f_0 f_1 \cdots f_m] \tag{8.17}$$

and the backward impulse response can be written as

$$h_b = [b_0\, b_1\, b_2 \ldots b_m]. \tag{8.18}$$

The polynomial form of the impulse response can be calculated as

$$h_{ij}(D) = \frac{h_f(D)}{h_b(D)} \rightarrow h_{ij}(D) = \frac{f_0 + f_1 D + f_2 D^2 + \ldots + f_m D^m}{b_0 + b_1 D + b_2 D^2 + b_m D^m}. \tag{8.19}$$

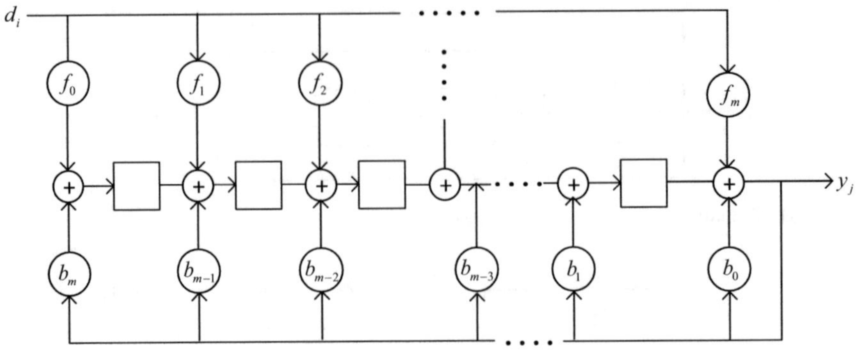


**Fig. 8.54** Convolutional encoder in observable canonical form

The convolutional encoder structure shown in Fig. 8.51 is called controllable canonical form of the convolutional encoder.

Another general structure of the convolutional encoder called observable canonical form is shown in Fig. 8.54.

For Fig. 8.54, the forward and backward impulse responses are the same as that of Eqs. (8.17) and (8.18). The polynomial form of the impulse response can be calculated using Eq. (8.19).

**Example 8.17** Find the impulse responses of the convolutional encoder shown in Fig. 8.55.

**Solution 8.17** The impulse response between $d$ and $y_1$ is $h_{11} = [1]$. To find the impulse response between $d$ and $y_2$, we can consider the forward and backward paths separately and find the corresponding impulse responses in polynomial form and take the division of them.

The forward path and its impulse response calculation are depicted in Fig. 8.56. The forward impulse response can be expressed in polynomial form as

$$h_f(D) = 1 + D + D^2 + D^3.$$

The backward path is depicted in Fig. 8.57, and while labeling the lines, we consider $y_2$ as input and end of the backward path as output node, and those lines having direct connection to the output node are labeled as "1," and the others are labeled as "0."

The backward impulse response can be written as $h_f = [1\ 0\ 1\ 1]$ which can be expressed in polynomial form as

$$h_b(D) = 1 + D^2 + D^3.$$

The polynomial form of the impulse response between $d$ and $y_2$ is calculated as

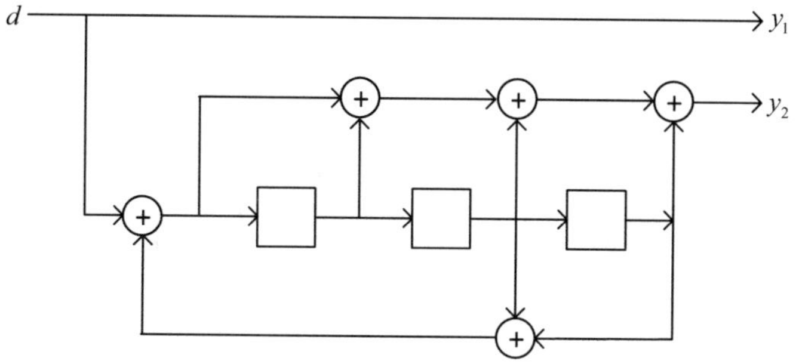


**Fig. 8.55** Convolutional encoder for Example 8.17

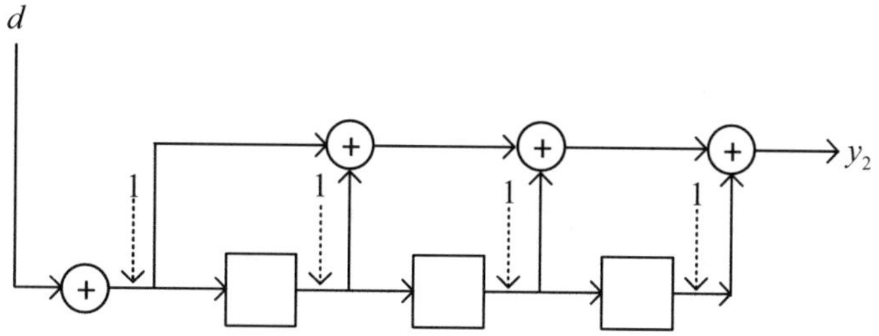


**Fig. 8.56** Impulse response of forward path

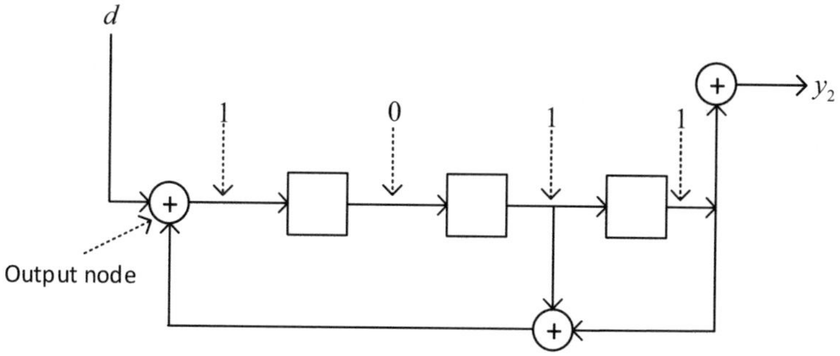


**Fig. 8.57** Impulse response of backward path

$$h(D) = \frac{h_f(D)}{h_b(D)} \rightarrow h(D) = \frac{1 + D + D^2 + D^3}{1 + D^2 + D^3}.$$

**Second Method**

The impulse response between $x$ and $y_2$ can also be calculated using the transfer function calculation approach as explained in Sect. 8.2.3. For this purpose, we first label the $x[n]$ and $a[n]$ signals as depicted in Fig. 8.58.

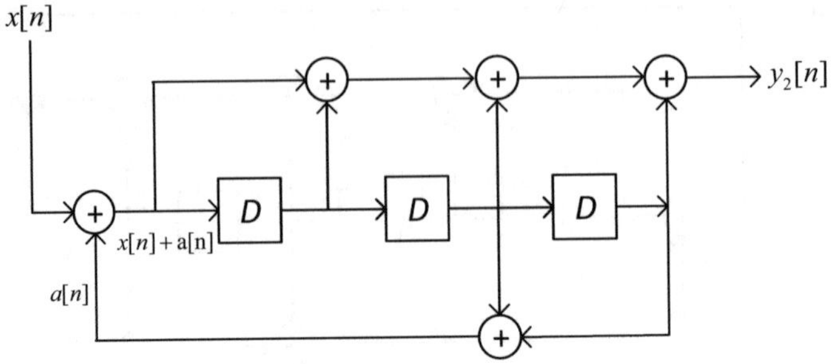


**Fig. 8.58** Convolutional encoder with delay elements

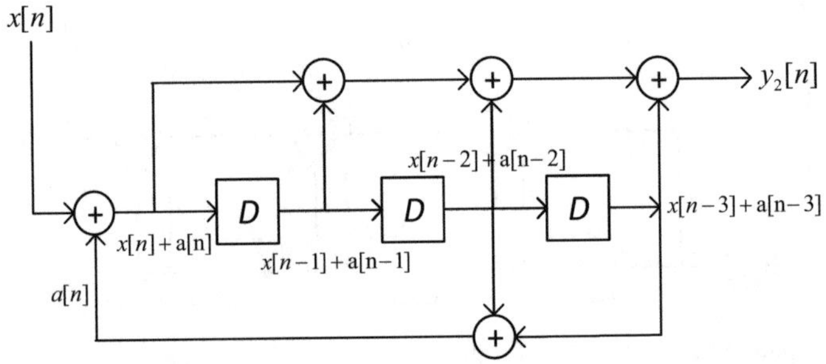


**Fig. 8.59** The outputs of the delay elements

Considering the memory cells as delay elements and binary vectors as digital sequences, we can draw the signal flow graph between $x$ and $y_2$ as in Fig. 8.59.

Using Fig. 8.59, we can write

$$y_2[n] = x[n] + a[n] + x[n-1] + a[n-1] + x[n-2] + a[n-2] \\ + x[n-3] + a[n-3] \tag{8.20}$$

and

$$a[n] = x[n-2] + a[n-2] + x[n-3] + a[n-3]. \tag{8.21}$$

From Eq. (8.20), we obtain

$$y_2(D) = x(D) + a(D) + Dx(D) + Da(D) + D^2x(D) + D^2a(D) + D^3x(D) \\ + D^3a(D) \rightarrow$$

$$y_2(D) = \left(1 + D + D^2 + D^3\right)x(D) + \left(1 + D + D^2 + D^3\right)a(D) \rightarrow$$

$$y_2(D) = \left(1 + D + D^2 + D^3\right)(x(D) + a(D)). \tag{8.22}$$

From Eq. (8.21), we get

$$a(D) = D^2x(D) + D^2a(D) + D^3x(D) + D^3a(D) \rightarrow$$
$$a(D) = \left(D^2 + D^3\right)x(D) + \left(D^2 + D^3\right)a(D) \rightarrow$$
$$a(D) = \frac{D^2 + D^3}{1 + D^2 + D^3}x(D). \tag{8.23}$$

When Eq. (8.23) is substituted into Eq. (8.22), we obtain

$$y_2(D) = \left(1 + D + D^2 + D^3\right)\left(x(D) + \frac{D^2 + D^3}{1 + D^2 + D^3}x(D)\right)$$

leading to

$$\frac{y_2(D)}{x(D)} = \frac{1 + D + D^2 + D^3}{1 + D^2 + D^3}$$

which is the polynomial form of impulse response between $x$ and $y_2$, i.e.,

$$h(D) = \frac{1 + D + D^2 + D^3}{1 + D^2 + D^3}.$$

**Example 8.18** Find the impulse response between $x$ and $y$ for the convolutional encoder shown in Fig. 8.60.

**Solution 8.18** We first label the $x[n]$ and $a[n]$ signals as depicted in Fig. 8.61.

Considering the memory cells as delay elements and binary vectors as digital sequences, we can draw the signal flow graph between $x$ and $y$ as depicted in Fig. 8.62.

Using Fig. 8.62, we can write that

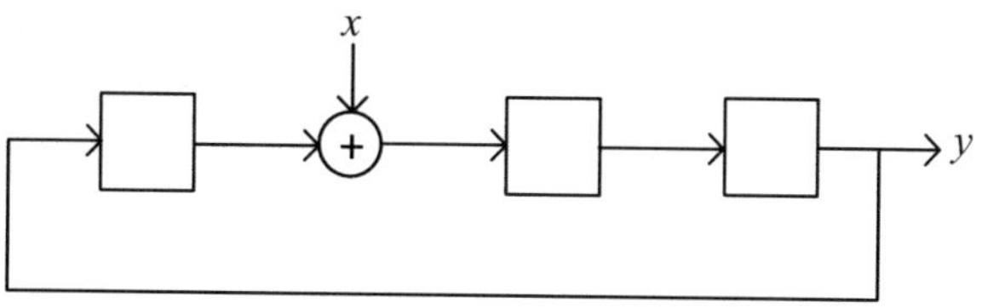


**Fig. 8.60** Convolutional encoder for Example 8.18

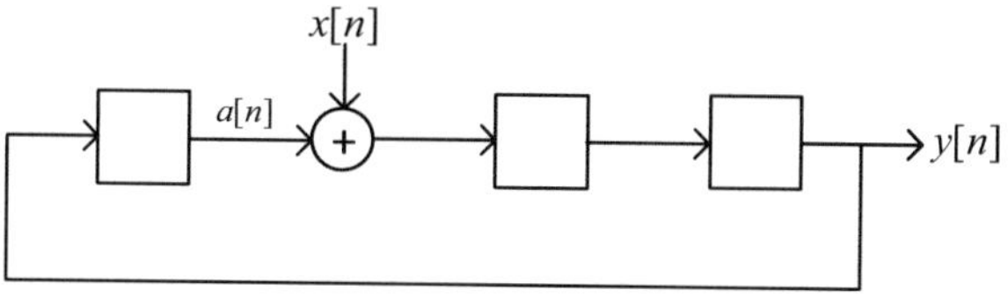


**Fig. 8.61** Labeling critical signals

$$y[n] = x[n-2] + a[n-2] \qquad a[n] = y[n-1]$$

from which we obtain

$$y[n] = x[n-2] + y[n-3]$$

where, taking the Z-transform of both sides, we get

$$Y(D) = D^2X(D) + D^3Y(D)$$

from which we obtain

$$\frac{Y(D)}{X(D)} = \frac{D^2}{1+D^3} \rightarrow h(D) = \frac{D^2}{1+D^3}$$

which is the impulse response between $x$ and $y$ in polynomial form.

**Exercises** Find the impulse responses of the convolutional encoders shown in Fig. 8.63.

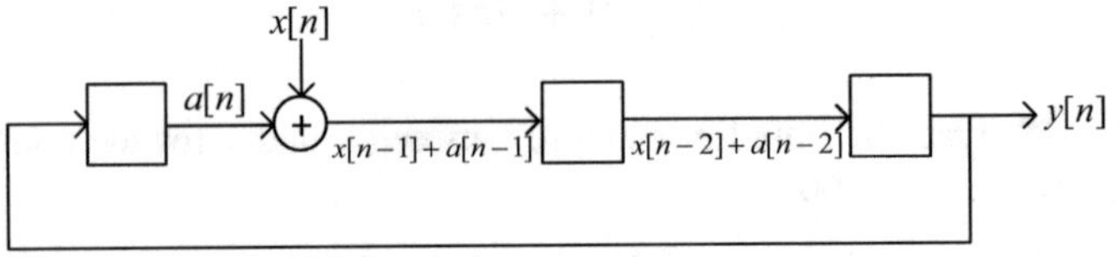


**Fig. 8.62** The outputs of the delay elements

**Fig. 8.63** Convolutional encoders for exercise

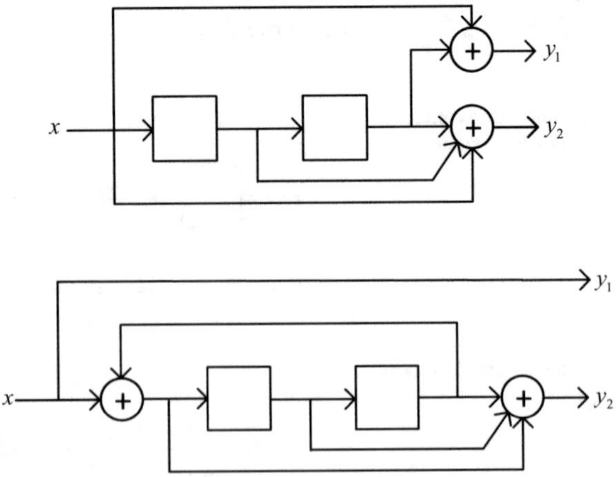

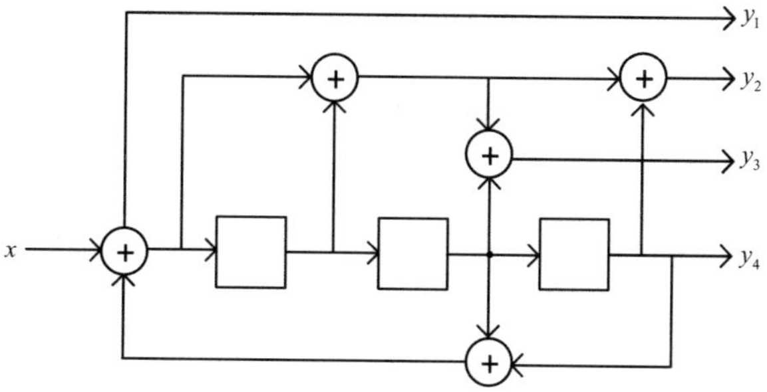


Fig. 8.64 Convolutional encoder for exercise

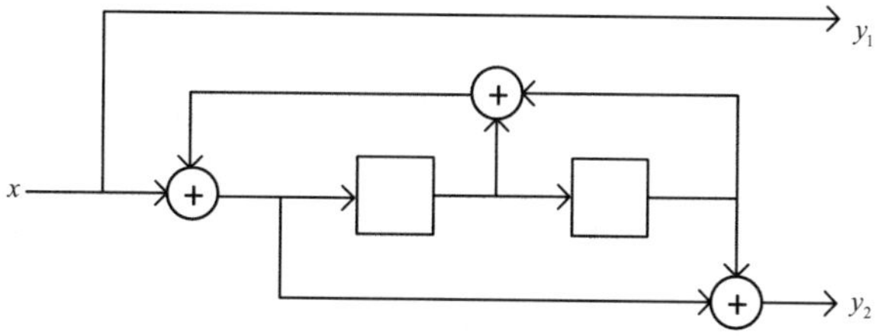


Fig. 8.65 Convolutional encoder for Example 8.19

**Exercises** Find the impulse responses of the convolutional encoder shown in Fig. 8.64.

**Example 8.19** The convolutional encoder shown in Fig. 8.65 is preferable in practical communication systems.

The generator matrix of the convolutional encoder shown in Fig. 8.65 can be found in polynomial form as

$$G(D) = (h_{11}(D)\ h_{12}(D)) \rightarrow G(D) = \left(1 \frac{1+D^2}{1+D+D^2}\right)$$

where expressing the impulse response $h_{12}(D)$ in octal form as

$$\frac{1+D^2}{1+D+D^2} \rightarrow \frac{101}{111} \rightarrow \left(\frac{5}{7}\right)_{\text{octal}}$$

we obtain the octal form of the generator matrix as

$$G = \left(1\ \frac{5}{7}\right)_{\text{octal}}.$$

## 8.6 Graphical Representation of Convolutional Codes

The operation of convolutional codes can be illustrated graphically. There are three types of graphical representation of convolutional codes, and these are tree diagram, state diagram, and trellis diagram representations. Tree diagram representation is not used due to its large size for even moderate code-word lengths. State diagrams are useful to understand the operation of the convolutional encoders, and trellis diagrams are used for the decoding operation of convolutional codes. They are employed for the utilization of Viterbi decoding algorithm. Now, let's explain these representations briefly.

### *8.6.1 State Diagram Representation*

The contents of the memory cells after each clock pulse are accepted as state. The state diagram illustrates the state change upon the application of a clock pulse and gives information about the output of the convolutional encoder for a specific state and external input.

**Example 8.20** Let's try to find the state diagram of the convolutional encoder shown in Fig. 8.66.

Since there are two cells in the convolutional encoder, the contents, i.e., outputs, of both memory cells can be one of the bit pairs 00, 01, 10, and 11. This means that there are 4 states of this convolutional encoder. Let's denote the pairs 00, 01, 10, and 11 by the symbols $S_0$, $S_1$, $S_2$, and $S_3$, respectively.

Initially the contents of the registers are all zeros as depicted in Fig. 8.67.

When 0 is applied as external input, i.e., when $x = 0$, the outputs of the convolutional encoder can be calculated as shown in Fig. 8.68.

And we can trace the behavior of the convolutional encoder by a state table or a state diagram as indicated in Fig. 8.69.

Upon the application of clock pulse, the next state can be calculated using Fig. 8.70 for the input $x = 0$ as 00, i.e., $S_0$. We can update the state table and diagram shown in Fig. 8.71 as in Fig. 8.72.

When 1 is applied as external input, i.e., for $x = 1$, the outputs of the convolutional encoder can be calculated as shown in Fig. 8.71 when present state is $S_0$.

The behavior of the convolutional encoder can be illustrated by a state table or a state diagram for present state $S_0$ and external input $x = 0$ as indicated in Fig. 8.72.

For present state $S_0$ and external input $x = 1$, we can find the next state upon the application of clock pulse using Fig. 8.71 as 10, i.e., $S_2$. We can update the state table and diagram shown in Fig. 8.72 as in Fig. 8.73.

For present state $S_1$ and external input $x = 0$, the output of the circuit can be found as $y_2y_1 = 00$ as illustrated in Fig. 8.74, and upon clock pulse, the next state happens to be $S_2 = 10$.

Fig. 8.66 Convolutional encoder for Example 8.20

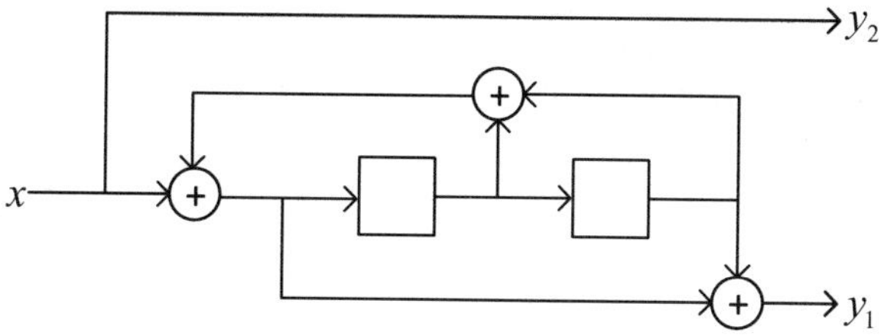


Fig. 8.67 Initial contents of the memory cells

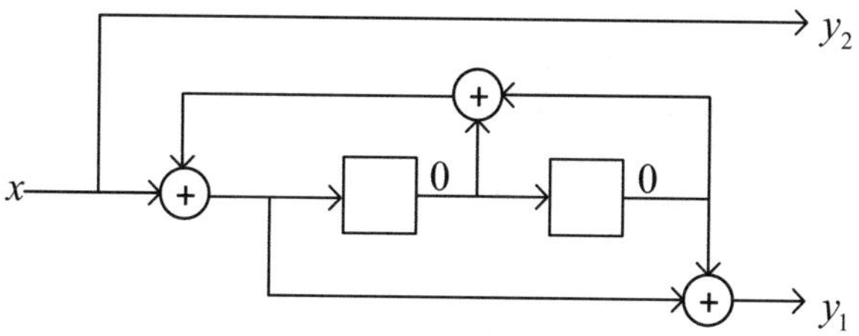


Fig. 8.68 Calculation of encoder's initial output

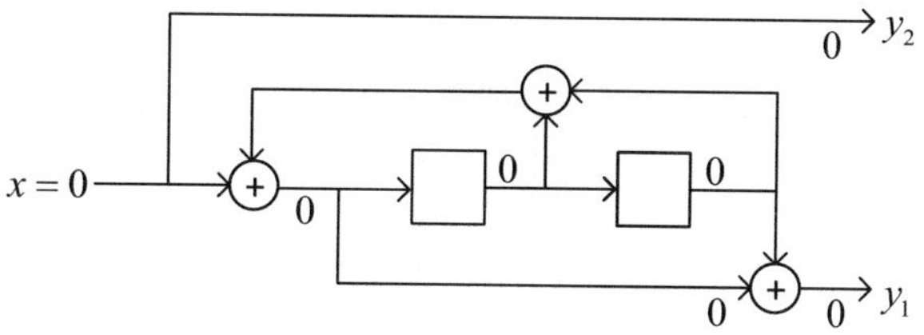


| Present State | Input | Output | Next State |
|---|---|---|---|
| $S_i$ | $x$ | $y_2y_1$ | $S_i$ |
| $S_0 = 00$ | 0 | 00 | |

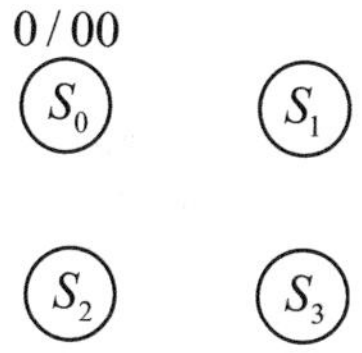


Fig. 8.69 State table and state diagram

| Present State | Input | Output | Next State |
|---|---|---|---|
| $S_i$ | $x$ | $y_2y_1$ | $S_i$ |
| $S_0 = 00$ | 0 | 00 | $S_0 = 00$ |

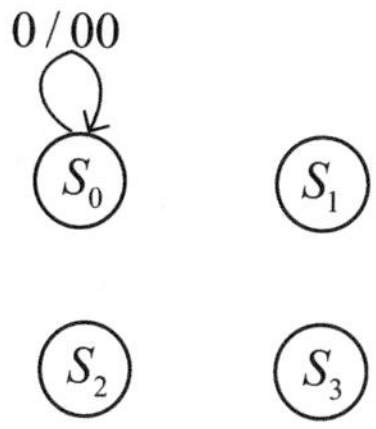


Fig. 8.70 State table and state diagram after first clock pulse

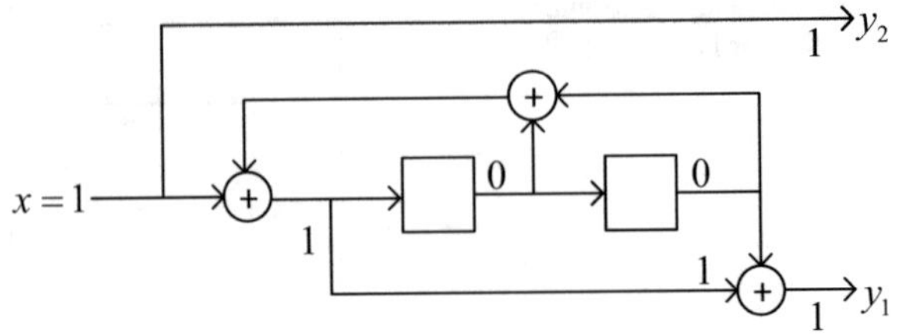


**Fig. 8.71** Calculation of the convolutional encoder output for state $S_0$ and input $x = 1$

| Present State $S_i$ | Input $x$ | Output $y_2y_1$ | Next State $S_i$ |
|---|---|---|---|
| $S_0 = 00$ | 0 | 00 | $S_0 = 00$ |
| $S_0 = 00$ | 1 | 11 | |

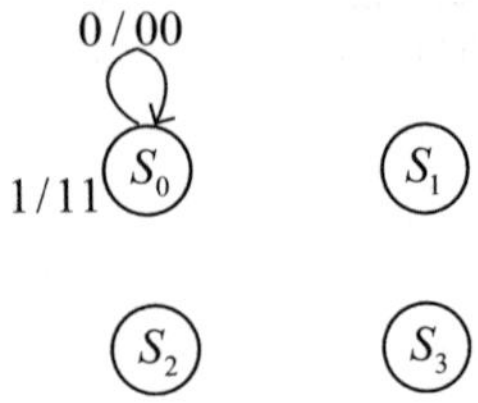


**Fig. 8.72** State table and state diagram construction

| Present State $S_i$ | Input $x$ | Output $y_2y_1$ | Next State $S_i$ |
|---|---|---|---|
| $S_0 = 00$ | 0 | 00 | $S_0 = 00$ |
| $S_0 = 00$ | 1 | 11 | $S_2 = 10$ |

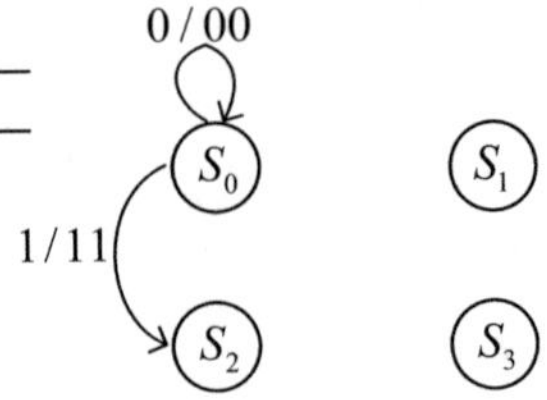


**Fig. 8.73** State table and state diagram after second clock pulse

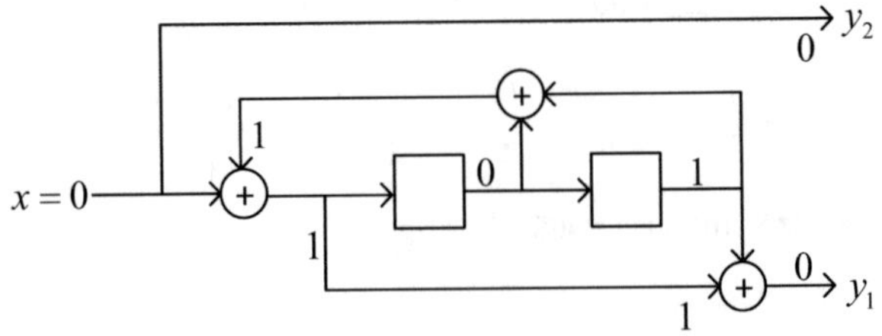


**Fig. 8.74** Calculation of convolutional encoder output and next state for present state $S_1$

The behavior of the circuit for the present state $S_1$ and external input $x = 0$ is expressed using state table and state diagram as in Fig. 8.75.

For present state $S_1$ and external input $x = 1$, the output of the circuit can be found as $y_2y_1 = 11$ as illustrated in Fig. 8.76, and upon clock pulse, the next state happens to be $S_0 = 00$.

The behavior of the circuit for the present state $S_1$ and external input $x = 1$ is expressed using state table and state diagram as in Fig. 8.77.

| Present State $S_i$ | Input $x$ | Output $y_2y_1$ | Next State $S_i$ |
|---|---|---|---|
| $S_0 = 00$ | 0 | 00 | $S_0 = 00$ |
| $S_0 = 00$ | 1 | 11 | $S_2 = 10$ |
| $S_1 = 01$ | 0 | 00 | $S_2 = 10$ |

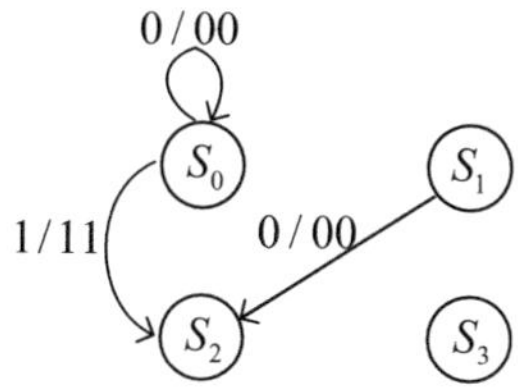


**Fig. 8.75** State table and state diagram after third clock pulse

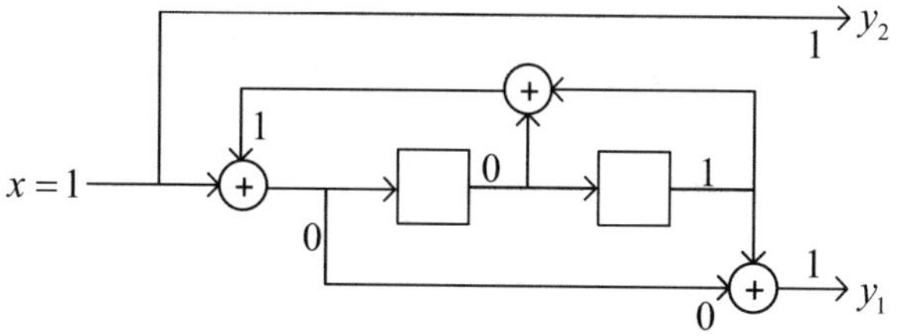


**Fig. 8.76** Encoder output and next state calculation for the present state $S_1$

| Present State $S_i$ | Input $x$ | Output $y_2y_1$ | Next State $S_i$ |
|---|---|---|---|
| $S_0 = 00$ | 0 | 00 | $S_0 = 00$ |
| $S_0 = 00$ | 1 | 11 | $S_2 = 10$ |
| $S_1 = 01$ | 0 | 10 | $S_2 = 10$ |
| $S_1 = 01$ | 1 | 11 | $S_0 = 00$ |

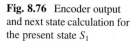

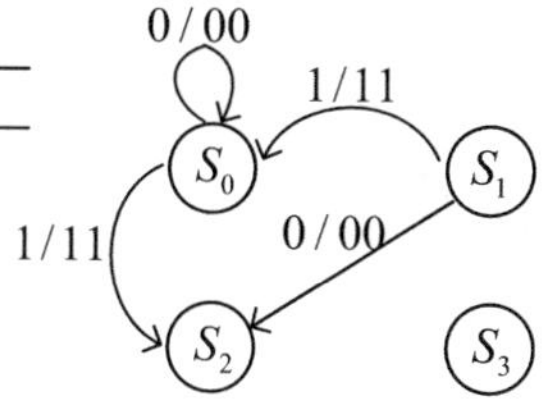


**Fig. 8.77** State table and state diagram after fourth clock pulse

For present state $S_2$ and external input $x = 0$, the output of the circuit can be found as $y_2y_1 = 01$ as illustrated in Fig. 8.78, and upon clock pulse, the next state happens to be $S_3 = 11$.

The behavior of the circuit for the present state $S_2$ and external input $x = 0$ is expressed using state table and state diagram as in Fig. 8.79.

For present state $S_2$ and external input $x = 1$, the output of the circuit can be found as $y_2y_1 = 10$ as illustrated in Fig. 8.80, and upon clock pulse, the next state happens to be $S_1 = 01$.

The behavior of the circuit for the present state $S_2$ and external input $x = 1$ is expressed using state table and state diagram as in Fig. 8.81.

For present state $S_3$ and external input $x = 0$, the output of the circuit can be found as $y_2y_1 = 01$ as illustrated in Fig. 8.82, and upon clock pulse, the next state happens to be $S_1 = 01$.

The behavior of the circuit for the present state $S_3$ and external input $x = 0$ is expressed using state table and state diagram as in Fig. 8.83.

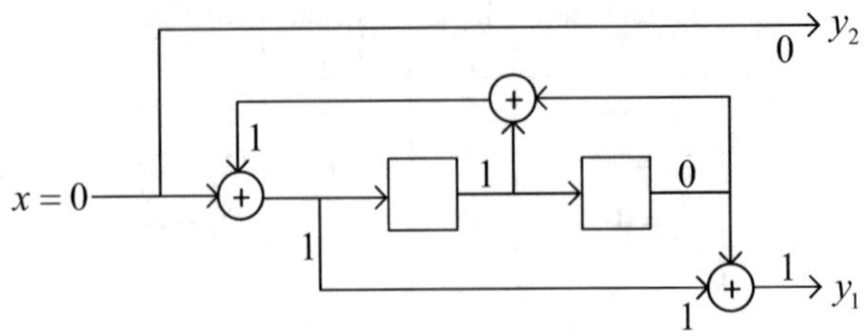


**Fig. 8.78** Calculation of encoder output for $x = 0$ when present state is $S_2$

| Present State $S_i$ | Input $x$ | Output $y_2y_1$ | Next State $S_i$ |
|---|---|---|---|
| $S_0 = 00$ | 0 | 00 | $S_0 = 00$ |
| $S_0 = 00$ | 1 | 11 | $S_2 = 10$ |
| $S_1 = 01$ | 0 | 10 | $S_2 = 10$ |
| $S_1 = 01$ | 1 | 11 | $S_0 = 00$ |
| $S_2 = 10$ | 0 | 01 | $S_3 = 11$ |

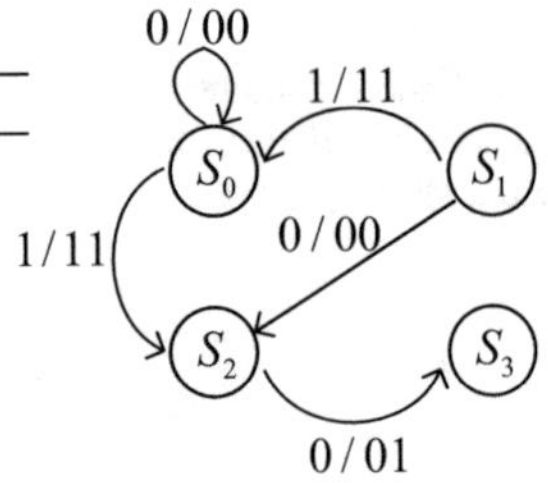


**Fig. 8.79** State table and state diagram including the states $S_0$, $S_1$, and $S_2$

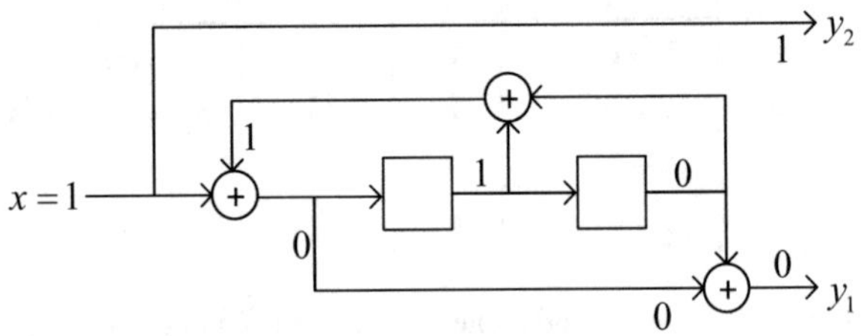


**Fig. 8.80** Calculation of encoder output and next state for $x = 1$ when present state is $S_2$

| $S_i$ | $x$ | $y_2y_1$ | $S_i$ |
|---|---|---|---|
| $S_0 = 00$ | 0 | 00 | $S_0 = 00$ |
| $S_0 = 00$ | 1 | 11 | $S_2 = 10$ |
| $S_1 = 01$ | 0 | 10 | $S_2 = 10$ |
| $S_1 = 01$ | 1 | 11 | $S_0 = 00$ |
| $S_2 = 10$ | 0 | 01 | $S_3 = 11$ |
| $S_2 = 10$ | 1 | 10 | $S_1 = 01$ |

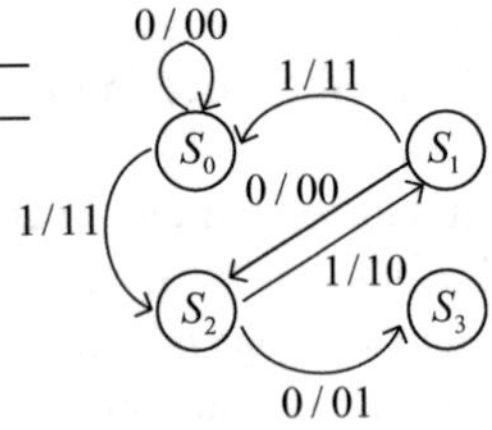


**Fig. 8.81** State table and state diagram including the transitions from $S_2$

For present state $S_3$ and external input $x = 1$, the output of the circuit can be found as $y_2y_1 = 10$ as illustrated in Fig. 8.84, and upon clock pulse, the next state happens to be $S_3 = 11$.

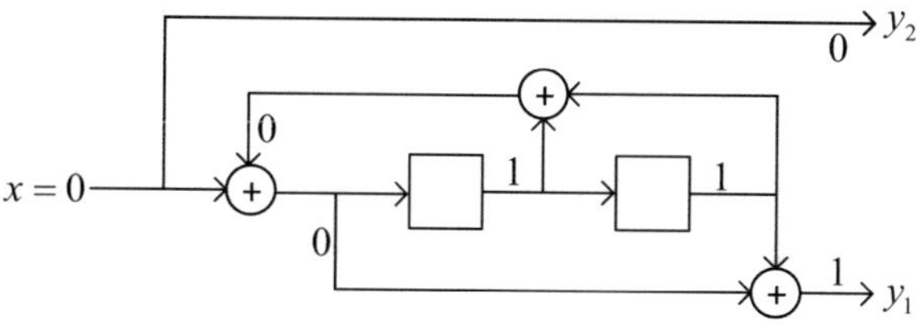


**Fig. 8.82** Calculation of next state and circuit outputs for present state 11 and input 0

| Present State | Input | Output | Next State |
|---|---|---|---|
| $S_i$ | $x$ | $y_2 y_1$ | $S_i$ |
| $S_0 = 00$ | 0 | 00 | $S_0 = 00$ |
| $S_0 = 00$ | 1 | 11 | $S_2 = 10$ |
| $S_1 = 01$ | 0 | 10 | $S_2 = 10$ |
| $S_1 = 01$ | 1 | 11 | $S_0 = 00$ |
| $S_2 = 10$ | 0 | 01 | $S_3 = 11$ |
| $S_2 = 10$ | 1 | 10 | $S_1 = 01$ |
| $S_3 = 11$ | 0 | 01 | $S_1 = 01$ |

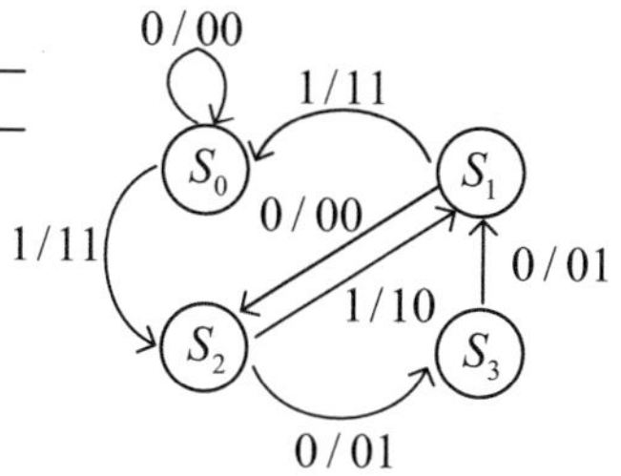


**Fig. 8.83** State table and state diagram with transitions from $S_3$ for $x = 0$

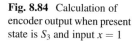

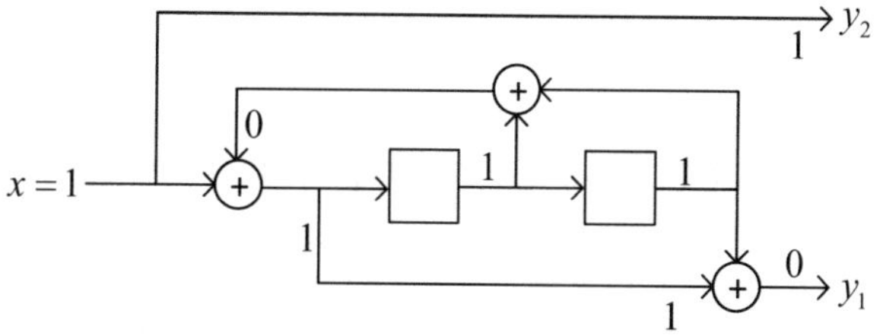


**Fig. 8.84** Calculation of encoder output when present state is $S_3$ and input $x = 1$

The behavior of the circuit for the present state $S_3$ and external input $x = 1$ is expressed using state table and state diagram as in Fig. 8.85.

The state diagram shown in Fig. 8.85 expresses the behavior of the circuit, and it can be used to find the code-word for any input sequence.

**Example 8.21** Find the output of the convolutional encoder shown in Fig. 8.86 for the input stream $x = 1010110$.

**Solution 8.21** Using the state diagram given in Fig. 8.85, we can calculate the output of the convolutional encoder for the given input sequence as

$$\boldsymbol{c} = [11\ 01\ 01\ 01\ 11\ 11\ 01].$$

| Present State | Input | Output | Next State |
|---|---|---|---|
| $S_i$ | $x$ | $y_2y_1$ | $S_i$ |
| $S_0 = 00$ | 0 | 00 | $S_0 = 00$ |
| $S_0 = 00$ | 1 | 11 | $S_2 = 10$ |
| $S_1 = 01$ | 0 | 10 | $S_2 = 10$ |
| $S_1 = 01$ | 1 | 11 | $S_0 = 00$ |
| $S_2 = 10$ | 0 | 01 | $S_3 = 11$ |
| $S_2 = 10$ | 1 | 10 | $S_1 = 01$ |
| $S_3 = 11$ | 0 | 01 | $S_1 = 01$ |
| $S_3 = 11$ | 1 | 01 | $S_3 = 11$ |

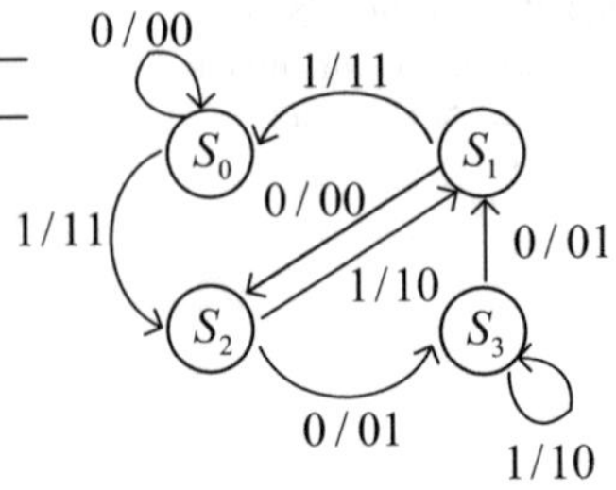


**Fig. 8.85** State table and state diagram including transitions from $S_3$ for $x = 1$

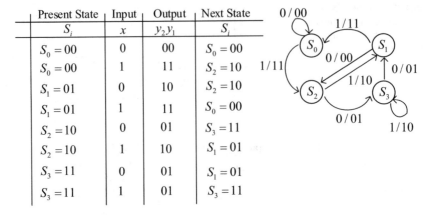

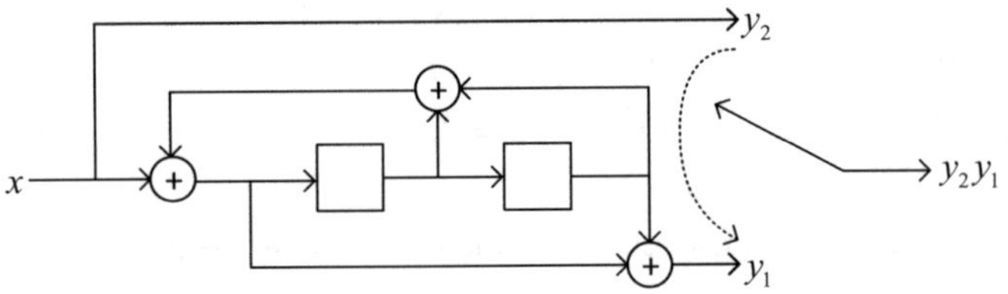


**Fig. 8.86** Convolutional encoder for Example 8.21

## 8.6.2 *Trellis Diagram Representation of Convolutional Codes*

State diagram of a convolutional encoder does not contain time information. The trellis diagram representation of a convolutional encoder contains all the information of the state diagram; in addition, it also contains the time information. The state diagram of the convolutional encoder shown in Fig. 8.87 can be obtained as in Fig. 8.88.

We can convert the state diagram in Fig. 8.88 to a trellis diagram as follows. There are four states in the state diagram, and initially, we are at the state $S_0$. This information can be illustrated as in Fig. 8.89.

According to the state diagram in Fig. 8.89, a transition from state $S_0$ to itself occurs for input 0 producing the output 00, and a transition from state $S_0$ to $S_2$ occurs for input 1 producing the output 10. We can illustrate this information as in Fig. 8.90.

At time instant $t = 1$, the possible transitions from state $S_0$ with the input/outputs label along the transition lines are depicted in Fig. 8.91.

For time instant $t = 1$, adding the possible transition for state $S_2$, we update the trellis diagram as in Fig. 8.92.

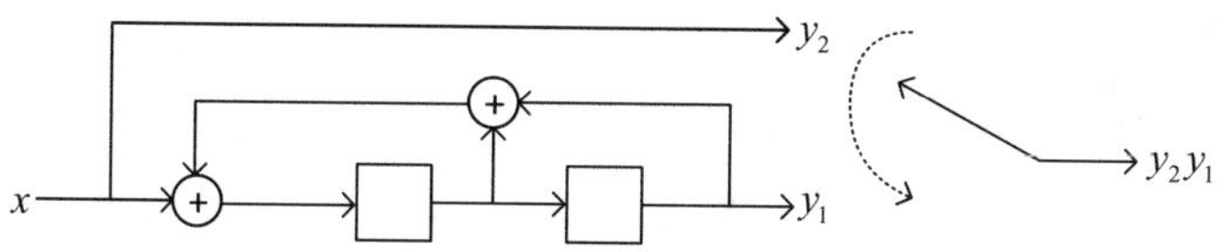


Fig. 8.87 A convolutional encoder

Fig. 8.88 The state diagram of the convolutional encoder in Fig. 8.87

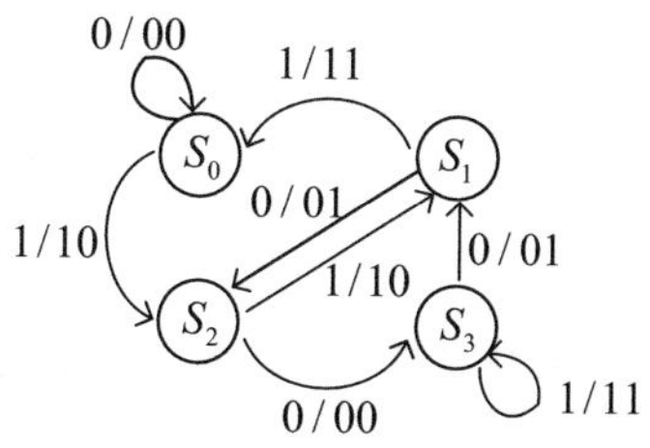


Fig. 8.89 Trellis diagram in its initial form

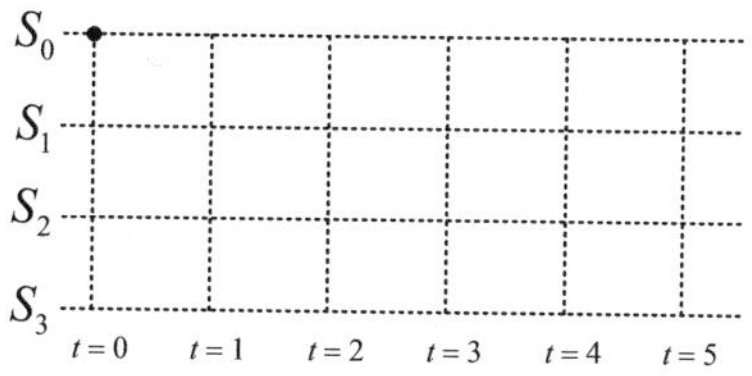


Fig. 8.90 Transitions from state $S_0$ at time instant $t = 0$

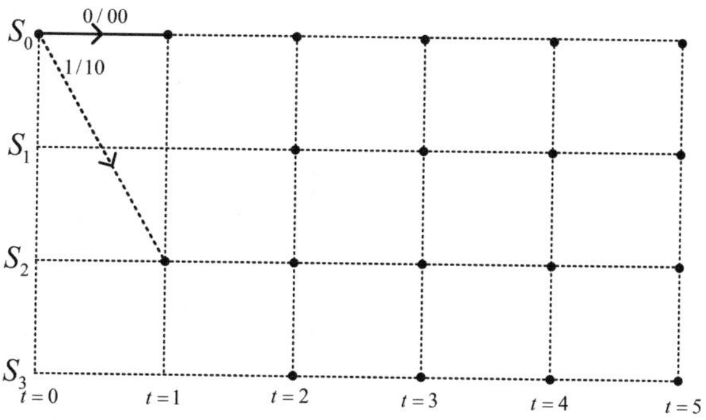


For time instant $t = 1$, the present state can be one of the states $S_0$, $S_1$, $S_2$, and $S_3$. We first draw the transition line for the states $S_0$ and $S_2$ as in Fig. 8.93.

If we add the transition lines for the states $S_1$ and $S_2$ to the trellis graph of Fig. 8.93, we obtain the trellis diagram of Fig. 8.94.

The transitions in the third section of the trellis diagram are repeated in the other sections as depicted in Fig. 8.95.

**Example 8.22** A convolutional encoder is given in Fig. 8.96.

For the given convolutional encoder:

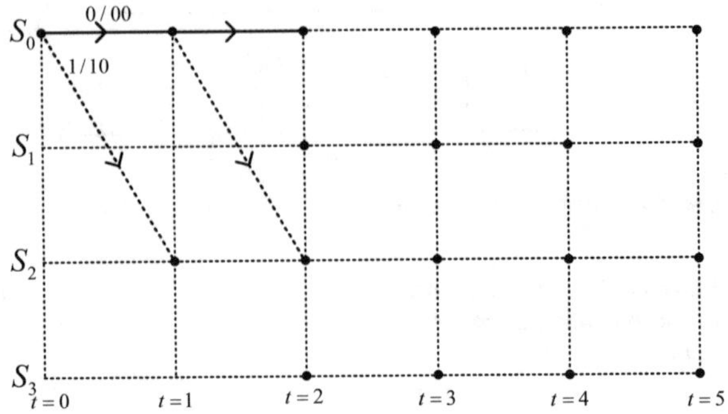


Fig. 8.91 Trellis diagram containing transitions from state $S_0$ at time instant $t = 1$

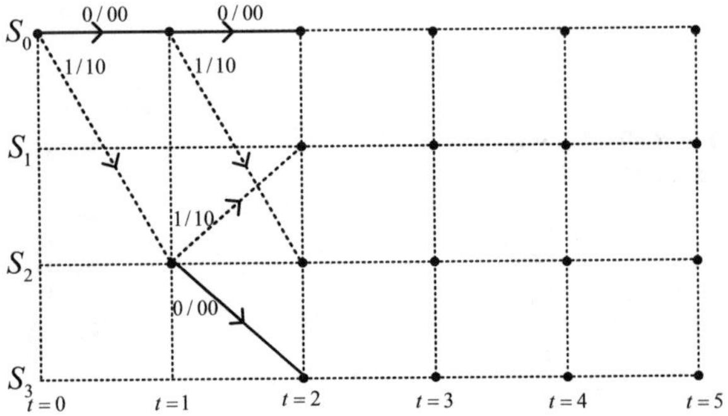


Fig. 8.92 Trellis diagram containing transitions from state $S_2$ at time instant $t = 1$

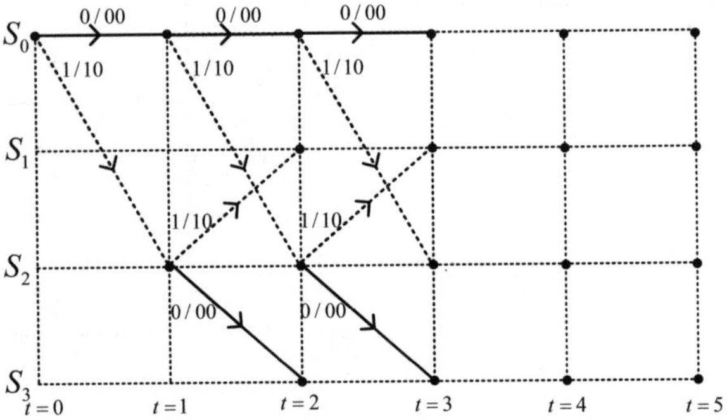


Fig. 8.93 Trellis diagram containing transitions for $t = 2$

(a) Obtain the state transition diagram.
(b) Obtain the trellis diagram.
(c) Encode the information sequence $\boldsymbol{d} = [1\ 0\ 1\ 1\ 1]$, and show the encoding path on the trellis diagram.

**Solution 8.22** Considering the operation of the circuit, we can obtain its state diagram as in Fig. 8.97 where the states $S_0$, $S_1$, $S_2$, and $S_3$ represent the binary sequences 00, 01, 10, and 11, respectively, and these binary sequences are the contents of the memory cells.

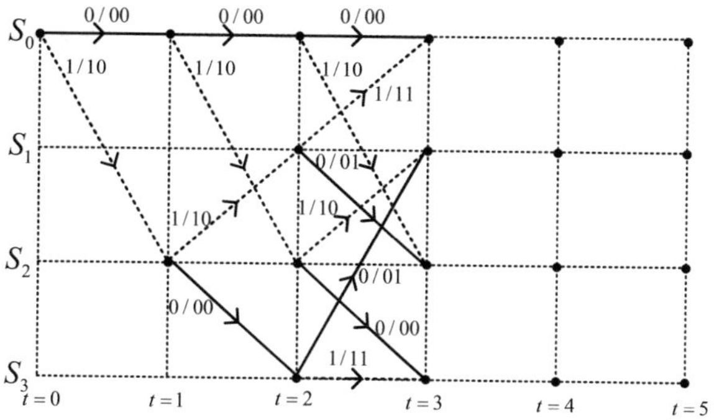


**Fig. 8.94** Trellis diagram containing all transitions for $t = 2$

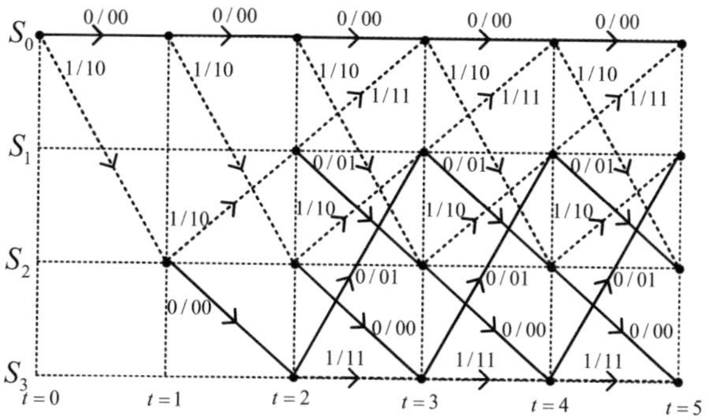


**Fig. 8.95** Complete trellis diagram

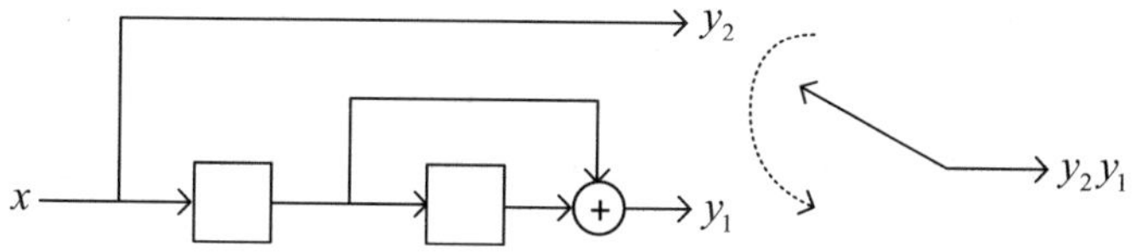


**Fig. 8.96** Convolutional encoder for Example 8.22

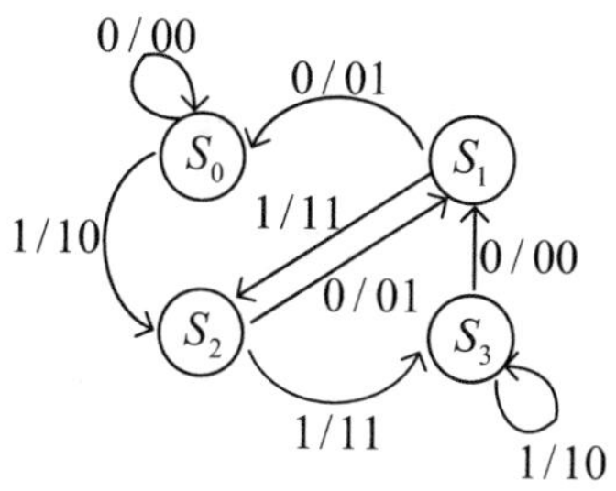


**Fig. 8.97** State diagram of the convolutional encoder in Fig. 8.96

Fig. 8.98 The first three stages of the trellis diagram

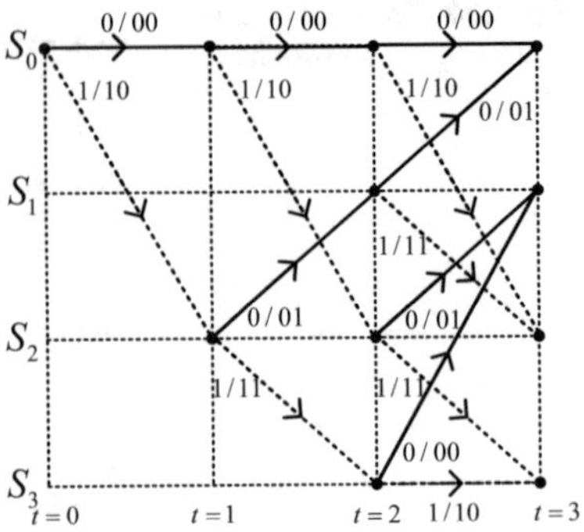


Fig. 8.99 The trellis diagram with five stages

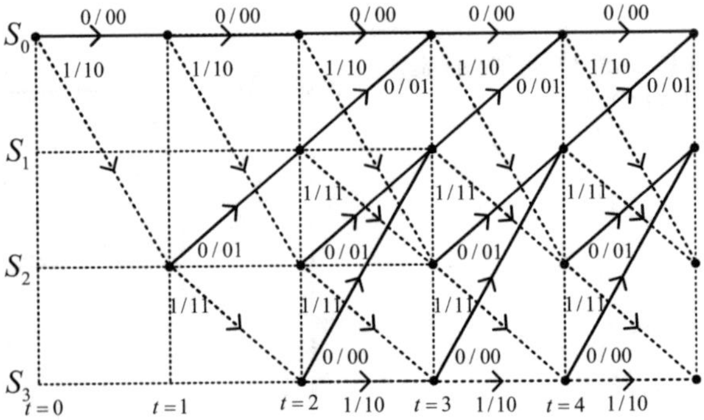


The first three stages of the trellis graph can be drawn using the state diagram as in Fig. 8.98.

Since the information word $\boldsymbol{d} = [1\ 0\ 1\ 1\ 1]$ contains 5 bits, we need five stages on the trellis graph. The trellis graph with five stages is depicted in Fig. 8.99.

The information vector $\boldsymbol{d} = [1\ 0\ 1\ 1\ 1]$ can be encoded using the trellis diagram starting with the state $S_0$ and making the state transitions considering the information bits as in Fig. 8.100.

When the output bits along the red path of Fig. 8.100 are concatenated, we get the convolutional encoder output, i.e., the code-word, as $\boldsymbol{c} = [10\ 01\ 11\ 11\ 10]$.

## 8.7 Decoding of Convolutional Codes

Convolutional codes are decoded using the Viterbi algorithm. The Viterbi algorithm finds the best path for on trellis diagram for a received sequence, and the best path corresponds to the transmitted convolutional code-word.

Let's now explain the hard Viterbi decoding algorithm for convolutional codes. Assume that we have a convolutional encoder with rate 1/2, and encode a data-word $\boldsymbol{d}$ and transmit it. For every data bit, the code-word $\boldsymbol{c}$ contains two code bits. The transmitted code-word may incur some bit errors at the receiver side.

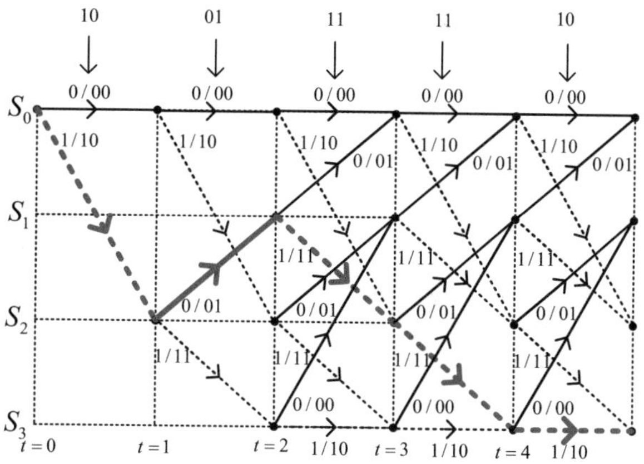


**Fig. 8.100** Encoding operation using the trellis diagram

## 8.7.1 Viterbi Algorithm

The Viterbi decoding algorithm can be outlined as follows:

(a) Divide the received bit sequence into bit pairs.
(b) Write the bit pairs along the top of the trellis graph.
(c) Calculate the branch metric of every transition.
(d) Calculate the state metrics starting from the first stage till the last stage, and while calculating the state metrics, follow the surviving path rules.
(e) After calculating the state metrics of the end-states, decide on the largest end-state metric.
(f) By back propagating from the end-state having the largest state metric to the starting state, decide on the surviving path.

Let's explain the terms branch metric, state metric, and surviving path concepts of the Viterbi algorithm.

**Branch Metric and Surviving Path**

Brach metric is the Hamming distance between the transition output and the corresponding bit pair of the code-word.

A state of the trellis can be reached from a number of states. Assuming that there are two incoming states for the destination state, the state metric of the destination state is calculated as

$$S_d = \min\left(S_a + BM_{ad}, S_b + BM_{bd}\right) \tag{8.24}$$

where $S_a$ and $S_b$ are the state metrics of the incoming states $A$ and $B$, and $S_d$ is the state metric of the destination state. The calculation of the state metric of the destination is illustrated in Fig. 8.101.

It is clear from Fig. 8.101 that only one of the branch metrics of two incoming transitions is used for the calculation of destination state metric. The chosen incoming path is called surviving path. For instance, if

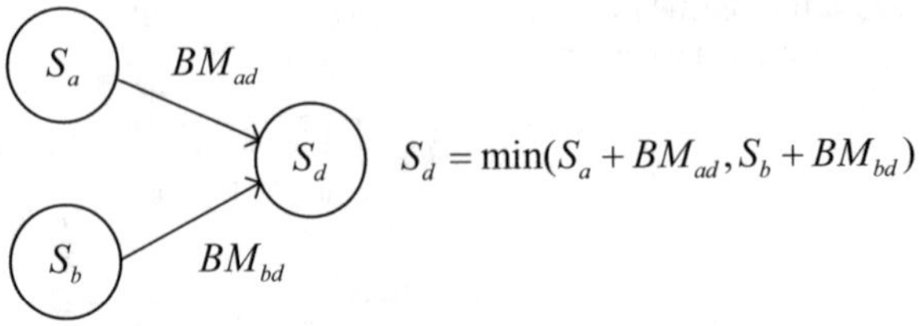


**Fig. 8.101** State metric calculation

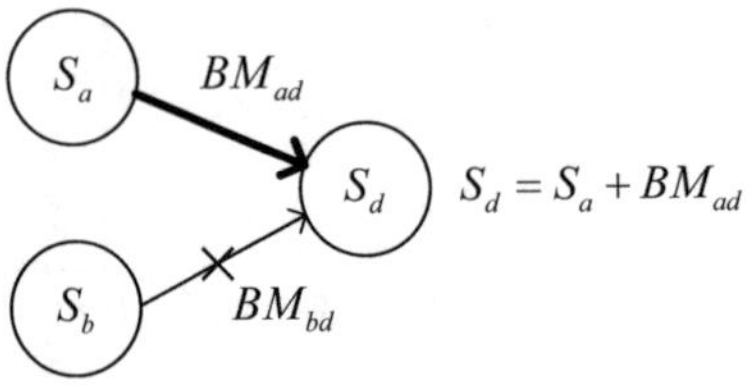


**Fig. 8.102** Deciding on the winner path

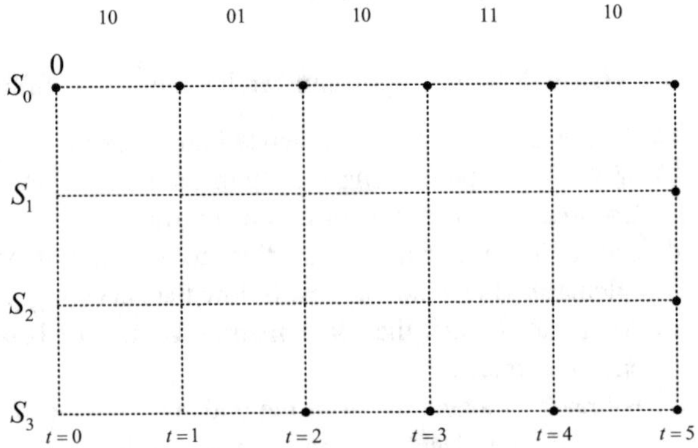


**Fig. 8.103** Trellis diagram with initial state metric

$$S_a + BM_{ad} < S_b + BM_{bd} \tag{8.25}$$

then the state metric of the destination node and the surviving path can be depicted in bold as in Fig. 8.102.

Now, let's solve an example to illustrate the concept of Viterbi decoding of convolutional codes.

**Example 8.23** In Example 8.22, when the data-word $\boldsymbol{d} = [1\ 0\ 1\ 1\ 1]$ is encoded, we get the code-word $\boldsymbol{c} = [10\ 01\ 11\ 11\ 10]$. Assume that $\boldsymbol{c}$ is transmitted, and the received word includes a single-bit error, and it is given as $\boldsymbol{r} = [10\ 01\ 10^{*}\ 11\ 10]$ where $*$ indicates the erroneous bit.

Using the received word and employing the Viterbi algorithm, we want to determine the transmitted code-word and hence correct the transmission error.

The Viterbi decoding operation can be outlined as follows:

Step 1: First, we divide the received word into bit pairs and place them to the top of the trellis and initialize the state metric of the starting state $S_0$ to 0 as in Fig. 8.103.

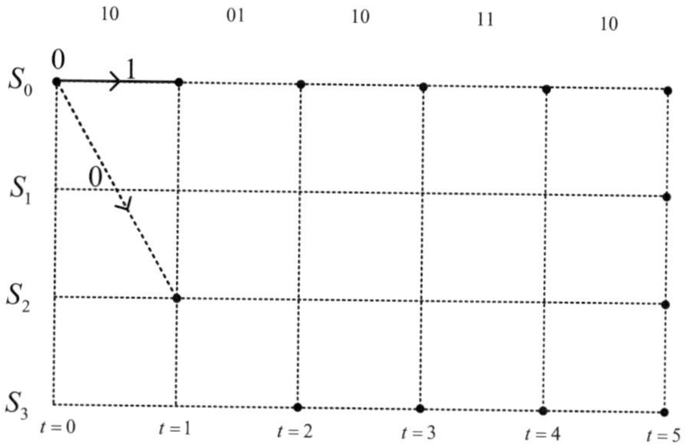


**Fig. 8.104** Branch metrics of the first stage

Step 2: From state $S_0$, we have two transitions, one to the $S_0$ and the other to the $S_2$. The branch metrics for these transitions are calculated using the first bit pair at the top of the trellis and the convolutional encoder outputs for these two transitions as

$$\begin{aligned} BM_{00}^{t=0} &= d_H(00, 10) \rightarrow BM_{00}^{t=0} = 1 \\ BM_{02}^{t=0} &= d_H(10, 10) \rightarrow BM_{02}^{t=0} = 0 \end{aligned}$$

and these branch metrics are depicted in Fig. 8.104.

Step 3: It is seen from Fig. 8.105 that, at time instant $t = 1$, the states $S_0$ and $S_2$ have single incoming transitions, and for this reason, state metrics of $S_0$ and $S_2$ at $t = 1$ can be calculated as

$$\begin{aligned} SM_0^{t=1} &= SM_0^{t=0} + BM_{00}^{t=0} \rightarrow SM_0^{t=1} = 0 + 1 \rightarrow SM_0^{t=1} = 1 \\ SM_2^{t=1} &= SM_0^{t=0} + BM_{02}^{t=0} \rightarrow SM_2^{t=1} = 0 + 0 \rightarrow SM_2^{t=1} = 0 \end{aligned}$$

and these state metrics are depicted in Fig. 8.105.

Step 4: For time instant $t = 1$, we have the transitions $S_0 \rightarrow S_0$, $S_0 \rightarrow S_2$ and $S_1 \rightarrow S_1$, $S_2 \rightarrow S_3$. The branch metrics for these transitions are calculated using the first bit pair at the top of the trellis and the convolutional encoder outputs for these two transitions as

$$\begin{aligned} BM_{00}^{t=1} &= d_H(00, 10) \rightarrow BM_{00}^{t=0} = 1 \\ BM_{02}^{t=1} &= d_H(10, 10) \rightarrow BM_{02}^{t=0} = 0 \\ BM_{21}^{t=1} &= d_H(01, 10) \rightarrow BM_{21}^{t=0} = 2 \\ BM_{22}^{t=1} &= d_H(11, 10) \rightarrow BM_{22}^{t=0} = 1 \end{aligned}$$

and these branch metrics are depicted in Fig. 8.106.

**Fig. 8.105** State metrics for $t = 1$

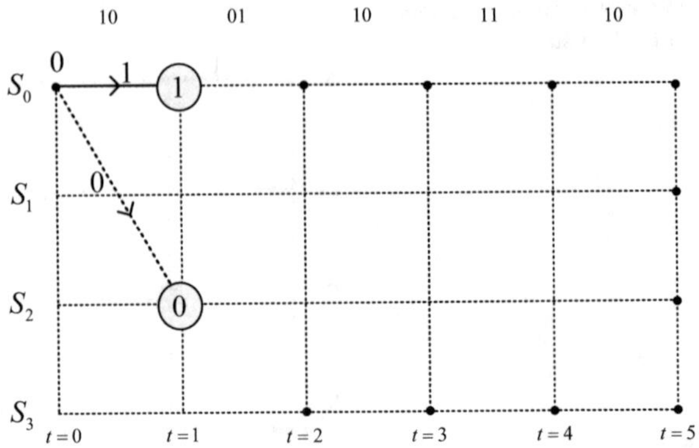


**Fig. 8.106** Branch metrics of the second stage

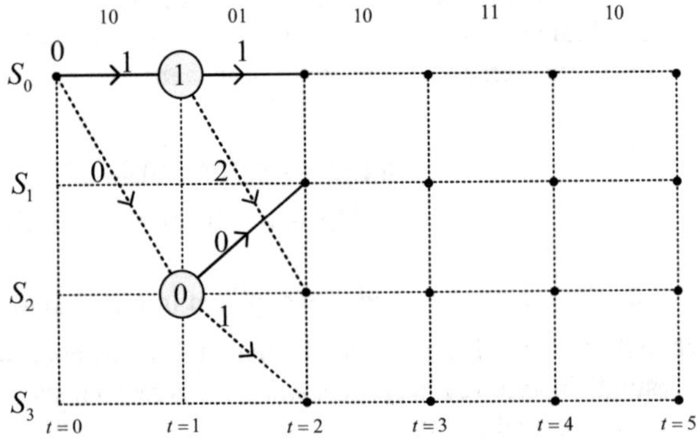


Step 5: For the time instant $t = 2$, all the states have single incoming transitions, and for this reason, state metrics can be calculated as

$$SM_0^{t=2} = SM_0^{t=1} + BM_{00}^{t=1} \rightarrow SM_0^{t=2} = 1 + 1 \rightarrow SM_0^{t=1} = 2$$
$$SM_1^{t=2} = SM_2^{t=1} + BM_{21}^{t=1} \rightarrow SM_1^{t=2} = 0 + 0 \rightarrow SM_2^{t=1} = 0$$
$$SM_2^{t=2} = SM_0^{t=1} + BM_{02}^{t=1} \rightarrow SM_2^{t=2} = 1 + 2 \rightarrow SM_2^{t=1} = 3$$
$$SM_3^{t=2} = SM_2^{t=1} + BM_{23}^{t=1} \rightarrow SM_3^{t=2} = 0 + 1 \rightarrow SM_2^{t=1} = 1$$

and these state metrics are depicted in Fig. 8.107.

Step 6: For time instant $t = 2$, we have the transitions

$$S_0 \rightarrow S_0 \quad S_0 \rightarrow S_2$$
$$S_1 \rightarrow S_1 \quad S_2 \rightarrow S_3$$
$$S_2 \rightarrow S_1 \quad S_2 \rightarrow S_3$$
$$S_3 \rightarrow S_1 \quad S_3 \rightarrow S_3$$

**Fig. 8.107** State metrics for $t = 2$

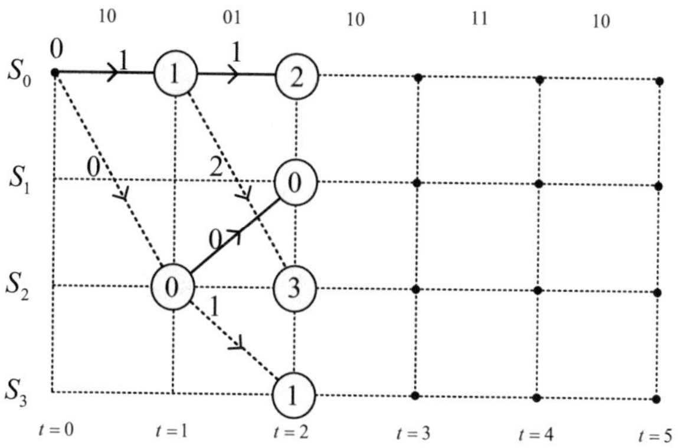


**Fig. 8.108** Branch metrics of the third stage

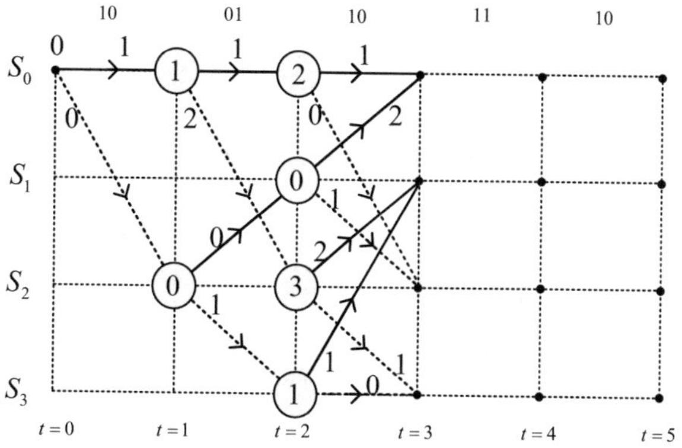


and the branch metrics for these transitions are calculated using the first bit pair at the top of the trellis and the convolutional encoder outputs for these two transitions as

$$BM_{00}^{t=2} = d_H(00, 10) \rightarrow BM_{00}^{t=2} = 1$$

$$BM_{02}^{t=2} = d_H(10, 10) \rightarrow BM_{02}^{t=2} = 0$$

$$BM_{10}^{t=2} = d_H(01, 10) \rightarrow BM_{10}^{t=2} = 2$$

$$BM_{12}^{t=2} = d_H(11, 10) \rightarrow BM_{12}^{t=2} = 1$$

$$BM_{21}^{t=2} = d_H(01, 10) \rightarrow BM_{21}^{t=2} = 2$$

$$BM_{23}^{t=2} = d_H(11, 10) \rightarrow BM_{22}^{t=2} = 1$$

$$BM_{31}^{t=2} = d_H(00, 10) \rightarrow BM_{21}^{t=2} = 1$$

$$BM_{33}^{t=2} = d_H(10, 10) \rightarrow BM_{22}^{t=2} = 0$$

and these branch metrics are depicted in Fig. 8.108.

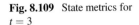

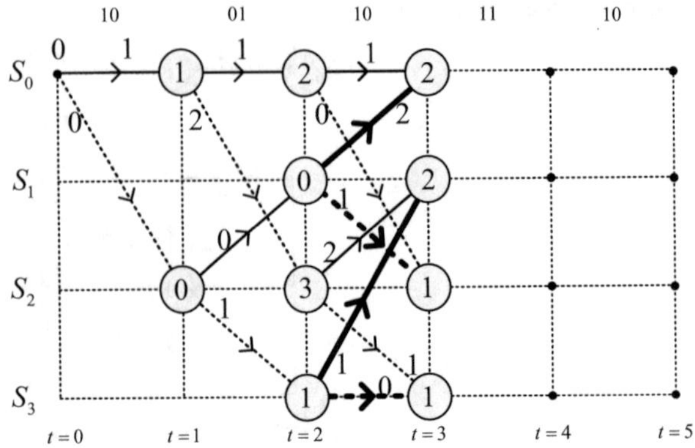


**Fig. 8.109** State metrics for $t = 3$

Step 7: For the time instant $t = 3$, all the states have two incoming transitions, and from these two incoming transitions, the one that makes the state metric smaller is chosen as the winner path, and the state metrics are calculated using the winner paths. With this information in mind, we can calculate the state metrics as

$$SM_0^{t=3} = \min\left(SM_0^{t=2} + BM_{00}^{t=2}, SM_1^{t=2} + BM_{10}^{t=2}\right) \rightarrow SM_0^{t=3} = \min(2+1, 0+2)$$
$$\rightarrow SM_0^{t=3} = 2$$

$$SM_1^{t=3} = \min\left(SM_2^{t=2} + BM_{21}^{t=2}, SM_3^{t=2} + BM_{31}^{t=2}\right) \rightarrow SM_1^{t=3} = \min(3+2, 1+1)$$
$$\rightarrow SM_1^{t=3} = 2$$

$$SM_2^{t=3} = \min\left(SM_0^{t=2} + BM_{02}^{t=2}, SM_1^{t=2} + BM_{12}^{t=2}\right) \rightarrow SM_2^{t=3} = \min(2+0, 0+1)$$
$$\rightarrow SM_2^{t=3} = 1$$

$$SM_3^{t=3} = \min\left(SM_2^{t=2} + BM_{23}^{t=2}, SM_3^{t=2} + BM_{33}^{t=2}\right) \rightarrow SM_3^{t=3} = \min(3+1, 1+0)$$
$$\rightarrow SM_3^{t=3} = 1$$

and these state metrics are depicted in Fig. 8.109 along with the winner paths.

Step 8: Proceeding in a similar manner, first, the calculation of the branch metrics for $t = 3$ and then the calculation of the state metrics for $t = 4$ are illustrated in Figs. 8.110 and 8.111.

Step 9: Like in the previous step, first, the calculation of the branch metrics for $t = 4$ and then the calculation of the state metrics for $t = 5$ are illustrated in Figs. 8.86 and 8.112.

If the state metrics calculated for both incoming transitions are the same, in this case, we choose one of the incoming transitions as the winner path. This is the case while calculating the state metric of the state $S_0$ at $t = 4$ (Fig. 8.113).

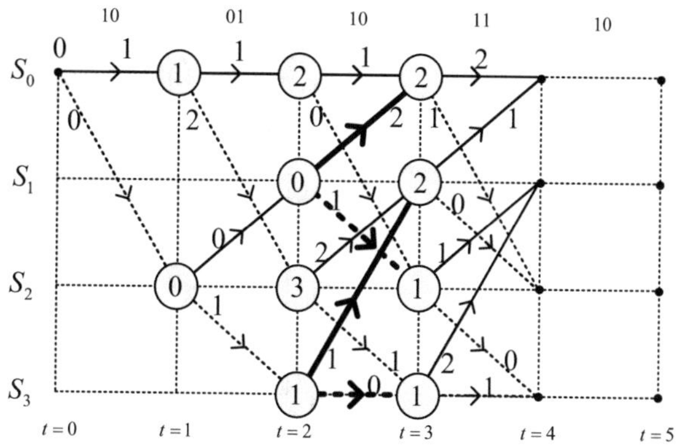


**Fig. 8.110** Branch metrics of the fourth stage

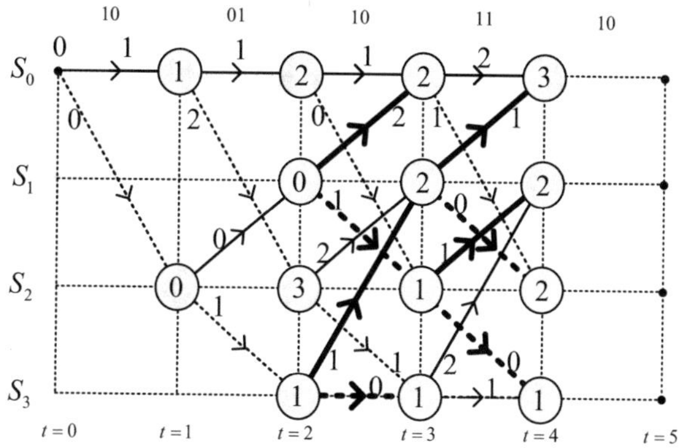


**Fig. 8.111** State metrics for $t = 4$

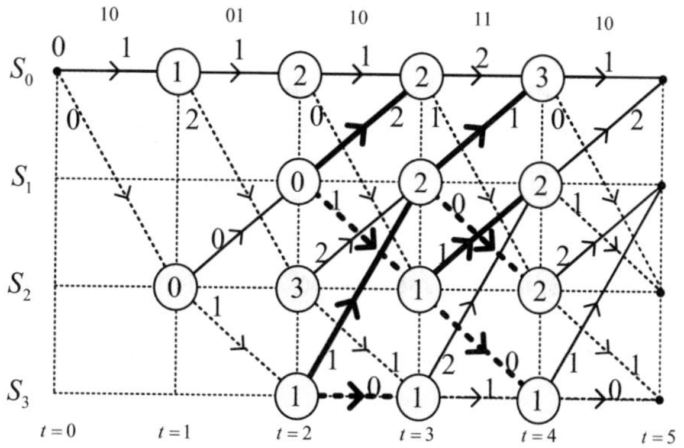


**Fig. 8.112** Branch metrics of the fifth stage

Step 10: For $t = 5$, the minimum state metric is decided as 1 which belongs to the state $S_3$, and by back propagation, we determine the survivor path as in Fig. 8.114. When the survivor path of Fig. 8.114 is inspected, we see that the survivor path shown in shadowed corresponds to the input sequence $\boldsymbol{d} = [1\ 0\ 1\ 1\ 1]$ which is correct.

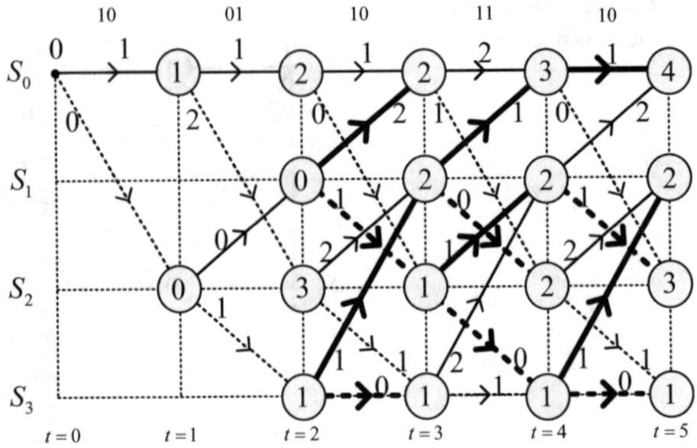


**Fig. 8.113** State metrics for $t = 5$

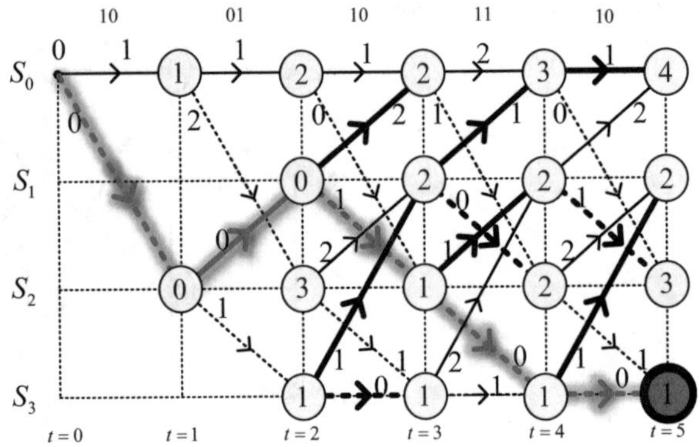


**Fig. 8.114** Determination of survivor path by back propagation

# Problems

1. Find the impulse responses of the convolutional encoder shown in Fig. P8.1, and find the output of the encoder for the input sequence $\boldsymbol{d} = [11010]$.

2. Find the impulse responses of the convolutional encoder shown in Fig. P8.2.

3. Find the transfer functions of the convolutional encoder shown in Fig. P8.3.

4. Find the generator matrix of the convolutional encoder shown in Fig. P8.4.

5. Find the transfer function of the convolutional encoder shown in Fig. P8.5.

6. Find the transfer function of the convolutional encoder shown in Fig. P8.6.

7. Find the generator matrix of the convolutional encoder shown in Fig. P8.7 in polynomial form.

8. Find the transfer function of the convolutional encoder shown in Fig. P8.8.

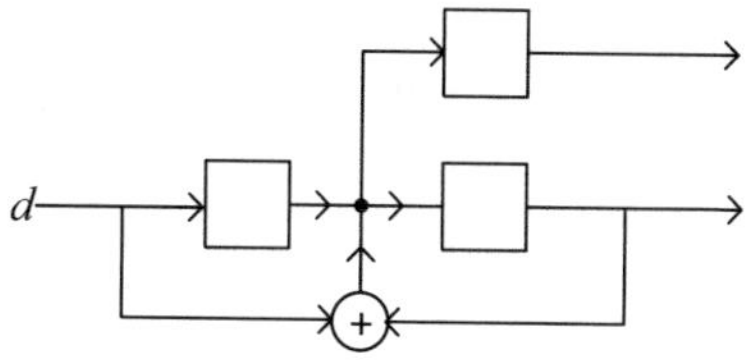


**Fig. P8.1** Convolutional encoder for Problem 1

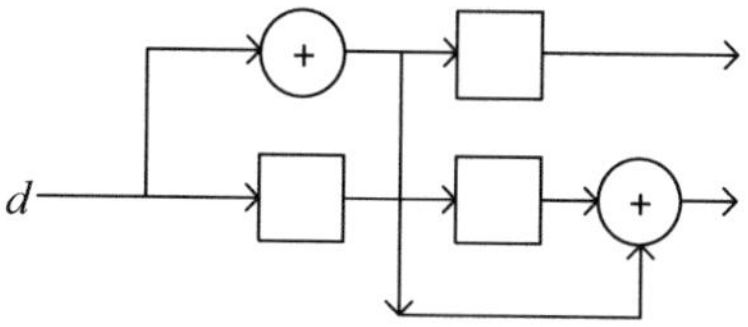


**Fig. P8.2** Convolutional encoder for Problem 2

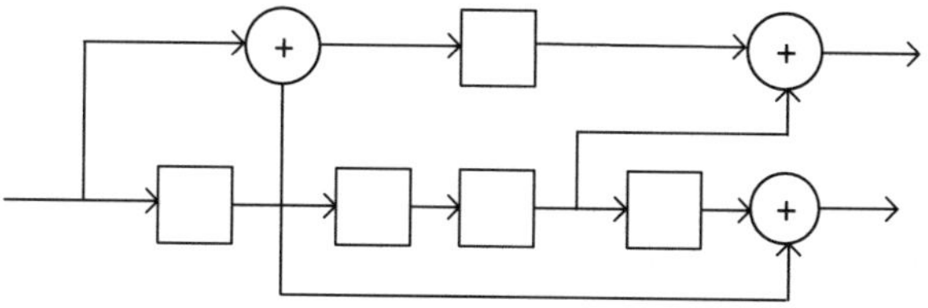

**Fig. P8.3** Convolutional encoder for Problem 3

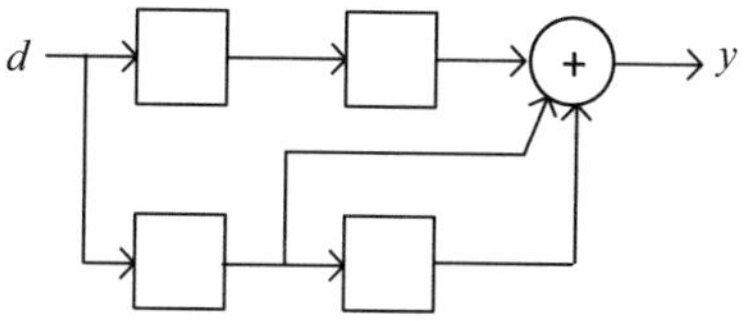


**Fig. P8.4** Convolutional encoder for Problem 4

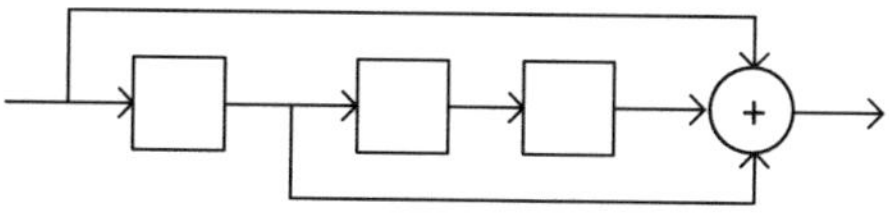

**Fig. P8.5** Convolutional encoder for Problem 5

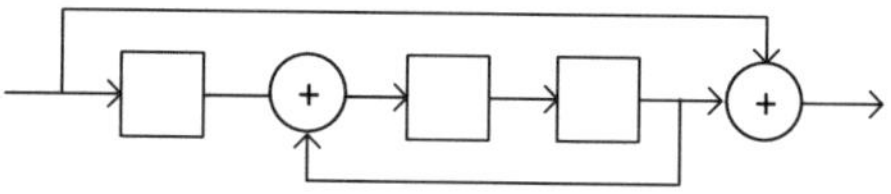

**Fig. P8.6** Convolutional encoder for Problem 6

9. Find the transfer function of the convolutional encoder shown in Fig. P8.9.

10. Find the generator matrix of the convolutional encoder shown in Fig. P8.10 in polynomial form.

11. Draw the state diagram of the convolutional encoder shown in Fig. P8.11.

12. Draw the trellis diagram of the convolutional encoder shown in Fig. P8.12.

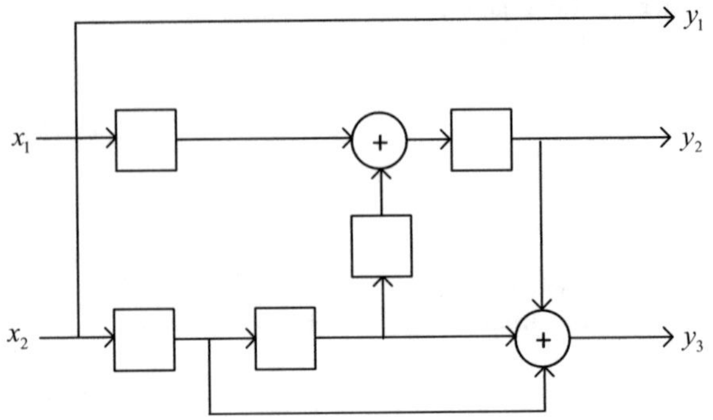


**Fig. P8.7** Convolutional encoder for Problem 7

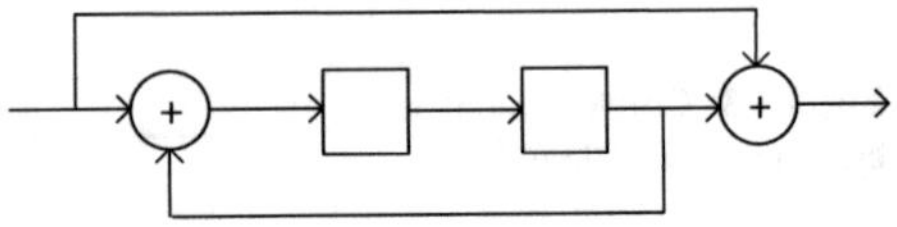

**Fig. P8.8** Convolutional encoder for Problem 8

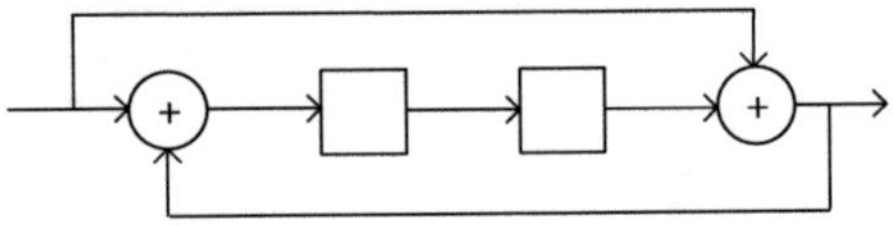

**Fig. P8.9** Convolutional encoder for Problem 9

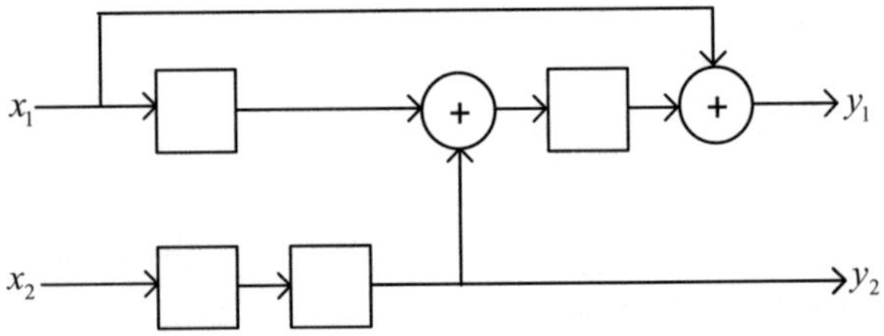


**Fig. P8.10** Convolutional encoder for Problem 10

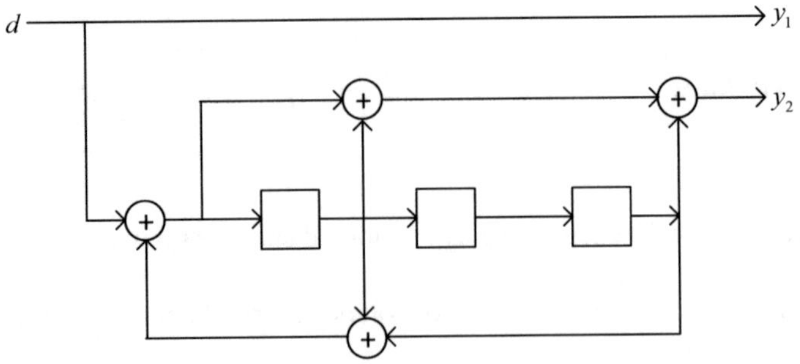


**Fig. P8.11** Convolutional encoder for Problem 11

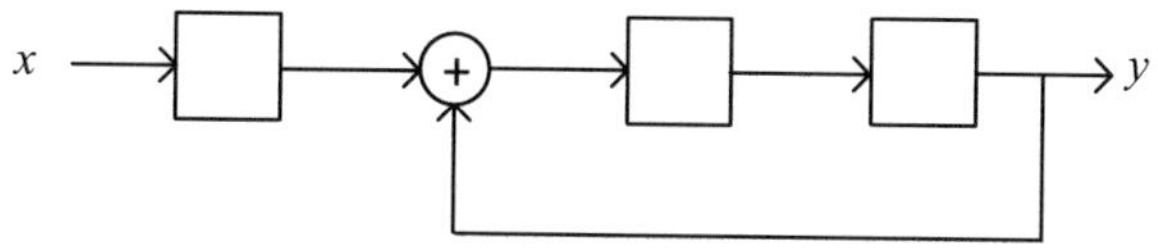


**Fig. P8.12** Convolutional encoder for Problem 12

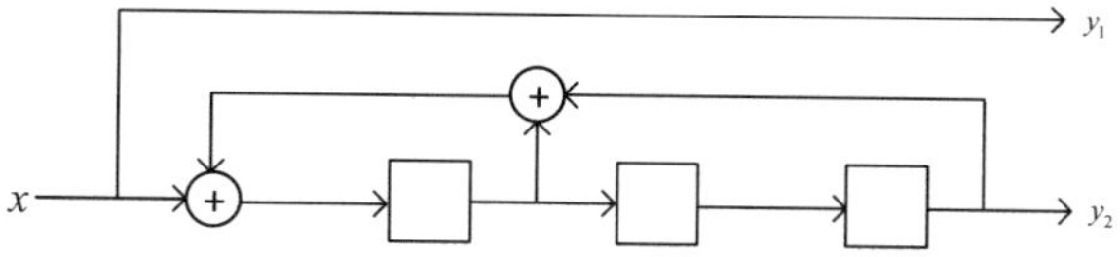


**Fig. P8.13** Convolutional encoder for Problem 13

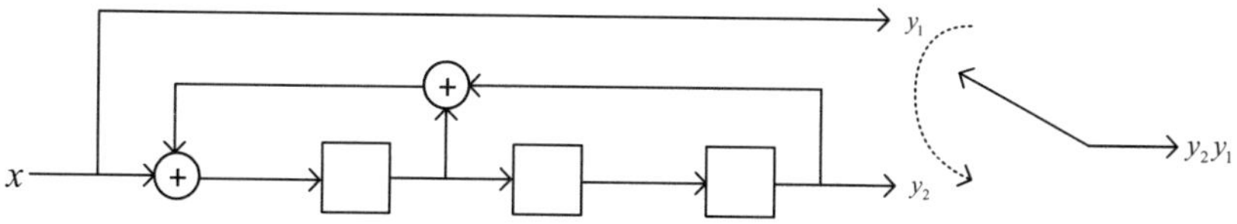


**Fig. P8.14** Convolutional encoder for Problem 14

13. Considering the convolutional encoder given in Fig. P8.13:
    (a) Obtain the state diagram.
    (b) Draw the trellis diagram.
    (c) Encode the data-word $\boldsymbol{d} = [101101]$ using the state diagram, and let $\boldsymbol{c}$ be the code-word obtained after encoding operation. Decode $\boldsymbol{c}$ using the Viterbi decoding algorithm.

14. Considering the convolutional encoder given in Fig. P8.14:
    (a) Obtain the state diagram.
    (b) Draw the trellis diagram.
    (c) Encode the data-word $\boldsymbol{d} = [11011]$ using the state diagram, and let $\boldsymbol{c}$ be the code-word obtained after encoding operation. Decode $\boldsymbol{c}$ using the Viterbi decoding algorithm.

# References

S. Gravano, *Introduction to Error Control Codes* (Oxford University Press, Oxford, 2001)
S. Li, D.J. Costello Jr., *Error Control Coding* (Prentice Hall, Englewood Cliffs, 2004)
S.B. Wicker, *Error Control Systems for Digital Communication and Storage* (Prentice Hall, Englewood Cliffs, 1995)

# Index




O. Gazi, *Forward Error Correction via Channel Coding*,
https://doi.org/10.1007/978-3-030-33380-5

Printed in the United States
By Bookmasters